Mathematical Thought
from Ancient to Modern Times

Mathematical Thought from Ancient to Modern Times

Volume 2

MORRIS KLINE

New York   Oxford   OXFORD UNIVERSITY PRESS

Oxford University Press

Oxford New York Toronto
Delhi Bombay Calcutta Madras Karachi
Petaling Jaya Singapore Hong Kong Tokyo
Nairobi Dar es Salaam Cape Town
Melbourne Auckland

and associated companies in
Berlin Ibadan

First published in 1972, in one volume, by Oxford University Press, Inc.,
200 Madison Avenue, New York, New York 10016

First issued as an Oxford University Press paperback, 1990

Oxford is a registered trademark of Oxford University Press

Library of Congress Cataloging-in-Publication Data
Kline, Morris, 1908–
Mathematical thought from ancient to modern times / Morris Kline.
p. cm. Includes bibliographical references.
ISBN 0-19-506136-5 (PBK) (v. 2)
1. Mathematics—History.   I. Title.
QA21.K516 1990 510′.9—dc20 89-25520

Printed in the United States of America

To my wife, Helen Mann Kline

Preface to the Three-Volume Paperback Edition of Mathematical Thought

The reception accorded the original edition of this book is most gratifying. I am flattered, if not a penny richer for it, by a pirated Chinese translation. Even more satisfying is a forthcoming authorized Spanish translation.

This work is part of my long-time efforts to humanize the subject of mathematics. At the very beginning of my career I banded with a few colleagues to produce a freshman text that departed from the traditional dry-as-dust mathematics textbook. Later, I wrote a calculus text with the same end in view. While I was directing a research group in electromagnetic theory and doing research myself, I still made time to write *Mathematics In Western Culture*, which is partly history and partly an exploration of the influence of mathematics upon philosophy, religion, literature, art, music, economic theory, and political thought. More recently I have written with the general reader in mind a book on the philosophical foundations of mathematics and a book on the underlying mathematical structure of a good deal of science, most especially cosmogony and physics.

I hope that students, teachers, as well as the general reader will profit from this more affordable and accessible three-volume paperback edition of *Mathematical Thought*. I wish to acknowledge the helpful suggestions made by Harold Edwards, Donald Gillis, and Robert Schlapp among others. My very special thanks go to Fred Pohle for his time, interest, and generosity. Having over the years taught a course based on this book, he saw a need for a multi-volume paperback version and provided the impetus for this edition. Beyond this he gave unstintingly of his time and knowledge in helping me correct errors. I am truly in his debt, as I am to my wife Helen, who undertook much of the work involved in preparing this edition.

Preface

If we wish to foresee the future of mathematics our proper course is to study the history and present condition of the science.
                                                    HENRI POINCARÉ

This book treats the major mathematical creations and developments from ancient times through the first few decades of the twentieth century. It aims to present the central ideas, with particular emphasis on those currents of activity that have loomed largest in the main periods of the life of mathematics and have been influential in promoting and shaping subsequent mathematical activity. The very concept of mathematics, the changes in that concept in different periods, and the mathematicians' own understanding of what they were achieving have also been vital concerns.

This work must be regarded as a survey of the history. When one considers that Euler's works fill some seventy volumes, Cauchy's twenty-six volumes, and Gauss's twelve volumes, one can readily appreciate that a one-volume work cannot present a full account. Some chapters of this work present only samples of what has been created in the areas involved, though I trust that these samples are the most representative ones. Moreover, in citing theorems or results, I have often omitted minor conditions required for strict correctness in order to keep the main ideas in focus. Restricted as this work may be, I believe that some perspective on the entire history has been presented.

The book's organization emphasizes the leading mathematical themes rather than the men. Every branch of mathematics bears the stamp of its founders, and great men have played decisive roles in determining the course of mathematics. But it is their ideas that have been featured; biography is entirely subordinate. In this respect, I have followed the advice of Pascal: "When we cite authors we cite their demonstrations, not their names."

To achieve coherence, particularly in the period after 1700, I have treated each development at that stage where it became mature, prominent, and influential in the mathematical realm. Thus non-Euclidean geometry is presented in the nineteenth century even though the history of the efforts to replace or prove the Euclidean parallel axiom date from Euclid's time onward. Of course, many topics recur at various periods.

To keep the material within bounds I have ignored several civilizations such as the Chinese,[1] Japanese, and Mayan because their work had no material impact on the main line of mathematical thought. Also some developments in mathematics, such as the theory of probability and the calculus of finite differences, which are important today, did not play major roles during the period covered and have accordingly received very little attention. The vast expansion of the last few decades has obliged me to include only those creations of the twentieth century that became significant in that period. To continue into the twentieth century the extensions of such subjects as ordinary differential equations or the calculus of variations would call for highly specialized material of interest only to research men in those fields and would have added inordinately to the size of the work. Beyond these considerations, the importance of many of the more recent developments cannot be evaluated objectively at this time. The history of mathematics teaches us that many subjects which aroused tremendous enthusiasm and engaged the attention of the best mathematicians ultimately faded into oblivion. One has but to recall Cayley's dictum that projective geometry is all geometry, and Sylvester's assertion that the theory of algebraic invariants summed up all that is valuable in mathematics. Indeed one of the interesting questions that the history answers is what survives in mathematics. History makes its own and sounder evaluations.

Readers of even a basic account of the dozens of major developments cannot be expected to know the substance of all these developments. Hence except for some very elementary areas the contents of the subjects whose history is being treated are also described, thus fusing exposition with history. These explanations of the various creations may not clarify them completely but should give some idea of their nature. Consequently this book may serve to some extent as a historical introduction to mathematics. This approach is certainly one of the best ways to acquire understanding and appreciation.

I hope that this work will be helpful to professional and prospective mathematicians. The professional man is obliged today to devote so much of his time and energy to his specialty that he has little opportunity to familiarize himself with the history of his subject. Yet this background is important. The roots of the present lie deep in the past and almost nothing in that past is irrelevant to the man who seeks to understand how the present came to be what it is. Moreover, mathematics, despite the proliferation into hundreds of branches, is a unity and has its major problems and goals. Unless the various specialties contribute to the heart of mathematics they are likely to be

1. A fine account of the history of Chinese mathematics is available in Joseph Needham's *Science and Civilization in China*, Cambridge University Press, 1959, Vol. 3, pp. 1–168.

sterile. Perhaps the surest way to combat the dangers which beset our fragmented subject is to acquire some knowledge of the past achievements, traditions, and objectives of mathematics so that one can direct his research into fruitful channels. As Hilbert put it, "Mathematics is an organism for whose vital strength the indissoluble union of the parts is a necessary condition."

For students of mathematics this work may have other values. The usual courses present segments of mathematics that seem to have little relationship to each other. The history may give perspective on the entire subject and relate the subject matter of the courses not only to each other but also to the main body of mathematical thought.

The usual courses in mathematics are also deceptive in a basic respect. They give an organized logical presentation which leaves the impression that mathematicians go from theorem to theorem almost naturally, that mathematicians can master any difficulty, and that the subjects are completely thrashed out and settled. The succession of theorems overwhelms the student, especially if he is just learning the subject.

The history, by contrast, teaches us that the development of a subject is made bit by bit with results coming from various directions. We learn, too, that often decades and even hundreds of years of effort were required before significant steps could be made. In place of the impression that the subjects are completely thrashed out one finds that what is attained is often but a start, that many gaps have to be filled, or that the really important extensions remain to be created.

The polished presentations in the courses fail to show the struggles of the creative process, the frustrations, and the long arduous road mathematicians must travel to attain a sizable structure. Once aware of this, the student will not only gain insight but derive courage to pursue tenaciously his own problems and not be dismayed by the incompleteness or deficiencies in his own work. Indeed the account of how mathematicians stumbled, groped their way through obscurities, and arrived piecemeal at their results should give heart to any tyro in research.

To cover the large area which this work comprises I have tried to select the most reliable sources. In the pre-calculus period these sources, such as T. L. Heath's *A History of Greek Mathematics*, are admittedly secondary, though I have not relied on just one such source. For the subsequent development it has usually been possible to go directly to the original papers, which fortunately can be found in the journals or in the collected works of the prominent mathematicians. I have also been aided by numerous accounts and surveys of research, some in fact to be found in the collected works. I have tried to give references for all of the major results; but to do so for all assertions would have meant a mass of references and the consumption of space that is better devoted to the account itself.

The sources have been indicated in the bibliographies of the various chapters. The interested reader can obtain much more information from these sources than I have extracted. These bibliographies also contain many references which should not and did not serve as sources. However, they have been included either because they offer additional information, because the level of presentation may be helpful to some readers, or because they may be more accessible than the original sources.

I wish to express thanks to my colleagues Martin Burrow, Bruce Chandler, Martin Davis, Donald Ludwig, Wilhelm Magnus, Carlos Moreno, Harold N. Shapiro, and Marvin Tretkoff, who answered numerous questions, read many chapters, and gave valuable criticisms. I am especially indebted to my wife Helen for her critical editing of the manuscript, extensive checking of names, dates, and sources, and most careful reading of the galleys and page proofs. Mrs. Eleanore M. Gross, who did the bulk of the typing, was enormously helpful. To the staff of Oxford University Press, I wish to express my gratitude for their scrupulous production of this work.

New York   M. K.
May 1972

Contents

Publisher's Note

to this Three-Volume Paperback Edition

Mathematical Thought from Ancient to Modern Times was first published by Oxford University Press as a one-volume cloth edition. In publishing this three-volume paperback edition we have retained the same pagination as the cloth in order to maintain consistency within the Index, Subject Index, and Notes. These volumes are paginated consecutively and, for the reader's convenience, both Indexes appear at the end of each volume.

Mathematical Thought
from Ancient to Modern Times

18
Mathematics as of 1700

> Those few things having been considered, the whole matter is
> reduced to pure geometry, which is the one aim of physics
> and mechanics.   G. W. LEIBNIZ

1. *The Transformation of Mathematics*

At the opening of the seventeenth century Galileo still found it necessary to
argue with the past. By the end of the century, mathematics had undergone
such extensive and radical changes that no one could fail to recognize the
arrival of a new era.

The European mathematicians produced far more between about 1550
and 1700 than the Greeks had in roughly ten centuries. This is readily
explained by the fact that, whereas mathematics in Greece was pursued by
only a handful of men, in Europe the spread of education, though by no
means universal, promoted the development of mathematicians in England,
France, Germany, Holland, and Italy. The invention of printing gave wide
access not only to the Greek works but to the results of the Europeans themselves, which, now readily available, served to stimulate new thoughts.

But the genius of the century is not evidenced solely by the expansion
of activity. The variety of new fields opened up in this brief period is impressive. The rise of algebra as a science (because the use of literal coefficients
permitted a measure of proof) as well as the vast expansion of its methods
and theory, the beginnings of projective geometry and the theory of probability, analytic geometry, the function concept, and above all the calculus
were major innovations, each destined to dwarf the one extensive accomplishment of the Greeks—Euclidean geometry.

Beyond the quantitative expansion and the new avenues of exploration
was the complete reversal of the roles of algebra and geometry. The Greeks
had favored geometry because it was the only way they could achieve rigor;
and even in the seventeenth century, mathematicians felt obliged to justify
algebraic methods with geometrical proofs. One could say that up to 1600
the body of mathematics was geometrical, with some algebraic and trigonometric appendages. After the work of Descartes, Fermat, and Wallis, algebra

became not only an effective methodology for its own ends but also the superior approach to the solution of geometric problems. The greater effectiveness of analytical methods in the calculus decided the competition, and algebra became the dominant substance of mathematics.

It was Wallis and Newton who saw clearly that algebra provided the superior methodology. Unlike Descartes, who regarded algebra as just technique, Wallis and Newton realized that it was vital subject matter. The work of Desargues, Pascal, and La Hire was depreciated and forgotten, and the geometric methods of Cavalieri, Gregory of Saint Vincent, Huygens, and Barrow were superseded. Pure geometry was eclipsed for about a hundred years, becoming at best an interpretation of algebra and a guide to algebraic thinking through coordinate geometry. It is true that excessive reverence for Newton's geometrical work in the *Principia*, reinforced by the enmity against the Continental mathematicians engendered by the dispute between Newton and Leibniz, caused the English mathematicians to persist in the geometrical development of the calculus. But their contributions were trivial compared to what the Continentals were able to achieve using the analytical approach. What was evident by 1700 was explicitly stated by no less an authority than Euler, who, in his *Introductio in Analysin Infinitorum* (1748), praises algebra as far superior to the synthetic methods of the Greeks.

It was with great reluctance that mathematicians abandoned the geometric approach. According to Henry Pemberton (1694–1771), who edited the third edition of Newton's *Principia*, Newton not only constantly expressed great admiration for the geometers of Greece but censured himself for not following them more closely than he did. In a letter to David Gregory (1661–1708), a nephew of James Gregory, Newton remarked that "algebra is the analysis of the bunglers in mathematics." But his own *Arithmetica Universalis* of 1707 did as much as any single work to establish the supremacy of algebra. Here he set up arithmetic and algebra as the basic science, allowing geometry only where it made demonstrations easier. Leibniz, too, noted the growing dominance of algebra and felt obliged to say, in an unpublished essay,[1] "Often the geometers could demonstrate in a few words what is very lengthy in the calculus . . . the view of algebra is assured, but it is not better."

Another, more subtle, change in the nature of mathematics had been unconsciously accepted by the masters. Up to 1550 the concepts of mathematics were immediate idealizations of or abstractions from experience. By that time negative and irrational numbers had made their appearance and were gradually gaining acceptance. When, in addition, complex numbers, an extensive algebra employing literal coefficients, and the notions of derivative and integral entered mathematics, the subject became dominated by

1. Couturat, L.: *Opuscules et fragments inédits de Leibniz*, 1903, reprinted by Georg Olms, 1961, p. 181.

concepts derived from the recesses of human minds. The notion of an instantaneous rate of change, in particular, though of course having some intuitive base in the physical phenomenon of velocity, is nevertheless far more of an intellectual construct and is also an entirely different contribution qualitatively than the mathematical triangle. Beyond these ideas, infinitely large quantities, which the Greeks had studiously avoided, and infinitely small ones, which the Greeks had skillfully circumvented, had to be contended with.

In other words, mathematicians were contributing concepts, rather than abstracting ideas from the real world. Nevertheless, these concepts were useful in physical investigations because (with the exception of complex numbers, which had yet to prove their worth) they had some ties to physical reality. Of course the Europeans were uneasy about the new types of numbers and the calculus notions without really discerning the cause of their concern. Yet as these concepts proved more and more useful in applications, they were at first grudgingly and later passively accepted. Familiarity bred not contempt but acceptability and even naturalness. After 1700, more and more notions, further removed from nature and springing full-blown from human minds, were to enter mathematics and be accepted with fewer qualms. For the genesis of its ideas mathematics gradually turned from the sensory to the intellectual faculties.

The incorporation of the calculus into the body of mathematics effected another change, in the very concept of mathematics, that subverted the ideal fashioned by the classical Greeks. We have already noted that the rise of algebra and the calculus introduced the problem of the logical foundations of these portions of mathematics and that this problem was not resolved. Throughout the century some mathematicians were upset by the abandonment of proof in the deductive sense, but their protests were drowned in the expanding content and use of algebra and the calculus; by the end of the century mathematicians had virtually dropped the requirement of clearly defined concepts and deductive proof. Rigorous axiomatic construction gave way to induction from particular examples, intuitive insights, loose geometrical evidence, and physical arguments. Since deductive proof had been the distinguishing feature of mathematics, the mathematicians were thus abandoning the hallmark of their subject.

In retrospect it is easy to see why they were forced into this position. As long as mathematicians derived their concepts from immediate experience, it was feasible to define the concepts and select the necessary axioms—though, at that, the logical basis for the theory of the integers that Euclid presented in Books VII to IX of the *Elements* was woefully deficient. But as they introduced concepts that no longer idealized immediate experiences, such as the irrational, negative, and complex numbers and the derivative and integral, they failed to recognize that these concepts were different in character, and so failed to realize that a basis for the axiomatic development other than self-evident truths was needed. It is true that the new concepts were far more subtle than the old ones; and the proper axiomatic basis, as we now know, could not have been readily erected.

How could critical mathematicians, well-versed in Greek mathematics, have been content to operate on a heuristic basis? They were concerned with major and in some cases pressing problems of science, and the mathematics they employed handled these problems. Rather than seeking full comprehension of the new creations or trying to erect the requisite deductive structure, they salved their consciences with their successes. An occasional recourse to philosophical or mystical doctrines succeeded in cloaking some difficulties so that they were no longer visible.

One new goal in particular characterizes the mathematics of the seventeenth and succeeding centuries—generality of methods and results. We have already noted the value placed on generality of method by Vieta, in his introduction of literal coefficients; by the projective geometers; by Fermat and Descartes in the exploration of curves; and by Newton and Leibniz in the treatment of functions. As far as generality of results is concerned, the accomplishments were limited. Many were just affirmations, such as that every polynomial equation of the nth degree has n roots, or that every equation of the second degree in x and y is a conic. Mathematical methods and notation were still too limited to permit the establishment of general results, but this became a goal of the mathematical efforts.

2. *Mathematics and Science*

Since classical Greek times, mathematics had been valued primarily for its role in the investigation of nature. Astronomy and music were constantly linked to mathematics, and mechanics and optics were certainly mathematical. However, the relationship of mathematics to science was altered in several ways by the work of the seventeenth century. First of all, because science, which was expanding enormously, had been directed by Galileo to use quantitative axioms and mathematical deduction (Chap. 16, sec. 3), the mathematical activity that was directly inspired by science became dominant.

Further, Galileo's injunction to seek mathematical description rather than causal explanation led to the acceptance of such a concept as the force of gravitation. This force and the laws of motion were the entire basis for Newton's system of mechanics. Since the only sure knowledge of gravitation was mathematical, mathematics became the substance of scientific theories. The insurgent seventeenth century found a qualitative world whose study was aided by mathematical abstractions and bequeathed a mathematical, quantitative world that subsumed under its mathematical laws the concreteness of the physical world.

Third, while the Greeks had employed mathematics freely in their science, there was, as long as the Euclidean basis sufficed for mathematics, a sharp distinction between it and science. Both Plato and Aristotle distinguished the two (Chap. 3, sec. 10 and Chap. 7, sec. 3), albeit in different ways; and Archimedes is especially clear about what is established mathematically and what is known physically. However, as the province of mathematics expanded, and mathematicians not only relied upon physical meanings to understand their concepts but accepted mathematical arguments because they gave sound physical results, the boundary between mathematics and science became blurred. Paradoxically, as science began to rely more and more upon mathematics to produce its physical conclusions, mathematics began to rely more and more upon scientific results to justify its own procedures.

The upshot of this interdependence was a virtual fusion of mathematics and vast areas of science. The compass of mathematics, as understood in the seventeenth century, may be seen from the *Cursus seu Mundus Mathematicus* (The Course or the World of Mathematics) by Claude-François Milliet Deschales (1621–78), published in 1674 and in an enlarged edition in 1690. Aside from arithmetic, trigonometry, and logarithms, he treats practical geometry, mechanics, statics, geography, magnetism, civil engineering, carpentry, stonecutting, military construction, hydrostatics, fluid flow, hydraulics, ship construction, optics, perspective, music, the design of firearms and cannons, the astrolabe, sundials, astronomy, calendar-reckoning, and horoscopy. Finally, he includes algebra, the theory of indivisibles, the theory of conics, and special curves such as the quadratrix and the spiral. This work was popular and esteemed. Though in the inclusion of some topics it reflects Renaissance interests, on the whole it presents a reasonable picture of the seventeenth- and even the eighteenth-century world of mathematics.

One might expect that the mathematicians would have been concerned to preserve the identity of their subject. But beyond the fact that they were obliged to depend upon physical meanings and results to defend their arguments, the greatest of the seventeenth- (and eighteenth-) century contributors to mathematics were either primarily scientists or at least equally concerned with both fields. Descartes, Huygens, and Newton, for example, were far greater physicists than mathematicians. Pascal, Fermat, and Leibniz were active in physics. In fact, it would be difficult to name an outstanding mathematician of the century who did not take a keen interest in science. As a consequence these men did not wish or seek to make any distinctions between the two fields. Descartes says in his *Rules for the Direction of the Mind* that mathematics is the science of order and measure and includes, besides algebra and geometry, astronomy, music, optics, and mechanics. Newton says in the *Principia*: "In mathematics we are to investigate the quantities of forces with their proportion consequent upon any conditions supposed; then, when we enter upon physics, we compare those proportions with the phenomena of Nature. . . ." Here physics refers to experimentation and observation. Newton's mathematics would be regarded as mathematical physics today.

3. *Communication Among Mathematicians*

Up to about 1550, mathematics was created by individuals or small groups headed by one or two prominent leaders. The results were communicated orally and occasionally written up in texts—which, however, were manuscripts. Since copies had also to be made by hand, they were scarce. By the seventeenth century printed books had become somewhat common, though even this improvement did not spread knowledge as widely as might be thought. Because the market for advanced mathematics was small, publishers had to charge high prices. Good printers were scarce. Publication was often followed by attacks on the authors from none-too-scrupulous opponents; it was not hard for such critics to find grounds for attack, especially because algebra and the calculus were not at all well grounded logically. Books in any case were not usually the answer for new creations because significant results did not warrant a book-sized publication.

As a consequence many mathematicians confined themselves to writing letters to friends in which they related their discoveries. Fearing that the letters would reach men who might take advantage of such unofficial documents, the writers often put their results in ciphers or anagrams, which they could then decode when challenged.

As more men began to participate in mathematical creation, the desire for exchange of information and for the stimulus of meeting people with the same intellectual interests resulted in the formation of scientific societies or academies. In 1601 the Accademia dei Lincei (Academy of the Lynx-like) was founded in Rome by young noblemen; it lasted thirty years. Galileo became a member in 1611. Another Italian society, the Accademia del Cimento (Academy of Experiments) was founded in Florence in 1657 as a formal organization of men who had been meeting in a laboratory founded by two members of the Medici family about ten years earlier. This academy included Vincenzo Viviani (1622–1703) and Torricelli, both pupils of Galileo, among its members. Unfortunately, the society was disbanded in 1667. In France, Desargues, Descartes, Gassendi, Fermat, and Pascal, among others, met privately under the leadership of Mersenne from 1630 on. This informal group was chartered by Louis XIV as the Académie Royale des Sciences in 1666, and its members were supported by the king. Paralleling what happened in France, an English group centered about John Wallis began in 1645 to hold meetings in Gresham College, London. These men emphasized mathematics and astronomy. The group was given a formal charter by

Charles II in 1662 and adopted the name of the Royal Society of London for the Promotion of Natural Knowledge. This society was concerned with putting mathematics and science to use and regarded dyeing, coinage, gunnery, the refinement of metals, and population statistics as subjects of interest. Finally, the Berlin Academy of Sciences, which Liebniz had advocated for some years, was opened in 1700 with Leibniz as its first president. In Russia, Peter the Great founded the Academy of Sciences of St. Petersburg in 1724.

The academies were important, not only in making possible direct contact and exchange of ideas, but because they also supported journals. The first of the scientific journals, though not sponsored by an academy, was the *Journal de Sçavans* or *Journal des Savants*, which began publication in 1665. This journal and the *Philosophical Transactions of the Royal Society*, which began publication in the same year, were the first journals to include mathematical and scientific articles. The French Académie des Sciences initiated the publication *Histoire de l'Académie Royale des Sciences avec les Mémoires de Mathématique et de Physique*. It also published the *Mémoires de Mathématique et de Physique Présentés à l'Académie Royale des Sciences par Divers Sçavans et Lus dans ses Assemblées*, also known as the *Mémoires des Savants Etrangers*. Another of the early scientific journals, the *Acta Eruditorum*, was begun in 1682 and, because it was published in Latin, soon acquired international readership. The Berlin Academy of Sciences sponsored the *Histoire de l'Académie Royale des Sciences et Belles-lettres* (whose title for some years was the *Miscellanea Berolinensia*).

The academies and their journals opened new outlets for scientific communication; these and later journals became the accepted medium for publication of new research. The academies furthered research in that most of them supported scientists. For example, Euler was supported by the Berlin Academy from 1741 to 1766 and Lagrange from 1766 to 1787. The St. Petersburg Academy supported Daniel and Nicholas Bernoulli at various times, and Euler from 1727 to 1741 and again from 1766 to his death in 1783. The founding of academies by the European governments marks also the official entry of governments into the area of science and the support of science. The usefulness of science had received recognition.

The institutions that a modern person would expect to play the major role in the creation and dissemination of knowledge—the universities—were ineffective. They were conservative and dogmatic, controlled by the official religions of the respective countries and very slow to incorporate new knowledge. On the whole they taught just a modicum of arithmetic, algebra, and geometry. Though there were some mathematicians at Cambridge University in the sixteenth century, from 1600 to 1630 there were none. In fact, in England during the early seventeenth century, mathematics did not form part of the curriculum. It was regarded as devilry. Wallis, who was born in 1616, says of the education common during his childhood, "Mathematics at that time with us was scarce looked on as academical but rather mechanical—as the business of tradesmen." He did attend Cambridge University and study mathematics there, but learned far more from independent study. Though prepared to be a professor of mathematics, he left Cambridge "because that study had died out there and no career was open to a teacher of that subject."

Professorships in mathematics were founded first at Oxford in 1619 and later at Cambridge. Prior to that, there had been only lecturers of low status. The Lucasian professorship at Cambridge, which Barrow was the first to hold, was founded in 1663. Wallis himself became a professor at Oxford in 1649 and held the chair until 1702. One obstacle to the enlistment of able professors was that they had to take holy orders, though exceptions were made, as in the case of Newton. The British universities generally (including also London, Glasgow, and Edinburgh) had roughly the same history: from about 1650 to 1750, they were somewhat active and then declined in activity until about 1825.

The French universities of the seventeenth and eighteenth centuries were inactive in mathematics. Not until the end of the eighteenth century, when Napoleon founded first-class technical schools, did they make any contribution. At German universities too, the mathematical activity of those two centuries was at a low level. Leibniz was isolated and, as we noted earlier, he railed against the teachings of the universities. The University of Göttingen was founded in 1731, but rose only slowly to any position of importance until Gauss became a professor there. The university centers of Geneva and Basel in Switzerland were exceptions in the period we are surveying; they could boast of the Bernoullis, Hermann, and others. The Italian universities were of some importance in the seventeenth century but lost ground in the eighteenth. When one notes that Pascal, Fermat, Descartes, Huygens, and Leibniz never taught at any university and that though Kepler and Galileo did teach for a while, they were court mathematicians for most of their lives, one sees how relatively unimportant the universities were.

4. *The Prospects for the Eighteenth Century*

The enormous seventeenth-century advances in algebra, analytic geometry, and the calculus; the heavy involvement of mathematics in science, which provided deep and intriguing problems; the excitement generated by Newton's astonishing successes in celestial mechanics; and the improvement in communications provided by the academies and journals all pointed to additional major developments and served to create immense exuberance about the future of mathematics.

There were obstacles to be overcome. The doubts as to the soundness of the calculus, the estrangement of the English and Continental mathematicians, the low state of the existing educational institutions, and the uncertainties of the support for careers in mathematics gave pause to young or would-be mathematicians. Nevertheless, the enthusiasm of the mathematicians was almost unbounded. They had glimpses of a promised land and were eager to press forward. They were, moreover, able to work in an atmosphere far more suitable for creation than at any time since 300 B.C. Classical Greek geometry had not only imposed restrictions on the domain of mathematics but had impressed a level of rigor for acceptable mathematics that hampered creativity. The seventeenth-century men had broken both of these bonds. Progress in mathematics almost demands a complete disregard of logical scruples; and, fortunately, the mathematicians now dared to place their confidence in intuitions and physical insights.

Bibliography

Hahn, Roger: *The Anatomy of a Scientific Institution: The Paris Academy of Sciences, 1666–1803*, University of California Press, 1971.

Hall, A. Rupert: *The Scientific Revolution, 1500–1800*, Longmans, Green, 1954, Chap. 7.

Hall, A. Rupert, and Marie Boas: *The Correspondence of Henry Oldenburg*, 4 vols., University of Wisconsin Press, 1968.

Hartley, Sir Harold: *The Royal Society: Its Origins and Founders*, The Royal Society, 1960.

Ornstein, M.: *The Role of Scientific Societies in the Seventeenth Century*, University of Chicago Press, 1938.

Purver, Margery: *The Royal Society, Concept and Creation*, Massachusetts Institute of Technology Press, 1967.

Wolf, Abraham: *A History of Science, Technology and Philosophy in the 16th and 17th Centuries*, 2nd ed., George Allen and Unwin, 1950, Chap. 4.

19
Calculus in the Eighteenth Century

> And therefore, whether the mathematicians of the present age
> act like men of science, in taking so much more pains to apply
> their principles than to understand them.   BISHOP BERKELEY

1. *Introduction*

The greatest achievement of the seventeenth century was the calculus. From
this source there stemmed major new branches of mathematics: differential
equations, infinite series, differential geometry, the calculus of variations,
functions of complex variables, and many others. Indeed, the beginnings of
some of these subjects were already present in the works of Newton and
Leibniz. The eighteenth century was devoted largely to the development of
some of these branches of analysis. But before this could be accomplished, the
calculus itself had to be extended. Newton and Leibniz had created basic
methods, but much remained to be done. Many new functions of one
variable and functions of two or more variables had either to be recognized
explicitly or created; the techniques of differentiation and integration had to
be extended to some of the existing functions and to others yet to be introduced; and the logical foundation of the calculus was still missing. The first
goal was to expand the subject matter of the calculus and this is the subject
of the present chapter and the next one.

The eighteenth-century men did extend the calculus and founded new
branches of analysis, though encountering in the process all the pangs,
errors, incompleteness, and confusion of the creative process. The mathematicians produced a purely formal treatment of calculus and the ensuing
branches of analysis. Their technical skill was unsurpassed; it was guided,
however, not by sharp mathematical thinking but by intuitive and physical
insights. These formal efforts withstood the test of subsequent critical
examination and produced great lines of thought. The conquest of new
domains of mathematics proceeds somewhat as do military conquests. Bold
dashes into enemy territory capture strongholds. These incursions must then
be followed up and supported by broader, more thorough and more cautious
operations to secure what has been only tentatively and insecurely grasped.

It will be helpful, in appreciating the work and arguments of the eighteenth-century thinkers, to keep in mind that they did not distinguish between algebra and analysis. Because they did not appreciate the need for the limit concept and because they failed to recognize the problems introduced by the use of infinite series, they naively regarded the calculus as an extension of algebra.

The key figure in eighteenth-century mathematics and the dominant theoretical physicist of the century, the man who should be ranked with Archimedes, Newton, and Gauss, is Leonhard Euler (1707–83). Born near Basel to a preacher, who wanted him to study theology, Leonhard entered the university at Basel and completed his work at the age of fifteen. While at Basel he learned mathematics from John Bernoulli. He decided to pursue the subject and began to publish papers at eighteen. At nineteen he won a prize from the French Académie des Sciences for a work on the masting of ships. Through the younger Bernoullis, Nicholas (1695–1726) and Daniel (1700–82), sons of John, Euler in 1733 secured an appointment at the St. Petersburg Academy in Russia. He started as an assistant to Daniel Bernoulli but soon succeeded him as a professor. Though Euler passed some painful years (1733–41) under the autocratic government, he did an amazing amount of research, the results of which appeared in papers published by the St. Petersburg Academy. He also helped the Russian government on many physical problems. In 1741, at the call of Frederick the Great, he went to Berlin, where he remained until 1766. During this period, he gave lessons to the princess of Anhalt-Dessau, niece of the king of Prussia. These lessons, on a variety of subjects—mathematics, astronomy, physics, philosophy, and religion—were later published as the *Letters to a German Princess* and are still read with pleasure. At the request of Frederick the Great, Euler worked on state problems of insurance and the design of canals and waterworks. Even during his twenty-five years in Berlin he sent hundreds of papers to the St. Petersburg Academy and advised it on its affairs.

In 1766, at the request of Catherine the Great, Euler returned to Russia, although fearing the effect on his weakened sight (he had lost the sight of one eye in 1735) of the rigors of the climate there. He became, in effect, blind shortly after returning to Russia, and during the last seventeen years of his life was totally blind. Nevertheless, these years were no less fruitful than the preceding ones. Euler had a prodigious memory and knew by heart the formulas of trigonometry and analysis and the first six powers of the first 100 prime numbers, to say nothing of innumerable poems and the entire *Æneid*. His memory was so phenomenal that he could carry out in his head numerical calculations that competent mathematicians had difficulty doing on paper.

Euler's mathematical productivity is incredible. His major mathematical fields were the calculus, differential equations, analytic and differential geometry of curves and surfaces, the theory of numbers, series, and the calculus of variations. This mathematics he applied to the entire domain of physics. He created analytical mechanics (as opposed to the older geometrical mechanics) and the subject of rigid body mechanics. He calculated the perturbative effect of celestial bodies on the orbit of a planet and the paths of projectiles in resisting media. His theory of the tides and work on the design and sailing of ships aided navigation. In this area his *Scientia Navalis* (1749) and *Théorie complète de la construction et de la manœuvre des vaisseaux* (1773) are outstanding. He investigated the bending of beams and calculated the safety load of a column. In acoustics he studied the propagation of sound and musical consonance and dissonance. His three volumes on optical instruments contributed to the design of telescopes and microscopes. He was the first to treat the vibrations of light analytically and to deduce the equation of motion taking into account the dependence on the elasticity and density of the ether, and he obtained many results on the refraction and dispersion of light. In the subject of light he was the only physicist of the eighteenth century who favored the wave as opposed to the particle theory. The fundamental differential equations for the motion of an ideal fluid are his; and he applied them to the flow of blood in the human body. In the theory of heat, he (and Daniel Bernoulli) regarded heat as an oscillation of molecules, and his *Essay on Fire* (1738) won a prize. Chemistry, geography, and cartography also interested him, and he made a map of Russia. The applications were said to be an excuse for his mathematical investigations; but there can be no doubt that he liked both.

Euler wrote texts on mechanics, algebra, mathematical analysis, analytic and differential geometry, and the calculus of variations that were standard works for a hundred years and more afterward. The ones that will concern us in this chapter are the two-volume *Introductio in Analysin Infinitorum* (1748), the first connected presentation of the calculus and elementary analysis; the more comprehensive *Institutiones Calculi Differentialis* (1755); and the three-volume *Institutiones Calculi Integralis* (1768–70); all are landmarks. All of Euler's books contained some highly original features. His mechanics, as noted, was based on analytical rather than geometrical methods. He gave the first significant treatment of the calculus of variations. Beyond texts he published original research papers of high quality at the rate of about eight hundred pages a year during most of the years of his life. The quality of these papers may be judged from the fact that he won so many prizes for them that these awards became an almost regular addition to his income. Some of the books and four hundred of his research papers were written after he became totally blind. A current edition of his collected works, when completed, will contain seventy-four volumes.

Unlike Descartes or Newton before him or Cauchy after him, Euler did not open up new branches of mathematics. But no one was so prolific or could so cleverly handle mathematics; no one could muster and utilize the resources of algebra, geometry, and analysis to produce so many admirable results. Euler was superbly inventive in methodology and a skilled technician. One finds his name in all branches of mathematics: there are formulas of Euler, polynomials of Euler, Euler constants, Eulerian integrals, and Euler lines.

One might suspect that such a volume of activity could be carried on only at the expense of all other interests. But Euler married and fathered thirteen children. Always attentive to his family and its welfare he instructed his children and his grandchildren, constructed scientific games for them, and spent evenings reading the Bible to them. He also loved to express himself on matters of philosophy; but here he exhibited his only weakness, for which he was often chided by Voltaire. One day he was forced to confess that he had never studied any philosophy and regretted having believed that one could understand that subject without studying it. But Euler's spirit for philosophic disputes remained undampened and he continued to engage in them. He even enjoyed the sharp criticism he provoked from Voltaire.

Surrounded by universal respect—well merited by the nobility of his character—he could at the end of his life consider as his pupils all the mathematicians of Europe. On September 7, 1783, after having discussed the topics of the day, the Montgolfiers,[1] and the discovery of Uranus, according to the oft-cited words of J. A. N. C. de Condorcet, "He ceased to calculate and to live."

2. The Function Concept

As we have seen, the concept of a function and the simpler algebraic and transcendental functions were introduced and used during the seventeenth century. As Leibniz, James and John Bernoulli, L'Hospital, Huygens, and Pierre Varignon (1654–1722) took up problems such as the motion of a pendulum, the shape of a rope suspended from two fixed points, motion along curved paths, motion with fixed compass bearing on a sphere (the loxodrome), evolutes and involutes of curves, caustic curves arising in the reflection and refraction of light, and the path of a point on one curve that rolls on another, they not only employed the functions already known but arrived at more complicated forms of elementary functions. As a consequence of these researches and general work on the calculus, the elementary functions were fully recognized and developed practically in the manner in which we now have them. For example, the logarithm function, which originated

1. The Montgolfiers were two brothers who in 1783 first successfully made an ascent in a balloon filled with heated air.

as the relationship between terms in a geometric and arithmetic progression and was treated in the seventeenth century as the series obtained by integrating $1/(1 + x)$, (Chap. 17, sec. 2), was introduced on a new basis. The study of the exponential function by Wallis, Newton, Leibniz, and John Bernoulli showed that the logarithm function is the inverse of the exponential function whose properties are relatively simple. In 1742 William Jones (1675–1749) gave a systematic introduction to the logarithm function in this manner (Chap. 13, sec. 2). Euler in his *Introductio* defines the two functions as

$$e^x = \lim_{n \to \infty} \left(1 + \frac{x}{n}\right)^n, \qquad \log x = \lim_{n \to \infty} n(x^{1/n} - 1).$$

The mathematics of the trigonometric functions was also systematized. Newton and Leibniz gave series expansions for these functions. The development of the formulas for the functions of the sum and difference of two angles, that is $\sin (x + y)$, $\sin (x - y)$, and so forth, is due to a number of men, among them John Bernoulli and Thomas Fantet de Lagny (1660–1734); the latter wrote a paper in the *Mémoires* of the Paris Academy for 1703 on this subject. Frédéric-Christian Mayer (dates unknown), one of the first members of the St. Petersburg Academy of Sciences, then derived the common identities of analytical trigonometry on the basis of the sum and difference formulas.[2] Finally Euler, in a prize paper of 1748 on the subject of the inequalities in the motions of Jupiter and Saturn, gave the full systematic treatment of the trigonometric functions.[3] The periodicity of the trigonometric functions is clear in Euler's *Introductio* of 1748 wherein he also introduced the radian measure of angles.[4]

Study of the hyperbolic functions began when it was noticed that the area under a circle was given by $\int \sqrt{a^2 - x^2}\, dx$ whereas the area under the hyperbola is given by $\int \sqrt{x^2 - a^2}\, dx$. Since the two differ by a sign, and the area under a circle can be expressed by the trigonometric functions (let $x = a \sin \theta$), whereas the area under the hyperbola is related to the logarithm function, there should be a relation involving imaginary numbers between the trigonometric functions and the logarithm function. This idea was developed by many men (see sec. 3). Finally, J. H. Lambert studied the hyperbolic functions comprehensively.[5]

The concept of function had been formulated by John Bernoulli. Euler, at the very outset of the *Introductio*, defines a function as any analytical expression formed in any manner from a variable quantity and constants.

2. *Comm. Acad. Sci. Petrop.*, 2, 1727.
3. *Opera*, (2), 25, 45–157.
4. *Opera*, (1), 9, 217–39, 305–7.
5. *Hist. de l'Acad. de Berlin*, 24, 1768, 327–54, pub. 1770 = *Opera Math.*, 2, 245–69.

He includes polynomials, power series, and logarithmic and trigonometric expressions. He also defines a function of several variables. There follows the notion of an algebraic function, in which the operations on the independent variable involve only algebraic operations, which in turn are divided into two classes: the rational, involving only the four usual operations, and the irrational, involving roots. He then introduces the transcendental functions, namely, the trigonometric, the logarithmic, the exponential, variables to irrational powers, and some integrals.

The principal difference among functions, writes Euler, consists in the combination of variables and constants that compose them. Thus, he adds, the transcendental functions are distinguished from the algebraic functions because the former repeat an infinite number of times the combinations of the latter; that is, the transcendental functions could be given by infinite series. Euler and his contemporaries did not regard it as necessary to consider the validity of the expressions obtained by the unending application of the four rational operations.

Euler distinguished between explicit and implicit functions and between single-valued and multiple-valued functions, the latter being roots of higher-degree equations in two variables, the coefficients of which are functions of one variable. Here, he says, if a function, such as $\sqrt[3]{P}$, where P is a one-valued function, has real values for real values of the argument, then most often it can be included among the single-valued functions. From these definitions (which are not free of contradictions), Euler turns to rational integral functions or polynomials. Such functions with real coefficients, he affirms, can be decomposed into first and second degree factors with real coefficients (see sec. 4 and Chap. 25, sec. 2).

By a continuous function, Euler, like Leibniz and the other eighteenth-century writers, meant a function specified by an analytic formula; his word "continuous" really means "analytic" for us, except for an occasional discontinuity as in $y = 1/x$.[6] Other functions were recognized; the curves representing them were called "mechanical" or "freely drawn."

Euler's *Introductio* was the first work in which the function concept was made primary and used as a basis for ordering the material of the two volumes. Something of the spirit of this book may be gathered from Euler's remarks on the expansion of functions in power series.[7] He asserts that any function can be so expanded but then states that "if anyone doubts that every function can be so expanded then the doubt will be set aside by actually expanding functions. However in order that the present investigation extend over the widest possible domain, in addition to the positive integral powers

6. In Vol. 2, Chap. 1 of his *Introductio*, Euler introduces "discontinuous" or mixed functions which require different analytic expressions in different domains of the independent variable. But the concept plays no role in the work.
7. *Opera*, (1), 8, Chap. 4, p. 74.

of z, terms with arbitrary exponents will be admitted. Then it is surely indisputable that every function can be expanded in the form $Az^\alpha + Bz^\beta + Cz^\gamma + Dz^\delta + \cdots$ in which the exponents $\alpha, \beta, \gamma, \delta, \cdots$ can be any numbers." For Euler the possibility of expanding all functions into series was confirmed by his own experience and the experience of all his contemporaries. And in fact, it was true in those days that all functions given by analytic expressions were developable in series.

Though a controversy about the notion of a function did arise in connection with the vibrating-string problem (see Chap. 22) and caused Euler to generalize his own notion of what a function was, the concept that dominated the eighteenth century was still the notion of a function given by a single analytic expression, finite or infinite. Thus Lagrange, in his *Théorie des fonctions analytiques* (1797), defined a function of one or several variables as any expression useful for calculation in which these variables enter in any manner whatsoever. In *Leçons sur le calcul des fonctions* (1806), he says that functions represent different operations that must be performed on known quantities to obtain the values of unknown quantities, and that the latter are properly only the last result of the calculation. In other words, a function is a combination of operations.

3. *The Technique of Integration and Complex Quantities*

The basic method for integrating even somewhat complicated algebraic functions and the transcendental functions—the technique introduced by Newton—was to represent the functions as series and integrate term by term. Little by little, the mathematicians developed the techniques of going from one closed form to another.

The eighteenth-century use of the integral concept was limited. Newton had utilized the derivative and the antiderivative or indefinite integral, whereas Leibniz had emphasized differentials and the summation of differentials. John Bernoulli, presumably following Leibniz, treated the integral as the inverse of the differential, so that if $dy = f'(x)\, dx$, then $y = f(x)$. That is, the Newtonian antiderivative was chosen as the integral, but differentials were used in place of Newton's derivative. According to Bernoulli, the object of the integral calculus was to find from a given relation among *differentials* of variables, the relation of the variables. Euler emphasized that the derivative is the ratio of the evanescent differentials and said that the integral calculus was concerned with finding the function itself. The summation concept was used by him only for the approximate evaluation of integrals. In fact, all of the eighteenth-century mathematicians treated the integral as the inverse of the derivative or of the differential dy. The existence of an integral was never questioned; it was, of course, found explicitly in most of the applications made in the eighteenth century, so that the question did not occur.

A few instances of the development of the technique of integration are worth noting. To evaluate

$$\int \frac{a^2\,dx}{a^2 - x^2}$$

James Bernoulli[8] had used the change of variable

$$x = a\,\frac{b^2 - t^2}{b^2 + t^2};$$

this converts the integral to the form

$$\int \frac{dt}{2at},$$

which is readily integrable as a logarithm function. John Bernoulli noticed in 1702 and published in the *Mémoires* of the Academy of Sciences of that year[9] the observation that

$$\frac{a^2}{a^2 - x^2} = \frac{a}{2}\left(\frac{1}{a + x} + \frac{1}{a - x}\right),$$

so that the integration can be performed at once. Thus the method of partial fractions was introduced. This method was also noted independently by Leibniz in the *Acta Eruditorum* of 1702.[10]

In correspondence between John Bernoulli and Leibniz, the method was applied to

$$\int \frac{dx}{ax^2 + bx + c}.$$

However, since the linear factors of $ax^2 + bx + c$ could be complex, the method of partial fractions led to integrals of the form

$$\int \frac{dx}{cx + d}$$

in which d at least was a complex number. Both Leibniz and John Bernoulli nevertheless integrated by using the logarithm rule and so involved the logarithms of complex numbers. Despite the confusion about complex numbers, neither hesitated to integrate in this manner. Leibniz said the presence of complex numbers did no harm.

John Bernoulli employed them repeatedly. In a paper published in 1702[11] he pointed out that, just as $adz/(b^2 - z^2)$ goes over by means of the

8. *Acta Erud.*, 1699 = *Opera*, 2, 868–70.
9. *Opera*, 1, 393–400.
10. *Math. Schriften*, 5, 350–66.
11. *Mém. de l'Acad. des Sci.*, Paris, 1702, 289 ff. = *Opera*, 1, 393–400.

substitution $z = b(t - 1)/(t + 1)$ into $adt/2bt$, so the differential

$$\frac{dz}{b^2 + z^2}$$

goes over by the substitution $z = \sqrt{-1}\, b(t - 1)/(t + 1)$ to

$$\frac{-dt}{\sqrt{-1}\, 2bt},$$

and the latter is the differential of the logarithm of an imaginary number. Since the original integral also leads to the arc tan function, Bernoulli had thus established a relation between the trigonometric and logarithmic functions.

However, these results soon raised lively discussions about the nature of the logarithms of negative and complex numbers. In his article of 1712[12] and in an exchange of letters with John Bernoulli during the years 1712–13, Leibniz affirmed that the logarithms of negative numbers are nonexistent (he said imaginary), while Bernoulli sought to prove that they must be real. Leibniz's argument was that positive logarithms are used for numbers greater than 1 and negative logarithms for numbers between 0 and 1. Hence there could be no logarithm for negative numbers. Moreover, if there were a logarithm for -1, then the logarithm of $\sqrt{-1}$ would be half of it; but surely there was no logarithm for $\sqrt{-1}$. That Liebniz could argue in this manner after having introduced the logarithms of complex numbers in integration is inexplicable. Bernoulli argued that since

(1) $$\frac{d(-x)}{-x} = \frac{dx}{x},$$

then $\log (-x) = \log x$; and since $\log 1 = 0$, so is $\log (-1)$. Leibniz countered that $d(\log x) = dx/x$ holds only for positive x. A second round of correspondence and disagreement took place between Euler and John Bernoulli during the years 1727–31. Bernoulli maintained his position, while Euler disagreed with it, though, at the time, he had no consistent position of his own.

The final clarification of what the logarithm of a complex number is became possible by virtue of related developments that are themselves significant and that led to the relationship between the exponential and the trigonometric functions. In 1714 Roger Cotes (1682–1716) published[13] a theorem on complex numbers, which, in modern notation, states that

(2) $$\sqrt{-1}\, \phi = \log_e (\cos \phi + \sqrt{-1} \sin \phi).$$

12. *Acta Erud.*, 1712, 167–69 = *Math. Schriften*, 5, 387–89.
13. *Phil. Trans.*, 29, 1714, 5–45.

In a letter to John Bernoulli of October 18, 1740, Euler stated that $y = 2 \cos x$ and $y = e^{\sqrt{-1}\,x} + e^{-\sqrt{-1}\,x}$ were both solutions of the same differential equation (which he recognized through series solutions), and so must be equal. He published this observation in 1743,[14] namely,

$$(3) \qquad \cos s = \frac{e^{\sqrt{-1}\,s} + e^{-\sqrt{-1}\,s}}{2}, \qquad \sin s = \frac{e^{\sqrt{-1}\,s} - e^{-\sqrt{-1}\,s}}{2\sqrt{-1}}.$$

In 1748 he rediscovered the result (2) of Cotes, which would also follow from (3).

While this development was taking place, Abraham de Moivre (1667–1754), who left France and settled in London when the Edict of Nantes protecting Huguenots was revoked, obtained, at least implicitly, the formula now named after him. In a note of 1722, which utilizes a result already published in 1707,[15] he says that one can obtain a relation between x and t, which represent the versines of two arcs (vers $\alpha = 1 - \cos \alpha$) that are in the ratio of 1 to n, by eliminating z from the two equations

$$1 - 2z^n + z^{2n} = -2z^n t \quad \text{and} \quad 1 - 2z + z^2 = -2zx.$$

In this result the de Moivre formula is implicit, for if one sets $x = 1 - \cos \phi$ and $t = 1 - \cos n\phi$, one can derive

$$(4) \qquad (\cos \phi \pm \sqrt{-1} \sin \phi)^n = \cos n\phi \pm \sqrt{-1} \sin n\phi.$$

For de Moivre n was an integer > 0. Actually, he never wrote the last result explicitly; the final formulation is due to Euler[16] and was generalized by him to all real n.

By 1747 Euler had enough experience with the relationship between exponentials, logarithms, and trigonometric functions to obtain the correct facts about the logarithms of complex numbers. In an article of 1749, entitled "De la controverse entre Mrs. [Messrs] Leibnitz et Bernoulli sur les logarithmes négatifs et imaginaires,"[17] Euler disagrees with Leibniz's counterargument that $d(\log x) = dx/x$ for positive x only. He says that Leibniz's objection, if correct, shatters the foundation of all analysis, namely, that the rules and operations apply no matter what the nature of the objects to which they are applied. He affirms that $d(\log x) = dx/x$ is correct for positive and negative x but adds that Bernoulli forgets that all one can conclude from (1) above is that $\log (-x)$ and $\log x$ differ by a constant. This constant must be $\log (-1)$ because $\log (-x) = \log (-1 \cdot x) = \log (-1) + \log x$. Hence, says Euler, Bernoulli has assumed, in effect, that $\log (-1) = 0$, but this must be proved. Bernoulli had given other arguments, which

14. *Miscellanea Berolinensia*, 7, 1743, 172–92 = *Opera*, (1), 14, 138–55.
15. *Phil. Trans.*, 25, 1707, 2368–71.
16. *Introductio*, Chap. 8.
17. *Hist. de l'Acad. de Berlin*, 5, 1749, 139–79, pub. 1751 = *Opera*, (1), 17, 195–232.

Euler also answers. For example, Bernoulli had argued that since $(-a)^2 = a^2$, then $\log (-a)^2 = \log a^2$ and so $2 \log (-a) = 2 \log a$ or $\log (-a) = \log a$. Euler counters that, since $(a\sqrt{-1})^4 = a^4$, then $\log a = \log (a\sqrt{-1}) = \log a + \log \sqrt{-1}$ and so in this case, presumably $\log \sqrt{-1}$ would have to be 0. But, Euler says, Bernoulli himself has proved in another connection that $\log \sqrt{-1} = \sqrt{-1}\, \pi/2$.

Leibniz had argued that since

(5) $$\log (1 + x) = x - \frac{1}{2} x^2 + \frac{1}{3} x^3 - \frac{1}{4} x^4 + \cdots$$

then for $x = -2$

$$\log (-1) = -2 - \frac{4}{2} - \frac{8}{3} \cdots,$$

from which one sees at least that $\log (-1)$ is not 0 (in fact, Leibniz had said that $\log (-1)$ is nonexistent). Euler's answer to this argument is that from

$$\frac{1}{1 + x} = 1 - x + x^2 - x^3 + x^4 \cdots,$$

for $x = -3$ one obtains

$$-\frac{1}{2} = 1 + 3 + 9 + 27 + \cdots$$

and for $x = 1$ one obtains

$$\frac{1}{2} = 1 - 1 + 1 - 1 + \cdots,$$

so that by adding the left and right sides

$$0 = 2 + 2 + 10 + 26 + \cdots.$$

Hence, says Euler, the argument from series proves nothing.

After refuting Leibniz and Bernoulli, Euler gives what is, by present standards, an incorrect argument. He writes

$$x = e^y = \left(1 + \frac{y}{i}\right)^i,$$

wherein i is an infinitely large number.[18] Then

$$x^{1/i} = 1 + \frac{y}{i}$$

18. In his earlier work Euler used i (the first letter of *infinitus*) for an infinitely large quantity. After 1777 he used i for $\sqrt{-1}$.

and so

$$y = i(x^{1/i} - 1).$$

Since $x^{1/i}$, "the root with infinitely large exponent i," has infinitely many complex values, y has such values, and since $y = \log x$, then so does $\log x$. In fact Euler now writes[19]

$$x = a + b\sqrt{-1} = c(\cos \phi + \sqrt{-1} \sin \phi).$$

Letting c be e^C he has

$$x = e^C(\cos \phi + i \sin \phi) = e^C e^{\sqrt{-1}(\phi \pm 2\lambda\pi)},$$

and so

(6) $$y = \log x = C + (\phi \pm 2\lambda\pi)\sqrt{-1}$$

where λ is a positive integer or zero. Thus, Euler affirms, for positive real numbers only one value of the logarithm is real and the others are all imaginary; but for negative real numbers and for imaginary numbers all values of the logarithm are imaginary. Despite this successful resolution of the problem, Euler's work was not accepted. D'Alembert advanced metaphysical, analytical, and geometrical arguments to show that $\log (-1) = 0$.

4. Elliptic Integrals

John Bernoulli, having succeeded in integrating some rational functions by the method of partial fractions, asserted in the *Acta Eruditorum* of 1702 that the integral of any rational function need not involve any other transcendental functions than trigonometric or logarithmic functions. Since the denominator of a rational function can be an nth degree polynomial in x, the correctness of the assertion depended on whether any polynomial with real coefficients can be expressed as a product of first and second degree factors with real coefficients. Leibniz in his paper in the *Acta* of 1702 thought this was not possible and gave the example of $x^4 + a^4$. He pointed out that

$$\begin{aligned} x^4 + a^4 &= (x^2 - a^2\sqrt{-1})(x^2 + a^2\sqrt{-1}) \\ &= (x + a\sqrt{\sqrt{-1}})(x - a\sqrt{\sqrt{-1}}) \\ &\times (x + a\sqrt{-\sqrt{-1}})(x - a\sqrt{-\sqrt{-1}}) \end{aligned}$$

and, he said, no two of these four factors multiplied together give a quadratic factor with real coefficients. Had he been able to express the square root of

19. $a + b\sqrt{-1} = \sqrt{a^2 + b^2}\left(\dfrac{a}{\sqrt{a^2 + b^2}} + \sqrt{-1}\,\dfrac{b}{\sqrt{a^2 + b^2}}\right) = c(\cos \phi + i \sin \phi)$.

$\sqrt{-1}$ and $-\sqrt{-1}$ as ordinary complex numbers he would have seen his error. Nicholas Bernoulli (1687–1759), a nephew of James and John, pointed out in the *Acta Eruditorum* of 1719 that $x^4 + a^4 = (a^2 + x^2)^2 - 2a^2x^2 = (a^2 + x^2 + ax\sqrt{2})(a^2 + x^2 - ax\sqrt{2})$; thus the function $1/(x^4 + a^4)$ could be integrated in terms of trigonometric and logarithmic functions.

The integration of irrational functions was also considered. James Bernoulli and Leibniz corresponded on this subject, because such integrands were being encountered frequently. James in 1694[20] was concerned with the elastica, the shape assumed by a thin rod when forces are applied to it—as, for example, at its ends. For one set of end-conditions he found that the equation of the curve is given by

$$dy = \frac{(x^2 + ab)\,dx}{\sqrt{a^4 - (x^2 + ab)^2}};$$

he could not integrate it in terms of the elementary functions. In connection with this work he introduced the lemniscate, whose rectangular coordinate equation is $(x^2 + y^2)^2 = a^2(x^2 - y^2)$ and whose polar coordinate equation is $r^2 = a^2 \cos 2\theta$. James tried to find the arc length, which from the vertex to an arbitrary point on the curve is given by

$$s = \int_0^r \frac{a^2}{\sqrt{a^4 - r^4}}\,dr,$$

and surmised that this integral, too, could not be integrated in terms of the elementary functions. Seventeenth-century attempts to rectify the ellipse, whose arc length is important for astronomy, led to the problem of evaluating

$$s = a \int_0^t \frac{(1 - k^2 t^2)\,dt}{\sqrt{(1 - t^2)(1 - k^2 t^2)}},$$

when the equation of the ellipse is taken as

$$\frac{x^2}{a^2} + \frac{y^2}{b^2} = 1$$

and in the integrand $k = (a^2 - b^2)/a^2$ and $t = x/a$. The problem of finding the period of a simple pendulum led to the integral

$$T = 4\sqrt{\frac{l}{g}} \int_0^{\pi/2} \frac{d\phi}{\sqrt{1 - k^2 \sin^2 \phi}}.$$

Such irrational integrands also occurred in finding the length of arc of a hyperbola, the trigonometric functions, and other curves. These integrals were already known by 1700; and others involving such integrands kept occurring throughout the eighteenth century. Thus Euler, in a definitive

20. *Acta Erud.*, 1694, 262–76 = *Opera*, 2, 576–600.

treatment of the elastica in the appendix to his 1744 book on the calculus of variations, obtained

$$dy = \frac{(\alpha + \beta x + \gamma x^2)\, dx}{\sqrt{a^4 - (\alpha + \beta x + \gamma x^2)^2}}$$

where the constants are for us immaterial. Like his predecessors, he resorted to series in order to obtain physical results.

The class of integrals comprised by the above examples is known as elliptic, the name coming from the problem of finding the length of arc of an ellipse. The eighteenth-century men did not know it, but these integrals cannot be evaluated in terms of the algebraic, circular, logarithmic, or exponential functions.[21]

The first investigations of elliptic integrals were directed not so much toward attempts to evaluate the integrals as toward the reduction of the more complicated ones to those arising in the rectification of ellipse and hyperbola. The reason for this approach is that from the geometrical point of view, which dominated at that time, the integrals for the elliptic and hyperbolic arcs seemed to be the simplest ones. A new point of view was opened up by the observation that the differential equation

(7) $$f(x)\, dx = \pm f(y)\, dy,$$

where $\int f(x)\, dx$ is a logarithmic function or an inverse trigonometric function, has as an integral an algebraic function of x and y; that is, in spite of the fact that it is impossible to find an algebraic integral of $f(x)\, dx$ itself, one can find an algebraic integral of the sum or difference of two such differentials. John Bernoulli then asked whether this property might not hold for other integrals than those of logarithms or inverse trigonometric functions.[22] He had discovered in 1698 that the difference of two arcs of the cubical parabola $(y = x^3)$ is integrable, a result he had obtained accidentally and regarded as most elegant. He then posed the more general problem of finding arcs of (higher) parabolas, ellipses, and hyperbolas whose sum or difference is equal to a rectilinear quantity, and affirmed that parabolic curves of the form $a^m y^p = b^n x^q$, $m + p = n + q$, are such that arcs of such curves, added or subtracted, equal a straight line. He gave no proof.

Count Giulio Carlo de' Toschi di Fagnano (1682–1766), an amateur mathematician, began in 1714 to take up these problems.[23] He considered the curves $y = (2/m + 2)x^{(m+2)/2}/a^{m/2}$ with m rational. For such curves it is rather straightforward to show (Fig. 19.1) that

$$\frac{m}{m+2} \cdot \int_{x_0}^{x_1} \frac{dx}{\sqrt{1 + (x/a)^m}} = \text{arc } PP_1 - (P_1 R_1 - PR)$$

21. This was proved by Liouville (*Jour. de l'Ecole Poly.*, 14, 1833, 124–93).
22. *Acta Erud.*, Oct. 1698, 462 ff. = *Opera*, 1, 249–53.
23. *Giornale dei Letterati d'Italia*, Vols. 19 ff.

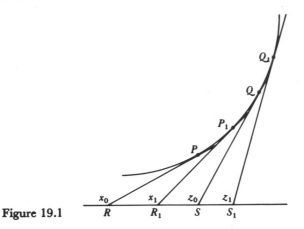

Figure 19.1

where x_0 and x_1 are the abscissas of R and R_1, and PR and P_1R_1 are the tangents at P and P_1 respectively. Likewise

$$\frac{m}{m+2}\int_{z_0}^{z_1}\frac{dz}{\sqrt{1+(z/a)^m}} = \text{arc } QQ_1 - (Q_1S_1 - QS).$$

If, therefore, for some relation of x to z we have

(8) $$\frac{dx}{\sqrt{1+(x/a)^m}} + \frac{dz}{\sqrt{1+(z/a)^m}} = 0,$$

then the sum of the two definite integrals would be 0, and we would have

(9) $$\text{arc } QQ_1 - \text{arc } PP_1 = (Q_1S_1 - QS) - (P_1R_1 - PR).$$

A solution of (8) for $m = 4$ is

(10) $$\frac{x}{a}\cdot\frac{z}{a} = 1.$$

Then for $m = 4$, on the curve $y = x^3/3a^2$, the difference of two arcs whose end-values x and z are related by (10) is expressible as a straight line segment. Fagnano also obtained integrals of (8) for $m = 6$ and $m = 3$.

Fagnano showed further that on the ellipse, as well as on the hyperbola, one can find infinitely many arcs whose difference can be expressed algebraically, even though the individual arcs cannot be rectified. Thus in 1716 he showed that the difference of any two elliptic arcs is algebraic. Analytically he had

(11) $$\frac{\sqrt{hx^2+l}}{\sqrt{fx^2+g}}\,dx + \frac{\sqrt{hz^2+l}}{\sqrt{fz^2+g}}\,dz = 0$$

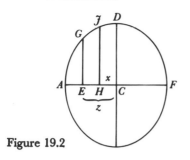

Figure 19.2

or, for brevity,

$$X \, dx + Z \, dz = 0$$

where h, l, f, g, x, and z satisfy the condition

(12) $$fhx^2 z^2 + flx^2 + flz^2 + gl = 0.$$

Fagnano showed that

(13) $$\int X \, dx + \int Z \, dz = -\frac{hxz}{\sqrt{-fl}}.$$

What this means geometrically is that if $2a$ is the minor axis FA of an ellipse (Fig. 19.2), $CH = x$, $CE = z$, JH is the ordinate at H, and GE, the ordinate at E, then

(14) $$\text{arc } JD + \text{arc } DG = \frac{-hxz}{2a^2} + C.$$

(To identify this with the integrals, let p be the parameter [latus rectum] of the ellipse, let $p - 2a = h$, $l = 2a^3$, $f = -2a$, $g = 2a^3$. Then z is $a\sqrt{2a^3 - 2ax^2}/\sqrt{2a^3 + hx^2}$.) When $x = 0$, arc JD vanishes and the algebraic term in (14) vanishes. By (12), $z = a$, and so arc DG becomes arc DA. This is the value of C. Then one can say

$$\text{arc } JD + \text{arc } GD = \frac{-hxz}{2a^2} + \text{arc } DA$$

or

$$\text{arc } JD - \text{arc } GA = \frac{-hxz}{2a^2}.$$

One result from this work,[24] still called Fagnano's theorem and obtained in 1716, states the following: Let

$$\frac{x^2}{a^2} + \frac{y^2}{b^2} = 1$$

24. *Opere*, 2, 287–92.

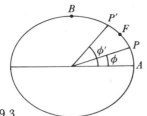

Figure 19.3

be the ellipse of eccentricity e and let $P(x, y)$ and $P'(x', y')$ be two points (Fig. 19.3) whose eccentric angles ϕ and ϕ' satisfy the condition

(15) $$\tan \phi \tan \phi' = \frac{b}{a}.$$

Then the theorem states that

(16) $$\text{arc } BP + \text{arc } BP' - \text{arc } BA = e^2xx'/a.$$

The points P and P' may coincide (while satisfying [15]). For this common position F, called Fagnano's point, he showed that

(17) $$\text{arc } BF - \text{arc } AF = a - b.$$

From 1714 on Fagnano also concerned himself with the rectification of the lemniscate by means of elliptic and hyperbolic arcs.

In 1717 and 1720 Fagnano integrated other combinations of differentials. Thus he showed that

(18) $$\frac{dx}{\sqrt{1 - x^4}} = \frac{dy}{\sqrt{1 - y^4}}$$

has the integral

(19) $$x = -\sqrt{\frac{1 - y^2}{1 + y^2}}$$

or

(20) $$x^2 + y^2 + x^2y^2 = 1.$$

One way of stating the result is: Between two integrals that express arcs of a lemniscate (for which $a = 1$), an algebraic relation exists, even though each integral separately is a transcendental function of a new kind.

Fagnano then proceeded to establish a number of similar relations in order to obtain special results about the lemniscate.[25] For example, he

25. *Giornale dei Letterati d'Italia*, 30, 1718, 87 ff. = *Opere*, 2, 304–13.

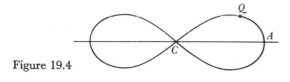

Figure 19.4

showed that if

(21)
$$\frac{dx}{\sqrt{1 - x^4}} = \frac{2\,dy}{\sqrt{1 - y^4}}$$

then

(22)
$$\frac{\sqrt{1 - y^4}}{y\sqrt{2}} = \frac{\sqrt{1 - x}}{\sqrt{1 + x}},$$

or by solving for x,

(23)
$$x = \frac{-1 + 2y^2 + y^4}{1 + 2y - y^4}.$$

From a variety of such results Fagnano showed how to find the points on the lemniscate (the values of r in $r^2 = a^2 \cos 2\theta$) that divide the quadrant, that is, the arc CQA in Figure 19.4, into n equal parts for certain values of n. He also showed how, given an arc CS, one can find the point I on this arc that divides CS into two equal parts. Further, he found the points on arc CQA which, when joined to C, divide the area under CQA and the horizontal axis into two, three, and five parts; and given the chords that divide this area into n equal parts, he found the chords bisecting each of these parts.

Thus Fagnano had done more than answer Bernoulli's question; he had shown that the same remarkable algebraic property that characterized the integrals representing logarithmic and inverse trigonometric functions held for at least certain classes of elliptic integrals.

About 1750 Euler noted Fagnano's work on the ellipse, hyperbola, and lemniscate and began a series of investigations of his own. In a paper, "Observationes de Comparatione Arcuum Curvarum Irrectificabilium,"[26] Euler, after repeating some of Fagnano's work, showed how to divide the area of a quadrant of the lemniscate into $n + 1$ parts if it has already been divided into n parts. He then points out that his and Fagnano's work has furnished some useful results on integration, for the equation (18) has, besides the obvious integral $x = y$, the additional particular integral

$$x = -\sqrt{(1 - y^2)/(1 + y^2)}.$$

26. *Novi Comm. Acad. Sci. Petrop.*, 6, 1756/7, 58–84, pub. 1761 = *Opera*, (1), 20, 80–107.

In his paper "De Integratione Æquationis Differentialis $m\,dx/\sqrt{1-x^4}$ $= n\,dy/\sqrt{1-y^4}$,"[27] Euler takes Fagnano's results as his point of departure. The integrals Fagnano had obtained for most of the differential equations he had considered were particular integrals. These were algebraic; but the complete integral might well be transcendental. Euler decided to look for complete integrals in algebraic form. He started with (18), but hoped to obtain the complete integral of

$$(24) \qquad \frac{m\,dx}{\sqrt{1-x^4}} = \frac{n\,dy}{\sqrt{1-y^4}}.$$

Here m/n is rational and the equation expresses the problem of finding two arcs of a lemniscate that have this ratio to each other. Euler says that he has been led by trial to believe that (24) has a complete integral that can be expressed algebraically when m/n is rational.

From the investigations of Fagnano it followed that equation (18) is satisfied by the particular integral (19) or (20). The integral of each side of (18) is an arc of a lemniscate with half-axis 1 and abscissa x, and the integration of the ordinary differential equation (18) amounts to finding two arcs that are of equal length. Euler had pointed out that $x = y$ is another particular integral of (18). The complete integral should be such that it reduces to each of these particular integrals for special values of the arbitrary constant. Guided by these facts, Euler found that the complete integral of (18) is

$$(25) \qquad x^2 + y^2 + c^2y^2x^2 = c^2 + 2xy\sqrt{1-c^4}$$

or

$$(26) \qquad x = \frac{y\sqrt{1-c^4} \pm c\sqrt{1-y^4}}{1 + c^2y^2}$$

where c is an arbitrary constant. Of course, given (25) we can readily verify that it is the complete integral of (18).

Implicit in the result (25) is what is often called Euler's addition theorem for these simple elliptic integrals. It is clear by straight differentiation that

$$(27) \qquad \int_0^x \frac{dx}{\sqrt{1-x^4}} = \int_0^y \frac{dx}{\sqrt{1-x^4}} + \int_0^c \frac{dx}{\sqrt{1-x^4}},$$

where c is a constant, is also a complete integral of (18). Hence x, y, and c must be related by (25). Thus the addition theorem says that if (27) holds for the elliptic integrals therein, then the upper limit x is an algebraic symmetric function, namely (26), of the arbitrarily chosen upper limits y and c of the other two integrals. The addition theorem applies to more general integrals, as we shall see.

27. *Novi Comm. Acad. Sci. Petrop.*, 6, 1756/7, 37–57, pub. 1761 = *Opera*, (1), 20, 58–79.

By utilizing these results (25) and (27), it is rather straightforward to show that if

$$(28) \qquad \int_0^y \frac{dx}{\sqrt{1 - x^4}} = n \int_0^x \frac{dx}{\sqrt{1 - x^4}},$$

then y is an algebraic function of x. This result is called Euler's multiplication theorem for the elliptic integral $\int_0^x dx/\sqrt{1 - x^4}$. From this result the complete integral of equation (28) follows; the important point is that it is an algebraic equation in x, y, and an arbitrary constant c. Euler shows how the complete integral might be obtained but does not give it explicitly.

In the same paper of 1756–57 and in Volume 7 of the same journal,[28] Euler tackled more general elliptic integrals. He gives the following result, which he says he obtained by trial and error. If one differentiates

$$(29) \quad \alpha + 2\beta(x + y) + \gamma(x^2 + y^2) + 2xy + 2\varepsilon xy(x + y) + \zeta x^2 y^2 = 0,$$

then the differential equation can be put in the form

$$(30) \qquad \frac{dx}{\sqrt{X}} + \frac{dy}{\sqrt{Y}} = 0$$

where X and Y are polynomials of the fourth degree of which four coefficients (the same in X and Y) can be expressed in terms of the five in (29) with the help of an arbitrary constant. Then (29) is the complete integral of (30), and when (30) is specialized to (18), then (29) becomes (25). Euler remarks that it is wonderful that, even though an integral of $dx/\sqrt{X}$ cannot be obtained in terms of circular or logarithmic functions, the equation (30) is satisfied by an algebraic relation. He then generalizes the results to

$$(31) \qquad \frac{m\, dx}{\sqrt{X}} = \frac{n\, dy}{\sqrt{Y}}, \qquad m/n \text{ rational,}$$

where X and Y are fourth degree polynomials with the same coefficients. The result is also in Euler's *Institutiones Calculi Integralis*,[29] where he shows what the results mean geometrically in terms of ellipse, hyperbola, and lemniscate.

From these results Euler was able to proceed with what is now called the addition theorem for elliptic integrals of the first kind. Consider the elliptic integral

$$(32) \qquad \int \frac{dx}{\sqrt{R(x)}}$$

28. *Novi Comm. Acad. Sci. Petrop.*, 7, 1758/9, 3–48, pub. 1761 = *Opera*, (1), 20, 153–200.
29. Vol. 1, Sec. 2, Chap. 6 = *Opera*, (1), 11, 391–423.

where $R(x) = Ax^4 + Bx^3 + Cx^2 + Dx + E$. Then the addition theorem states that the equation

$$(33) \qquad \frac{dx}{\sqrt{R(x)}} = \frac{dy}{\sqrt{R(y)}}$$

is satisfied by a certain definite algebraic equation in x and y such that y is rationally expressible in terms of x, the corresponding value of $\sqrt{R(x)}$, the arbitrary constants x_0 and y_0, and the corresponding values of $\sqrt{R(x_0)}$ and $\sqrt{R(y_0)}$. Also y takes on the arbitrarily preassigned value y_0 when x takes on the arbitrary value x_0.

This result leads readily to another theorem, which may be more enlightening. If the sum or the difference of two elliptic integrals of the form

$$(34) \qquad \int \frac{dx}{\sqrt{R(x)}}$$

is equated to a third integral of the same form, and if further the lower limit of integration and the coefficients under the radical are the same for all three integrals, then the upper limit of integration of the third integral is an algebraic function of the two other upper limits, the common lower limit, and the corresponding values of $\sqrt{R(x)}$ at the two upper limits and the lower limit.

Euler went further. Just as Fagnano's treatment of the difference of two lemniscate arcs led him to the general elliptic integral of the first kind, so Fagnano's treatment of the difference of two elliptic arcs (see [11]) led Euler to an addition theorem for integrals of a second kind.[30] He expressed regret that his methods were not extensible to higher roots than the square root or to radicands of higher than fourth degree. He also saw a great defect in his work in that he had not obtained his complete algebraic integrals by a general method of analysis; consequently his results were not naturally related to other parts of the calculus.

The definitive work on elliptic integrals was done by Adrien-Marie Legendre (1752–1833), a professor at the Ecole Militaire who served on several governmental committees; he later became an examiner of students at the Ecole Polytechnique. Up to his death in 1833 he never ceased to work with passion and regularity. His name lives in a great number of theorems, very varied, because he tackled most diverse questions. But he had neither the originality nor the profundity of Lagrange, Laplace, and Monge. Legendre's work gave birth to very important theories, but only after it was taken over by more powerful minds. He ranks just after the three contemporaries just mentioned.

30. *Novi Comm. Acad. Sci. Petrop.*, 7, 1758/9, 3–48, pub. 1761 = *Opera*, (1), 20, 153–200 and *Inst. Cal. Integ.*, 1, ¶645 = *Opera*, (1), 11, ¶645.

Euler's addition theorems were the main results of the theory of elliptic integrals when Legendre took up the subject in 1786. For four decades he was the only man who added investigations concerning these integrals to the literature. He devoted two basic papers to the subject,[31] then wrote the *Exercices de calcul intégral* (3 vols., 1811, 1817, 1826), the *Traité des fonctions elliptiques*[32] (2 vols., 1825–26), and three supplements giving accounts of the work of Abel and Jacobi in 1829 and 1832. Euler's results, like Fagnano's, were tied to geometrical considerations; Legendre concentrated on the analysis proper.

Legendre's chief result, which is in his *Traité*, was to show that the general elliptic integral

$$(35) \qquad \int \frac{P(x)}{\sqrt{R(x)}}\, dx$$

where $P(x)$ is any rational function of x and $R(x)$ is the usual general fourth degree polynomial, can be reduced to three types:

$$(36) \qquad \int \frac{dx}{\sqrt{1 - x^2}\sqrt{1 - l^2 x^2}},$$

$$(37) \qquad \int \frac{x^2\, dx}{\sqrt{1 - x^2}\sqrt{1 - l^2 x^2}},$$

$$(38) \qquad \int \frac{dx}{(x - a)\sqrt{1 - x^2}\sqrt{1 - l^2 x^2}}.$$

Legendre designated these three types as elliptic integrals of the first, second, and third kinds, respectively.

He also showed that by further transformations these three integrals can be reduced to the three forms

$$(39) \qquad F(k, \phi) = \int_0^\phi \frac{d\phi}{\sqrt{1 - k^2 \sin^2 \phi}}, \qquad 0 < k < 1$$

$$(40) \qquad E(k, \phi) = \int_0^\phi \sqrt{1 - k^2 \sin^2 \phi}\, d\phi, \qquad 0 < k < 1$$

$$(41) \qquad \pi(n, k, \phi) = \int_0^\phi \frac{d\phi}{(1 + n \sin^2 \phi)\sqrt{1 - k^2 \sin^2 \phi}}, \qquad 0 < k < 1$$

where n is any constant. In these forms one sees that the values of the integrals from $\phi = 0$ to $\phi = \pi/2$ are repeated but in reverse order from $\phi = \pi/2$ to

31. *Hist. de l'Acad. des Sci., Paris*, 1786, 616–43 and 644–83.
32. The use of the word "function" in this text is misleading. He studied elliptic integrals and at times those with variable upper limits. These are, of course, functions of the upper limits. But the term "elliptic functions" refers today to the functions introduced later by Abel and Jacobi.

$\phi = \pi$. The notation $\Delta(k, \phi)$ for $\sqrt{1 - k^2 \sin^2 \phi}$ was also introduced by Legendre.

These forms can be converted, by the change of variable $x = \sin \phi$, to the Jacobi forms

$$(42) \qquad F(k, x) = \int_0^x \frac{dx}{\sqrt{1 - x^2}\sqrt{1 - k^2 x^2}}$$

$$(43) \qquad E(k, x) = \int_0^x \sqrt{\frac{1 - k^2 x^2}{1 - x^2}}\, dx$$

$$(44) \qquad \pi(n, k, x) = \int_0^x \frac{dx}{(1 + nx^2)\sqrt{(1 - x^2)(1 - k^2 x^2)}} \, .$$

The quantity k is called the modulus of each elliptic integral. If the limits of integration are $\phi = \pi/2$ or $x = 1$, then the integrals are called complete; otherwise, incomplete.

There was much merit in Legendre's work on elliptic integrals. He drew many inferences, previously unstated, from the work of his predecessors and organized the mathematical subject; but he did not add any basic ideas, nor did he attain the new insight of Abel and Jacobi (Chap. 27, sec. 6) who inverted these integrals and so conceived of the elliptic *functions*. Legendre did get to know the work of Abel and Jacobi and praised them, with much humility and, probably, some bitterness. In devoting the supplements to his 1825 work to their new ideas, he understood very well that this material threw into the shade all he had done on the subject. He had overlooked one of the great discoveries of his epoch.

5. *Further Special Functions*

The elliptic indefinite integrals are new transcendental functions. As the analytical work of the eighteenth century developed, more transcendental functions were obtained. Of these the most important is the gamma function, which arose from work on two problems, interpolation theory and anti-differentiation. The problem of interpolation had been considered by James Stirling (1692–1770), Daniel Bernoulli, and Christian Goldbach (1690–1764). It was posed to Euler and he announced his solution in a letter of October 13, 1729, to Goldbach.[33] A second letter, of January 8, 1730, brought in the integration problem.[34] In 1731 Euler published results on both in a paper, "De Progressionibus. . . ."[35]

33. Fuss, *Correspondance*, 1, 3–7.
34. Fuss, *Correspondance*, 1, 11–18.
35. *Comm. Acad. Sci. Petrop.*, 5, 1730/1, 36–57, pub. 1738 = *Opera*, (1), 14, 1–24.

The interpolation problem was to give meaning to $n!$ for nonintegral values of n. Euler noted that

$$(45) \qquad n! = \left[\left(\frac{2}{1}\right)^n \frac{1}{n+1}\right]\left[\left(\frac{3}{2}\right)^n \frac{2}{n+2}\right]\left[\left(\frac{4}{3}\right)^n \frac{3}{n+3}\right]\cdots$$

$$= \prod_{k=1}^{\infty} \left(\frac{k+1}{k}\right)^n \frac{k}{k+n}.$$

The equation is seen to be formally correct if one cancels common factors in the infinite product. However, this analytic expression for $n!$, unlike the basic definition $n(n-1)\cdots 2\cdot 1$, has a meaning for all n except negative integral values. Euler noticed that for $n = 1/2$ the right side yields, after a bit of manipulation, the infinite product of Wallis

$$(46) \qquad \frac{\pi}{2} = \left(\frac{2\cdot 2}{1\cdot 3}\right)\left(\frac{4\cdot 4}{3\cdot 5}\right)\left(\frac{6\cdot 6}{5\cdot 7}\right)\left(\frac{8\cdot 8}{7\cdot 9}\right)\cdots.$$

In the notation $\Gamma(n+1) = n!$, introduced later by Legendre, Euler also showed that $\Gamma(n+1) = n\Gamma(n)$ and so obtained $\Gamma(3/2)$, $\Gamma(5/2)$, and so forth.

Euler could have used (45) as his generalization of the factorial concept. In fact, it is often introduced today in the equivalent form, which Euler also gave, namely,

$$(47) \qquad \lim_{m\to\infty} \frac{m!\,(m+1)^n}{(n+1)(n+2)\cdots(n+m)}.$$

But the connection with Wallis's result led Euler to take up an integral already considered by Wallis, namely,

$$(48) \qquad \int_0^1 x^e(1-x)^n\,dx$$

wherein for Euler e and n are now arbitrary. Euler evaluated this integral by expanding $(1-x)^n$ by the binomial theorem and obtained

$$(49) \quad \int_0^1 x^e(1-x)^n\,dx$$

$$= \frac{1}{e+1} - \frac{n}{1(e+2)} + \frac{n(n-1)}{1\cdot 2(e+3)} - \frac{n(n-1)(n-2)}{1\cdot 2\cdot 3(e+4)} + \cdots.$$

For $n = 0, 1, 2, 3, \ldots$ the sums on the right side are respectively

$$(50) \quad \frac{1}{e+1}, \frac{1}{(e+1)(e+2)}, \frac{1\cdot 2}{(e+1)(e+2)(e+3)},$$

$$\frac{1\cdot 2\cdot 3}{(e+1)(e+2)(e+3)(e+4)}, \ldots.$$

Hence for positive integral n Euler found that

$$(51) \qquad \int_0^1 x^e (1 - x)^n \, dx = \frac{n!}{(e + 1)(e + 2) \cdots (e + n + 1)}.$$

Euler now sought an expression for $n!$ with n arbitrary. By a series of transformations that would not be wholly acceptable to us today, Euler arrived at

$$(52) \qquad n! = \int_0^1 (-\log x)^n \, dx.$$

This integral has meaning for almost all arbitrary n. It is called the second Eulerian integral, or, as Legendre later called it, the gamma function, and is denoted by $\Gamma(n + 1)$. [Gauss let $\pi(n) = \Gamma(n + 1)$]. Later, in 1781 (pub. 1794), Euler gave the modern form, which is obtained from (52) by letting $t = -\log x$,

$$(53) \qquad \Gamma(n + 1) = \int_0^\infty x^n e^{-x} \, dx.$$

The integral (48), which Legendre called the first Eulerian integral, became standardized as the beta function,

$$(54) \qquad B(m, n) = \int_0^1 x^{m-1}(1 - x)^{n-1} \, dx.$$

Euler discovered the relation between the two integrals,[36] namely,

$$B(m, n) = \frac{\Gamma(m)\Gamma(n)}{\Gamma(m + n)}.$$

In his *Exercices de calcul intégral*, Legendre made a profound study of the Eulerian integrals and arrived at the duplication formula

$$(55) \qquad \Gamma(2x) = (2\pi)^{-1/2} \, 2^{2x-(1/2)} \Gamma(x) \Gamma\left(x + \frac{1}{2}\right).$$

Gauss studied the gamma function in his work on the hypergeometric function[37] and extended Legendre's result to what is called the multiplication formula:

$$(56) \quad \Gamma(nx) = (2\pi)^{(1-n)/2} \, n^{nx-(1/2)} \Gamma(x) \Gamma\left(x + \frac{1}{n}\right) \Gamma\left(x + \frac{2}{n}\right)$$

$$\cdots \Gamma\left(x + \frac{n-1}{n}\right).$$

36. *Novi Comm. Acad. Sci. Petrop.*, 16, 1771, 91–139, pub. 1772 = *Opera*, (1), 17, 316–57.
37. *Comm. Soc. Gott.*, II, 1813 = *Werke*, 3, 123–162, p. 149 in part.

6. *The Calculus of Functions of Several Variables*

The development of the calculus of functions of two and three variables took place early in the century. We shall note just a few of the details.

Though Newton derived from polynomial equations in x and y, that is $f(x, y) = 0$, expressions which we now obtain by partial differentiation of f with respect to x and y, this work was not published. James Bernoulli also used partial derivatives in his work on isoperimetric problems, as did Nicholas Bernoulli (1687–1759) in a paper in the *Acta Eruditorum* of 1720 on orthogonal trajectories. However, it was Alexis Fontaine des Bertins (1705–71), Euler, Clairaut, and d'Alembert who created the theory of partial derivatives.

The difference between an ordinary and a partial derivative was at first not explicitly recognized, and the same symbol d was used for both. Physical meaning dictated that in the case of functions of several independent variables the desired derivative was to take account of the change in one variable only.

The condition that $dz = p \, dx + q \, dy$, where p and q are functions of x and y, be an exact differential, that is, be obtainable from $z = f(x, y)$ by forming the differential $dz = (\partial f/\partial x) \, dx + (\partial f/\partial y) \, dy$ was obtained by Clairaut.[38] His result is that $p \, dx + q \, dy$ is an exact differential, that is, there is an f such that $\partial f/\partial x = p$ and $\partial f/\partial y = q$, if and only if $\partial p/\partial y = \partial q/\partial x$.

The major impetus to work with derivatives of functions of two or more variables came from the early work on partial differential equations. Thus a calculus of partial derivatives was supplied by Euler in a series of papers devoted to problems of hydrodynamics. In a paper of 1734[39] he shows that if $z = f(x, y)$ then

$$\frac{\partial^2 z}{\partial x \, \partial y} = \frac{\partial^2 z}{\partial y \, \partial x}.$$

In other papers written from 1748 to 1766, he treats change of variable, inversion of partial derivatives, and functional determinants. In works of 1744 and 1745 on dynamics, d'Alembert extended the calculus of partial derivatives.

Multiple integrals were really involved in Newton's work in the *Principia* on the gravitational attraction exerted by spheres and spherical shells on particles. However, Newton used geometrical arguments. In the eighteenth century Newton's work was cast in analytical form and extended. Multiple integrals appear in the first half of the century and were used to denote the solution of $\partial^2 z/\partial x \, \partial y = f(x, y)$. They were also used, for example, to determine the gravitational attraction exerted by a lamina on particles.

38. *Mém. de l'Acad. des Sci.*, Paris, 1739, 425–36, and 1740, 293–323.
39. *Comm. Acad. Sci. Petrop.*, 7, 1734/5, 174–93, pub. 1740 = *Opera*, (1), 22, 36–56.

Thus the attraction of an elliptical lamina of thickness δc on a point directly over the center and distance c units is a constant times the integral

$$\delta c \int\int \frac{c\ dx\ dy}{(c^2 + x^2 + y^2)^{3/2}}$$

taken over the ellipse $(x^2/a^2) + (y^2/b^2) = 1$. This was evaluated by Euler in 1738 by repeated integration.[40] He integrated with respect to y and used infinite series to expand the new integrand as a function of x.

By 1770 Euler did have a clear conception of the definite double integral over a bounded domain enclosed by arcs, and he gave the procedure for evaluating such integrals by repeated integration.[41] Lagrange, in his work on the attraction by ellipsoids of revolution,[42] expressed the attraction as a triple integral. Finding it difficult to evaluate in rectangular coordinates, he transformed to spherical coordinates. He introduced

$$x = a + r \sin \phi \cos \theta$$
$$y = b + r \sin \phi \sin \theta$$
$$z = c + r \cos \phi,$$

where a, b, and c are the coordinates of the new origin, θ is the longitude angle, ϕ the colatitude, and $0 \leq \phi \leq \pi$, $0 \leq \theta \leq 2\pi$. The essence of the transformation of the integral is to replace $dx\ dy\ dz$ by $r^2 \sin \theta\ d\theta\ d\phi\ dr$. Thus Lagrange began the subject of the transformation of multiple integrals. In fact, he gave the general method, though not very clearly. Laplace also gave the spherical coordinate transformation almost simultaneously.[43]

7. The Attempts to Supply Rigor in the Calculus

Accompanying the expansion of the concepts and techniques of the calculus were efforts to supply the missing foundations. The books on the calculus that appeared after Newton's and Leibniz's unsuccessful attempts to explain the concepts and justify the procedures tried to clear up the confusion but actually added to it.

Newton's approach to the calculus was potentially easier to rigorize than Leibniz's, though the latter's methodology was more fluid and more convenient for application. The English thought they could secure the rigor for both approaches by trying to tie them to Euclid's geometry. But they confused Newton's moments (his indivisible increments) with his fluxions, which dealt with continuous variables. The Continentals, following Leibniz, worked with differentials and tried to rigorize this concept. Differentials

40. *Comm. Acad. Sci. Petrop.*, 10, 1738, 102–15, pub. 1747 = *Opera*, (2), 6, 175–88.
41. *Novi Comm. Acad. Sci. Petrop.*, 14, 1769, 72–103, pub. 1770 = *Opera*, (1), 17, 289–315.
42. *Nouv. Mém. de l'Acad. de Berlin*, 1773, 121–48, pub. 1775 = *Œuvres*, 3, 619–58.
43. *Mém. des sav. étrangers*, 1772, 536–44, pub. 1776 = *Œuvres*, 8, 369–477.

were treated either as infinitesimals, that is, quantities not zero but not of any finite size, or sometimes, as zero.

Brook Taylor (1685–1731), who was secretary of the Royal Society from 1714 to 1718, in his *Methodus Incrementorum Directa et Inversa* (1715), sought to clarify the ideas of the calculus but limited himself to algebraic functions and algebraic differential equations. He thought he could always deal with finite increments but was vague on their transition to fluxions. Taylor's exposition, based on what we would call finite differences, failed to obtain many backers because it was arithmetical in nature when the British were trying to tie the calculus to geometry or to the physical notion of velocity.

Some idea of the obscurity of the eighteenth-century efforts and their lack of success may also be gained from Thomas Simpson's (1710–61) *A New Treatise on Fluxions* (1737). After some preliminary definitions, he defines a fluxion thus: "The magnitude by which any flowing quantity would be uniformly increased in a given portion of time with the generating celerity at any proposed position or instant (was [were] it from thence to continue invariable) is the fluxion of the said quantity at that position or instant." In our language Simpson is defining the derivative by saying that it is (dy/dt) Δt. Some authors gave up. The French mathematician Michel Rolle at one point taught that the calculus was a collection of ingenious fallacies.

The eighteenth century also witnessed new attacks on the calculus. The strongest was made by Bishop George Berkeley (1685–1753), who feared the growing threat to religion posed by mechanism and determinism. In 1734 he published *The Analyst, Or A Discourse Addressed to an Infidel Mathematician. Wherein It is examined whether the Object, Principles, and Inferences of the modern Analysis are more distinctly conceived, or more evidently deduced, than Religious Mysteries and Points of Faith.* "*First cast out the beam out of thine own Eye; and then shalt thou see clearly to cast out the mote out of thy brother's Eye.*" (The "infidel" was Edmond Halley.)[44]

Berkeley rightly pointed out that the mathematicians were proceeding inductively rather than deductively. Nor did they give the logic or reasons for their steps. He criticized many of Newton's arguments; for example, in the latter's *De Quadratura*, where he said he avoided the infinitely small, he gave x the increment denoted by o, expanded $(x + o)^n$, subtracted the x^n, divided by o to find the ratio of the increments of x^n and x, and then dropped terms involving o, thus obtaining the fluxion of x^n. Berkeley said that Newton first did give x an increment but then let it be zero. This, he said, is a defiance of the law of contradiction and the fluxion obtained was really $0/0$. Berkeley attacked also the method of differentials, as presented by l'Hospital and others on the Continent. The ratio of the differentials, he

44. George Berkeley: *The Works*, G. Bell and Sons, 1898, Vol. 3, 1–51.

said, should determine the secant and not the tangent; one undoes this error by neglecting higher differentials. Thus, "by virtue of a twofold mistake you arrive, though not at a science, yet at the truth," because errors were compensating for each other. He also picked on the second differential $d(dx)$ because it is the difference of a quantity dx that is itself the least discernible quantity. He says, "In every other science men prove their conclusions by their principles, and not their principles by the conclusions."

As for the derivative regarded as the ratio of the evanescent increments in y and x or dy and dx, these were "neither finite quantities, nor quantities infinitely small, nor yet nothing." These rates of change were but "the ghosts of departed quantities. Certainly ... he who can digest a second or third fluxion ... need not, methinks, be squeamish about any point in Divinity." He concluded that the principles of fluxions were no clearer than those of Christianity and denied that the object, principles, and inferences of the modern analysis were more distinctly conceived or more soundly deduced than religious mysteries and points of faith.

A reply to the *Analyst* was made by James Jurin (1684–1750). In 1734 he published *Geometry, No Friend to Infidelity*, in which he maintained that fluxions are clear to those versed in geometry. He then tried ineffectually to explain Newton's moments and fluxions. For example, Jurin defined the limit of a variable quantity as "some determinate quantity, to which the variable quantity is supposed continually to approach" and to come nearer than any given difference. However, then he added, "but never to go beyond it." This definition he applied to a variable ratio (the difference quotient). Berkeley's crushing answer, entitled *A Defense of Freethinking in Mathematics* (1735),[45] indicated that Jurin was trying to defend what he did not understand. Jurin replied, but did not clarify matters.

Benjamin Robins (1707–51) then entered the fray with articles and a book, *A Discourse Concerning the Nature and Certainty of Sir Isaac Newton's Method of Fluxions and of Prime and Ultimate Ratios* (1735). Robins neglected the moments of Newton's first paper but emphasized fluxions and prime and ultimate ratios. He defined a limit thus: "We define an ultimate magnitude to be a limit, to which a varying magnitude can approach within any degree of nearness whatever, though it can never be made absolutely equal to it." Fluxions he considered to be the right idea, and prime and ultimate ratios as only an explanation. He added that the method of fluxions is established without recourse to limits despite the fact that he gave explanations in terms of a variable approaching a limit. Infinitesimals he disavowed.

To answer Berkeley, Colin Maclaurin (1698–1746), in his *Treatise of Fluxions* (1742), attempted to establish the rigor of the calculus. It was a commendable effort but not correct. Like Newton, Maclaurin loved geom-

45. George Berkeley: *The Works*, G. Bell and Sons, 1898, Vol. 3, 53–89.

etry, and therefore tried to found the doctrine of fluxions on the geometry of the Greeks and the method of exhaustion, particularly as used by Archimedes. He hoped thereby to avoid the limit concept. His accomplishment was to use geometry so skillfully that he persuaded others to use it and neglect analysis.

The Continental mathematicians relied more upon the formal manipulation of algebraic expressions than on geometry. The most important representative of this approach is Euler, who rejected geometry as a basis for the calculus and tried to work purely formally with functions, that is, to argue from their algebraic (analytic) representation.

He denied the concept of an infinitesimal, a quantity less than any assignable magnitude and yet not 0. In his *Institutiones* of 1755 he argued,[46]

> There is no doubt that every quantity can be diminished to such an extent that it vanishes completely and disappears. But an infinitely small quantity is nothing other than a vanishing quantity and therefore the thing itself equals 0. It is in harmony also with that definition of infinitely small things, by which the things are said to be less than any assignable quantity; it certainly would have to be nothing; for unless it is equal to 0, an equal quantity can be assigned to it, which is contrary to the hypothesis.

Since Euler banished differentials he had to explain how dy/dx, which was $0/0$ for him, could equal a definite number. He does this as follows: Since for any number n, $n \cdot 0 = 0$, then $n = 0/0$. The derivative is just a convenient way of determining $0/0$. To justify dropping $(dx)^2$ in the presence of a dx Euler says $(dx)^2$ vanishes before dx does, so that, for example, the ratio of $dx + (dx)^2$ to dx is 1. He did accept ∞ as a number, for example, as the sum $1 + 2 + 3 + \cdots$. He also distinguished orders of ∞. Thus $a/0 = \infty$, but $a/(dx)^2$ is infinite to the second order, and so on.

Euler then proceeds to obtain the derivative of $y = x^2$. He gives x the increment ω; the increment of y is $\eta = 2x\omega + \omega^2$ and the ratio of η/ω is $2x + \omega$. He then says that this ratio approaches $2x$ the smaller ω is taken. However, he emphasizes that these differentials η and ω are absolutely zero and that nothing can be inferred from them other than their mutual ratio, which is in the end reduced to a finite quantity. Thus Euler accepts unqualifiedly that there exist quantities that are absolutely zero but whose ratios are finite numbers. There is more "reasoning" of this nature in Chapter 3 of the *Institutiones*, where he encourages the reader by remarking that the derivative does not hide so great a mystery as is commonly thought and which in the mind of many renders the calculus suspect.

As another example of Euler's reasoning, let us take his derivation of the

46. *Opera*, (1), 10, 69.

differential of $y = \log x$, in section 180 of his *Institutiones* (1755). Replacing x by $x + dx$ gives

$$dy = \log (x + dx) - \log x = \log \left(1 + \frac{dx}{x}\right).$$

Now he calls upon a result in Chapter 7 of Volume 1 of his *Introductio* (1748),

$$(57) \qquad \log_e (1 + z) = z - \frac{z^2}{2} + \frac{z^3}{3} - \frac{z^4}{4} + \cdots.$$

Replacing z by dx/x gives

$$dy = \frac{dx}{x} - \frac{dx^2}{2x^2} + \frac{dx^3}{3x^3} - \cdots.$$

Since all the terms beyond the first one are evanescent, we have

$$d(\log x) = \frac{dx}{x}.$$

We should keep in mind that Euler's texts were the standard of his day. What Euler did contribute in his formalistic approach was to free the calculus from geometry and base it on arithmetic and algebra. This step at least prepared the way for the ultimate justification of the calculus on the basis of the real number system.

Lagrange, in a paper of 1772[47] and in his *Théorie des fonctions analytiques*[48] made the most ambitious attempt to rebuild the foundations of the calculus. The subtitle of his book reveals his folly. It reads: "Containing the principal theorems of the differential calculus without the use of the infinitely small, or vanishing quantities, or limits and fluxions, and reduced to the art of algebraic analysis of finite quantities."

Lagrange criticizes Newton's approach by pointing out that, regarding the limiting ratio of arc to chord, Newton considers chord and arc equal not before or after vanishing, but *when* they vanish. As Lagrange correctly points out, "That method has the great inconvenience of considering quantities in the state in which they cease, so to speak, to be quantities; for though we can always properly conceive the ratios of two quantities as long as they remain finite, that ratio offers to the mind no clear and precise idea, as soon as its terms both become nothing at the same time." Maclaurin's *Treatise of Fluxions*, he says, shows how difficult it is to demonstrate the method of fluxions. He is equally dissatisfied with the little zeros (infinitesimals) of Leibniz and Bernoulli and with the absolute zeros of Euler, all of which, "although correct in reality are not sufficiently clear to serve as foundation of a science whose certitude should rest on its own evidence."

47. *Nouv. Mém. de l'Acad. de Berlin*, 1772, pub. 1774 = *Œuvres*, 3, 441–76.
48. 1797; 2nd ed., 1813 = *Œuvres*, 9.

Lagrange wished to give the calculus all the rigor of the demonstrations of the ancients and proposed to do this by reducing the calculus to algebra, which, as we noted above, included infinite series as extensions of polynomials. In fact, for Lagrange, function theory is the part of algebra concerned with the derivatives of functions. Specifically Lagrange proposed to use power series. With becoming modesty he remarks that it is strange that this method did not occur to Newton.

He now wants to use the fact that any function $f(x)$ can be expressed thus:

$$(58) \qquad f(x + h) = f(x) + ph + qh^2 + rh^3 + sh^4 + \cdots$$

wherein the coefficients, $p, q, r, \ldots$ involve x but are independent of h. However, he wishes to be sure before proceeding that such a power series expansion is always possible. Of course, he says, this is known through any number of familiar examples, but he does agree that there are exceptional cases. As exceptional cases Lagrange has in mind those in which some derivative of $f(x)$ becomes infinite and those in which the function and its derivatives become infinite. These exceptions happen only at isolated points; hence, with Lagrange they do not count. In a similar cavalier fashion he deals with a second difficulty. Lagrange and Euler accepted without question that a series expansion containing integral and fractional powers of h was surely possible, but Lagrange wished to eliminate the need for fractional powers. Fractional powers, he believed, could arise only if $f(x)$ contained radicals. But these, too, he dismisses as exceptional cases. Hence he is ready to proceed with (58).

By a somewhat involved but purely formal argument Lagrange concludes that we get $2q$ from p in the same way that we get p from $f(x)$, and a similar conclusion holds for the other coefficients $r, s, \ldots$ in (58). Hence if we let p be denoted by $f'(x)$ and designate by $f''(x)$ a function derived from $f'(x)$ as $f'(x)$ is derived from $f(x)$, then

$$p = f'(x), \qquad q = \frac{1}{2!} f''(x), \qquad r = \frac{1}{3!} f'''(x), \ldots,$$

and so (58) gives

$$f(x + h) = f(x) + hf'(x) + \frac{h^2}{2!} f''(x) + \cdots.$$

Lagrange now concludes that the final "expression has the advantage of showing how the terms of the series depend on each other, and especially how when one knows how to form the first derivative function, one can form all the derivative functions which enter the series." A little later he adds, "For one who knows the rudiments of the differential calculus it is clear that these derivative functions coincide with dy/dx, d^2y/dx^2, $\cdots$."

Lagrange has yet to show how one derives p or $f'(x)$ from $f(x)$. Here he uses (58) and neglects all terms after the second. Then $f(x + h) - f(x) = ph$. He divides by h and concludes that $p = f'(x)$.

Actually Lagrange's assumption that a function can be expanded in a power series is one weak point in the scheme. The criteria now known for such an expansion involve the existence of derivatives, and this is what Lagrange sought to avoid. His arguments to justify the power series only added to the confusion about which functions can be so expanded. Even if such an expansion is possible, Lagrange shows how to calculate the coefficients only if we can get the first one, that is, $f'(x)$; and here he does the same crude thing his predecessors did. Finally, the question of the convergence of the series (58) is really not discussed. He does show that for h small enough the last term kept is greater than what is neglected. He also gives in this book the Lagrange form of the remainder in a Taylor expansion (Chap. 20, sec. 7), but it plays no role in the above development. Despite all these weaknesses, Lagrange's approach to the calculus met great favor for quite some time. Later it was abandoned.

Lagrange believed he had dispensed with the limit concept. He did agree[49] that the calculus could be founded on a theory of limits but said the kind of metaphysics one must employ is foreign to the spirit of analysis. Despite the inadequacies of his approach, he did contribute, as did Euler, to divorcing the foundations of analysis from geometry and mechanics; in this his influence was decisive. While this separation is not pedagogically desirable, since it bars intuitive understanding, it did make clear that logically analysis must stand on its own feet.

Toward the end of the century the mathematician, soldier, and administrator Lazare N. M. Carnot (1753–1823) wrote a popular, widely sold book, *Réflexions sur la métaphysique du calcul infinitésimal* (1797), in which he sought to make the calculus precise. He tried to show that the logic rested on the method of exhaustion and that all the ways of treating the calculus were but simplifications or shortcuts whose logic could be supplied by founding them on that method. After much thought he, like Berkeley, ended up concluding that errors in the usual arguments of the calculus were compensating for each other.

Among the multitude of efforts to rigorize the calculus, a few were on the right track. The most notable of these were d'Alembert's and, earlier, Wallis's. D'Alembert believed that Newton had the right idea and that he himself was merely explaining Newton's meaning. In his article "Différentiel" in the famous *Encyclopédie ou Dictionnaire Raisonné des Sciences, des Arts, et des Métiers* (1751–80), he says, "Newton has never regarded the differential calculus as a calculus of infinitesimals, but as a method of prime and ulti-

49. *Œuvres*, 1, 325.

mate ratios, that is to say, a method of finding the limit of these ratios." But d'Alembert defined a differential as "an infinitely small quantity or at least smaller than any assignable magnitude." He did believe the calculus of Leibniz could be built up on three rules of differentials; however, he favored the derivative as a limit. In his pursuit of the use of limits he, like Euler, argued that $0/0$ may be equal to any quantity one wishes.

In another article, "Limite," he says: "The theory of limits is the true metaphysics of the calculus. . . . It is never a question of infinitesimal quantities in the differential calculus: it is uniquely a question of limits of finite quantities. Thus the metaphysics of the infinite and infinitely small quantities, larger or smaller than one another, is totally useless to the differential calculus." Infinitesimals were merely a manner of speaking that avoided the lengthier description in terms of limits. In fact, d'Alembert gave a good approximation to the correct definition of limit in terms of a variable quantity approaching a fixed quantity more closely than any given quantity, though here too he talks about the variable never reaching the limit. However, he did not give a formal exposition of the calculus that incorporated and utilized his basically correct views.

He, too, was vague on a number of points; for example, he defined the tangent to a curve as the limit of the secant when the two points of intersection become one. This vagueness, especially in his statement of the notion of limit, caused many to debate the question of whether a variable can reach its limit. Since there was no explicit, correct presentation, d'Alembert advised students of the calculus, "Persist and faith will come to you."

Sylvestre-François Lacroix (1765–1843), in the second edition (1810–1819) of his *Traité du calcul différentiel et du calcul intégral* had the idea more explicitly that the ratio of two quantities, each of which approaches 0, can approach a definite number as a limit. He gives the ratio $ax/(ax + x^2)$ and notes that this ratio is the same as $a/(a + x)$, and the latter approaches 1 as x approaches 0. Moreover, he points out that 1 is the limit even when x approaches 0 through negative values. However, he also speaks of the ratio of the limits when these are 0 and even uses the symbol $0/0$. He does introduce the differential dy of a function $y = f(x)$ in terms of the derivative; that is, $dy = f'(x)\, dx$. Hence if $y = ax^3$, $dy = 3ax^2\, dx$. He first used the term "differential coefficient" for the derivative; thus $3ax^2$ is the differential coefficient.

Almost every mathematician of the eighteenth century made some effort or at least a pronouncement on the logic of the calculus, and though one or two were on the right track, all the efforts were abortive. The distinction between a very large number and an infinite "number" was hardly made. It seemed clear that a theorem that held for any n must hold for n infinite. Likewise a difference quotient was replaced by a derivative, and a sum of a finite number of terms and an integral were hardly distinguished.

Mathematicians passed from one to the other freely. In 1755, in his *Institutiones*, Euler distinguished between the increment in a function and the differential of that function and between a summation and the integral, but these distinctions were not immediately taken up. All the efforts could be summed up in Voltaire's description of the calculus as "the art of numbering and measuring exactly a Thing whose Existence cannot be conceived."

In view of the almost complete absence of any foundations, how could the mathematicians proceed with the manipulation of the variety of functions? In addition to their great reliance upon physical and intuitive meanings, they did have a model in mind—the simpler algebraic functions, such as polynomials and rational functions. They carried over to all functions the properties they found in these explicit, concrete functions: continuity, the existence of isolated infinities and discontinuities, expansion in power series, and the existence of derivatives and integrals. But when they were obliged, largely through the work on the vibrating string, to broaden the concept of function, as Euler put it, to any freely drawn curves (Euler's mixed or irregular or discontinuous functions), they could no longer use the simpler functions as a guide. And when the logarithmic function had to be extended to negative and complex numbers, they really proceeded without any reliable basis at all; this is why arguments on such matters were common. The rigorization of the calculus was not achieved until the nineteenth century.

Bibliography

Bernoulli, James: *Opera*, 2 vols., 1744, reprinted by Birkhaüser, 1968.

Bernoulli, John: *Opera Omnia*, 4 vols., 1742, reprinted by Georg Olms, 1968.

Boyer, Carl B.: *The Concepts of the Calculus*, Dover (reprint), 1949, Chap. 4.

Brill, A. and M. Nöther: "Die Entwicklung der Theorie der algebraischen Funktionen in älterer and neuerer Zeit," *Jahres. der Deut. Math.-Verein.*, 3, 1892/3, 107–566.

Cajori, Florian: "History of the Exponential and Logarithmic Concepts," *Amer. Math. Monthly*, 20, 1913, 5–14, 35–47, 75–84, 107–17, 148–51, 173–82, 205–10.

————: *A History of the Conceptions of Limits and Fluxions in Great Britain from Newton to Woodhouse*, Open Court, 1919.

————: "The History of Notations of the Calculus," *Annals of Math.*, (2), 25, 1923, 1–46.

Cantor, Moritz: *Vorlesungen über Geschichte der Mathematik*, B. G. Teubner, 1898 and 1924, Vols. 3 and 4, relevant sections.

Davis, Philip J.: "Leonhard Euler's Integral: A Historical Profile of the Gamma Function," *Amer. Math. Monthly*, 66, 1959, 849–69.

Euler, Leonhard: *Opera Omnia*, B. G. Teubner and Orell Füssli, 1911–; see references to specific volumes in the chapter.

Fagnano, Giulio Carlo: *Opere matematiche*, 3 vols., Albrighi Segati, 1911.

Fuss, Paul H. von: *Correspondance mathématique et physique de quelques célèbres géomètres du XVIIIème siècle*, 2 vols., 1843, Johnson Reprint Corp., 1967.

Hofmann, Joseph E.: "Über Jakob Bernoullis Beiträge zur Infinitesimal-mathematik," *L'Enseignement Mathématique*, (2), 2, 61–171, 1956; also published separately by Institut de Mathématiques, Genève, 1957.

Mittag-Leffler, G.: "An Introduction to the Theory of Elliptic Functions," *Annals of Math.*, (2), 24, 1922–23, 271–351.

Montucla, J. F.: *Histoire des mathématiques*, A. Blanchard (reprint), 1960, Vol. 3, pp. 110–380.

Pierpont, James: "Mathematical Rigor, Past and Present," *Amer. Math. Soc. Bulletin*, 34, 1928, 23–53.

Struik, D. J.: *A Source Book in Mathematics, 1200–1800*, Harvard University Press, 1969, pp. 333–38, 341–51, 374–91.

20
Infinite Series

1. *Introduction*

Infinite series were in the eighteenth century and are still today considered an essential part of the calculus. Indeed, Newton considered series inseparable from his method of fluxions because the only way he could handle even slightly complicated algebraic functions and the transcendental functions was to expand them into infinite series and differentiate or integrate term by term. Leibniz in his first published papers of 1684 and 1686 also emphasized "general or indefinite equations." The Bernoullis, Euler, and their contemporaries relied heavily on the use of series. Only gradually, as we pointed out in the preceding chapter, did the mathematicians learn to work with the elementary functions in closed form, that is, as simple analytical expressions. Nevertheless, series were still the only representation for some functions and the most effective means of calculating the elementary transcendental functions.

The successes obtained by using infinite series became more numerous as the mathematicians gradually extended their discipline. The difficulties in the new concept were not recognized, at least for a while. Series were just infinite polynomials and appeared to be treatable as such. Moreover, it seemed clear, as Euler and Lagrange believed, that every function could be expressed as a series.

2. *Initial Work on Infinite Series*

Infinite series, usually in the form of infinite geometric progressions with common ratio less than 1, appear very early in mathematics. Aristotle[1] even recognized that such series have a sum. They appear sporadically among the later medieval mathematicians, who considered infinite series to calculate

---

1. *Physica*, Book III, Chap. 6, 206b, 3–33.

436

the distance traveled by moving bodies when the velocity changes from one period of time to another. Oresme, who had considered a few such series, even proved in a tract, *Quæstiones Super Geometriam Euclidis* (c. 1360), that the harmonic series

$$1 + \frac{1}{2} + \frac{1}{3} + \frac{1}{4} + \frac{1}{5} + \cdots$$

is divergent by the method used today, namely, to replace the series by the series of lesser terms

$$\frac{1}{2} + \frac{1}{2} + \left(\frac{1}{4} + \frac{1}{4}\right) + \left(\frac{1}{8} + \frac{1}{8} + \frac{1}{8} + \frac{1}{8}\right) + \cdots$$

and to note that the latter series diverges because we can obtain as many groups of terms each of magnitude 1/2 as we please. However, one must not conclude that Oresme or mathematicians in general began to distinguish convergent and divergent series.

In his *Varia Responsa* (1593, *Opera*, 347–435) Vieta gave the formula for the sum of an *infinite* geometric progression. He took from Euclid's *Elements* that the sum of n terms of $a_1 + a_2 + \cdots + a_n$ is given by

$$\frac{s_n - a_n}{s_n - a_1} = \frac{a_1}{a_2}.$$

Then if $a_1/a_2 > 1$, a_n approaches 0 as n becomes infinite, so that

$$s_\infty = \frac{a_1^2}{a_1 - a_2}.$$

In the middle of the seventeenth century Gregory of Saint Vincent, in his *Opus Geometricum* (1647), showed that the Achilles and the Tortoise paradox could be resolved by summing an infinite geometric series. The finiteness of the sum showed that Achilles would overtake the tortoise at a definite time and place. Gregory gave the first explicit statement that an infinite series represents a magnitude, namely, the sum of the series, which he called the limit of the series. He says the "terminus of a progression is the end of the series to which the progression does not attain, even if continued to infinity, but to which it can approach more closely than by any given interval." He made many other statements that are less accurate and less clear, but he did contribute to the subject and influenced many pupils.

Mercator and Newton (Chap. 17, sec. 2) found the series

$$\log (1 + x) = x - \frac{1}{2} x^2 + \frac{1}{3} x^3 + \cdots.$$

The observation was made that the series has an infinite value for $x = 2$, whereas, according to the left side, it should yield log 3. Wallis noted this

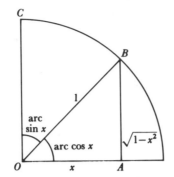

Figure 20.1

difficulty but could not explain it. Newton obtained many other series for algebraic and transcendental functions. Thus, to obtain the series for arc sin x, in 1666, he used the fact (Fig. 20.1) that the area $OBC = (1/2)$ arc sin x, so arc sin $x = \int_0^x \sqrt{1 - x^2}\, dx - x\sqrt{1 - x^2}/2$. He got the result by expanding the right side into series, integrating term by term, and combining the two series. He also obtained the series for arc tan x. In his *De Analysi* of 1669 he gave the series for sin x, cos x, arc sin x, and e^x. Some of these he got from others by inverting a series, that is, solving it for the independent variable in terms of the dependent variable. His method of doing this is crude and inductive. Nevertheless, Newton was immensely pleased with his derivation of so many series.

Collins received Newton's *De Analysi* in 1669 and communicated the results on series to James Gregory on December 24, 1670. Gregory answered (Turnbull, *Correspondence*, 1, 52–58 and 61–64) on February 15, 1671 that he had obtained other series, among them

$$\tan x = x + \frac{x^3}{3} + \frac{2}{15} x^5 + \frac{17}{315} x^7 + \cdots,$$

$$\sec x = 1 + \frac{x^2}{2} + \frac{5}{24} x^4 + \frac{61}{720} x^6 + \cdots.$$

His derivations of these series are not known. Leibniz, too, obtained series for sin x, cos x, and arc tan x, presumably independently, in 1673. The series for the transcendental functions were the most fertile method available in the early stages of the calculus for handling these functions and are a significant part of Newton's and Leibniz's work on the calculus.

These men and others who used the binomial theorem for fractional and negative exponents to obtain many of the series not only ignored the questions that arose from the use of series but had no proof of the binomial theorem. They also accepted unquestioningly that the series was equal to the function that was being expanded.

James Bernoulli in 1702[2] derived the series for sin x and cos x by using expressions he had derived for sin $n\alpha$ in terms of sin α and then letting α approach 0 while n becomes infinite, so that $n\alpha$ approaches x while n sin α, which equals $n\alpha$ sin α/α, also approaches x. Wallis had mentioned in the Latin edition of his *Algebra* (1693) that Newton had given these series again in 1676; Bernoulli noted this remark but failed to acknowledge Newton's priority. Moreover, de Moivre gave a proof of Newton's results in the *Philosophical Transactions* of 1698;[3] though Bernoulli used and referred to this journal in other work, he gave no indication that he was aware through this source of Newton's work.

One of the major uses of series beyond their service in differentiation and integration is to calculate special quantities, such as π and e, and the logarithmic and trigonometric functions. Newton, Leibniz, James Gregory, Cotes, Euler, and many others were interested in series for this purpose. However, some series converge so slowly that they are almost useless for calculation. Thus Leibniz in 1674[4] obtained the famous result

$$\frac{\pi}{4} = 1 - \frac{1}{3} + \frac{1}{5} - \frac{1}{7} + \cdots.$$

However, it would require about 100,000 terms to compute π, even to the accuracy obtained by Archimedes. Likewise, the series for log $(1 + x)$ converges very slowly, so that many terms have to be taken into account to achieve an accuracy of a few decimal places. This series was transformed in various ways to produce more rapidly converging series. Thus James Gregory (*Exercitationes Geometricae*, 1668) obtained

$$\frac{1}{2} \log \left(\frac{1 + z}{1 - z} \right) = z + \frac{1}{3} z^3 + \frac{1}{5} z^5 + \cdots,$$

which proved to be more useful for the calculation of logarithms. The problem of transforming a series into another that converges more rapidly was pursued by many men throughout the eighteenth century. One such transformation, due to Euler, is given in Section 4.

Still another use of series was initiated by Newton. Given the implicit function $f(x, y) = 0$, to work with y as a function of x one wants the explicit function. There may be several such explicit functions, as is evident in the trivial case of $x^2 + y^2 - 1 = 0$, which has two solutions $y = \pm \sqrt{1 - x^2}$, both emanating from the point $(1, 0)$. In this simple case the two solutions can be expressed in terms of closed analytic expressions. But generally each expression for y must be expressed as an infinite series in x. However, these series are not necessarily power series, particularly if the points at which the

2. *Opera*, 2, 921–29.
3. Vol. 20, 190–93.
4. *Math. Schriften*, 5, 88–92; also *Acta Erud.*, 1682 = *Math. Schriften*, 5, 118–22.

expansions are sought are singular points ($f_x = f_y = 0$). In his *Method of Fluxions* Newton published a scheme for determining the forms of the several series, one for each explicit solution. His method, which uses what is known as Newton's parallelogram, shows how to determine the first few exponents in a series of the form

$$y = a_1 x^m + a_2 x^{m+n} + a_3 x^{m+2n} + \cdots.$$

The coefficients of the series can then be determined by the method of undetermined coefficients. Actually Newton gave only specific examples, from which one must infer the method.

The problem of determining the exponents in each series is troublesome. Taylor, James Stirling, and Maclaurin gave rules; Maclaurin tried to extend and prove them but made no progress. A proof of Newton's method was given independently by Gabriel Cramer and Abraham G. Kästner (1719–1800).

3. *The Expansion of Functions*

One of the problems faced by mathematicians in the late seventeenth and eighteenth centuries was interpolation of table values. Greater accuracy of the interpolated values of the trigonometric, logarithmic, and nautical tables was necessary to keep pace with progress in navigation, astronomy, and geography. The common method of interpolation (the word is Wallis's) is called linear interpolation because it assumes that the function is a linear function of the independent variable in the interval between two known values. However, the functions in question are not linear; and the mathematicians realized that a better method of interpolation was needed.

The method we are about to describe was initiated by Briggs in his *Arithmetica Logarithmica* (1624), though the key formula was given by James Gregory in a letter to Collins (Turnbull, *Correspondence*, 1, 45–48) of November 23, 1670, and independently by Newton. Newton's work appears in Lemma 5 of Book III of the *Principia* and in the *Methodus Differentialis*, which, though published in 1711, was written by 1676. The method uses what are called finite differences and is the first major result in the calculus of finite differences.

Suppose $f(x)$ is a function whose values are known at a, $a + c$, $a + 2c$, $a + 3c, \ldots, a + nc$. Let

$$\Delta f(a) = f(a + c) - f(a),$$
$$\Delta f(a + c) = f(a + 2c) - f(a + c),$$
$$\Delta f(a + 2c) = f(a + 3c) - f(a + 2c),$$
$$\cdots\cdots\cdots\cdots\cdots\cdots\cdots\cdots\cdots\cdots\cdots$$

Further, let

$$\Delta^2 f(a) = \Delta f(a + c) - \Delta f(a)$$
$$\Delta^3 f(a) = \Delta^2 f(a + c) - \Delta^2 f(a),$$
$$\cdots\cdots\cdots\cdots\cdots\cdots\cdots\cdots\cdots\cdots\cdots\cdots$$

Then the Gregory-Newton formula states that

(1) $$f(a + h) = f(a) + \frac{h}{c} \Delta f(a) + \frac{\frac{h}{c}\left(\frac{h}{c} - 1\right)}{1 \cdot 2} \Delta^2 f(a) + \cdots.$$

Newton sketched a proof but Gregory did not.

To calculate a value of $f(x)$ at any value x between the known values, one simply gives h the value $x - a$. This calculated value is not necessarily the true value of the function; what the formula yields is the value of a polynomial in h that agrees with the true function at the special values a, $a + c, a + 2c, \ldots$.

The Gregory-Newton formula was also used to carry out approximate integration. Given a function, say $g(x)$, to be integrated, perhaps in order to find the area under the corresponding curve, one uses the values of $g(x)$ to obtain $g(a), g(a + c), g(a + 2c), \ldots$ and their differences and higher-order differences; these values are substituted in (1). Then (1) gives a polynomial approximation to $g(x)$, and, as Newton points out, since polynomials are readily integrated, one gets an approximation to the desired integral of $g(x)$.

Gregory also applied (1) to the function $(1 + d)^x$. He knew the value of this function at $x = 0, 1, 2, 3, \ldots$. Then $f(0) = 1$, $\Delta f(0) = d$, $\Delta^2 f(0) = d^2$, and so on. Thus by letting $a = 0$, $c = 1$, and $h = x - 0$ in (1), and using the values of $f(0), \Delta f(0), \ldots$ he got

(2) $$(1 + d)^x = 1 + dx + \frac{x(x - 1)}{1 \cdot 2} d^2 + \frac{x(x - 1)(x - 2)}{1 \cdot 2 \cdot 3} d^3 + \cdots.$$

Thus Gregory obtained the binomial expansion for general x.

The Gregory-Newton interpolation formula was used by Brook Taylor to develop the most powerful single method for expanding a function into an infinite series. The binomial theorem, division of the denominator of a rational function into the numerator, and the method of undetermined co-efficients are limited devices. In his *Methodus Incrementorum Directa et Inversa* (1715), the first publication in which he treated the calculus of finite differences, Taylor derived the theorem that still bears his name and which he had stated in 1712. Incidentally, he praises Newton but makes no mention of Leibniz's work of 1673 on finite differences, though Taylor knew this work. Taylor's theorem was known to James Gregory in 1670 and was discovered independently somewhat later by Leibniz; however, these two men did not publish it. John Bernoulli did publish practically the same result in the *Acta*

Eruditorum of 1694; and though Taylor knew this result he did not refer to it. His own "proof" was different. What he did amounts to letting c be Δx in the Gregory-Newton formula. Then, for example, the third term on the right side of (1) becomes

$$(3) \qquad \frac{h(h - \Delta x)}{1 \cdot 2} \frac{\Delta^2 f(a)}{\Delta x^2}.$$

Taylor concluded that when $\Delta x = 0$, this term becomes $h^2 f''(a)/2!$, and so the entire Gregory-Newton formula becomes

$$(4) \qquad f(a + h) = f(a) + f'(a)h + f''(a) \frac{h^2}{2!} + f'''(a) \frac{h^3}{3!} + \cdots.$$

Of course Taylor's method was not rigorous, nor did he consider the question of convergence.

Taylor's theorem for $a = 0$ is now called Maclaurin's theorem. Colin Maclaurin, who succeeded James Gregory as professor at Edinburgh, gave this special case in his *Treatise of Fluxions* (1742) and stated that it was but a special case of Taylor's result. However, historically it has been credited to Maclaurin as a separate theorem. Incidentally, Stirling gave this special case for algebraic functions in 1717 and for general functions in his *Methodus Differentialis* of 1730.

Maclaurin's proof of his result is by the method of undetermined coefficients. He proceeds as follows. Let

$$(5) \qquad f(z) = A + Bz + Cz^2 + Dz^3 + \cdots.$$

Then

$$f'(z) = B + 2Cz + 3Dz^2 + \cdots$$
$$f''(z) = 2C + 6Dz + \cdots$$
$$\dots\dots\dots\dots\dots\dots\dots\dots\dots$$

Let $z = 0$ in each equation and determine $A, B, C, \cdots$. He did not worry about convergence and proceeded to use the result.

4. *The Manipulation of Series*

James and John Bernoulli did a great deal of work with series. James wrote five papers between 1689 and 1704 that were published by his nephew Nicholas (1695–1726) (John's son) as a supplement to James's *Ars Conjectandi* (1713). Most of the work in these papers is devoted to the use of series representations of functions for the purposes of differentiating and integrating the functions and obtaining areas under curves and lengths of curves. While these applications were a substantial contribution to the calculus, there were no especially novel features. However, some of the methods he used to sum series are worth noting because they illustrate the nature of mathematical thought in the eighteenth century.

In the first paper (1689),[5] he starts with the series

(6) $$N = \frac{a}{c} + \frac{a}{2c} + \frac{a}{3c} + \cdots,$$

from which

(7) $$N - \frac{a}{c} = \frac{a}{2c} + \frac{a}{3c} + \frac{a}{4c} + \cdots.$$

He now subtracts (7) from (6); in this process each term on the right side of (7) is subtracted from the term above it. This yields

(8) $$\frac{a}{c} = \frac{a}{1 \cdot 2c} + \frac{a}{2 \cdot 3c} + \frac{a}{3 \cdot 4c} + \cdots.$$

This is a correct result but incorrectly derived, because the original series is divergent. James says that the procedure is questionable and should not be used without some circumspection.

He then considers the ordinary harmonic series and shows that its sum is infinite.[6] He considers the terms

(9) $$\frac{1}{n + 1} + \frac{1}{n + 2} + \cdots + \frac{1}{n^2}$$

and says this sum is larger than $(n^2 - n) \cdot (1/n^2)$ because there are $n^2 - n$ terms and each is at least as large as the last. But

$$(n^2 - n) \cdot \left(\frac{1}{n^2}\right) = 1 - \frac{1}{n}.$$

Hence if we add $1/n$ to (9)

$$\frac{1}{n} + \frac{1}{n + 1} + \frac{1}{n + 2} + \cdots + \frac{1}{n^2} > 1.$$

Thus, he says, we can go from one group of terms to another, each group having a sum greater than 1. Hence we can obtain a finite number of terms whose sum is as large as we please; and therefore the sum of the whole series must be infinite. Consequently, he also points out, the sum of an infinite series whose "last" term vanishes can be *infinite*; this is contrary to his earlier belief and the belief of many eighteenth-century mathematicians, including Lagrange.

5. *Opera*, 1, 375–402.
6. *Opera*, 1, 392.

John Bernoulli had previously given a different "proof" of the infinite sum of the harmonic series. It runs thus:

(10) $\dfrac{1}{2} + \dfrac{1}{3} + \dfrac{1}{4} + \dfrac{1}{5} + \cdots = \dfrac{1}{1\cdot 2} + \dfrac{2}{2\cdot 3} + \dfrac{3}{3\cdot 4} + \dfrac{4}{4\cdot 5} + \cdots$

$$= \left(\dfrac{1}{1\cdot 2} + \dfrac{1}{2\cdot 3} + \dfrac{1}{3\cdot 4} + \dfrac{1}{4\cdot 5} + \cdots\right) + \left(\dfrac{1}{2\cdot 3} + \dfrac{1}{3\cdot 4} + \dfrac{1}{4\cdot 5} + \cdots\right)$$

$$+ \left(\dfrac{1}{3\cdot 4} + \dfrac{1}{4\cdot 5} + \cdots\right) + \left(\dfrac{1}{4\cdot 5} + \cdots\right) + \cdots.$$

Now, using (8), wherein we let a and c be 1, we get from (10) that

$$\dfrac{1}{2} + \dfrac{1}{3} + \dfrac{1}{4} + \cdots =$$

$$1 + \left(1 - \dfrac{1}{2}\right) + \left(1 - \dfrac{1}{2} - \dfrac{1}{6}\right) + \left(1 - \dfrac{1}{2} - \dfrac{1}{6} - \dfrac{1}{12}\right) + \cdots$$

$$= 1 + \dfrac{1}{2} + \dfrac{1}{3} + \dfrac{1}{4} + \cdots.$$

If we let $A = 1/2 + 1/3 + 1/4 + 1/5 + \cdots$, we have proved that $A = 1 + A$. If A were finite, this result would be impossible.

In the succeeding four tracts on series James Bernoulli does many things so loosely that it is difficult to believe he had ever recognized the need for caution with infinite series. For example, in the second tract (1692),[7] he argues as follows: From the formula for a geometrical progression we have $1 + 1/2 + 1/4 + 1/8 + \cdots = 2$. Then by taking $1/3$ of both sides $1/3 + 1/6 + 1/12 + \cdots = 2/3$, and by taking $1/5$ of both sides of the original series $1/5 + 1/10 + 1/20 + \cdots = 2/5$, and so on. Hence the sum of the left sides, which is the entire harmonic series, equals the sum of the right sides, or

$$1 + \dfrac{1}{2} + \dfrac{1}{3} + \dfrac{1}{4} + \cdots = 2 + \dfrac{2}{3} + \dfrac{2}{5} + \cdots = 2\left(1 + \dfrac{1}{3} + \dfrac{1}{5} + \cdots\right).$$

Hence the sum of the odd terms is half the sum of the harmonic series. Then $1/2 + 1/4 + 1/6 + 1/8 + \cdots$ is also $1/2$ the harmonic series, so that $1 + 1/3 + 1/5 + \cdots = 1/2 + 1/4 + 1/6 + \cdots$.

In the third tract (1696)[8] he writes

$$\dfrac{l}{m+n} = \dfrac{l}{m}\left(1 + \dfrac{n}{m}\right)^{-1} = \dfrac{l}{m} - \dfrac{ln}{m^2} + \dfrac{ln^2}{m^3} - \cdots;$$

7. *Opera*, 1, 517–42.
8. *Opera*, 2, 745–64.

and when $n = m$,

(11)
$$\frac{l}{2m} = \frac{l}{m} - \frac{l}{m} + \frac{l}{m} - \cdots,$$

which he describes as a not inelegant paradox.

In the second paper on series he replaced the general term by a sum or difference of two other terms, and then performed other operations that lead to specific results. This replacement is correct for absolutely convergent series but not for conditionally convergent ones. Hence he got wrong results which he also described as paradoxes.

One of James's very interesting results deals with the series of reciprocals of the nth powers of the natural numbers, that is, with $1 + 1/2^n + 1/3^n + 1/4^n + \cdots$. James proved that the sum of the odd-numbered terms is to the sum of the even-numbered terms as $2^n - 1$ is to 1. This is correct for $n \geq 2$. However, James did not hesitate to apply it for the case $n = 1$ and $n = 1/2$. This last result he found paradoxical.

Another result on series due to James says that the sum of the series $1 + 1/\sqrt{2} + 1/\sqrt{3} + \cdots$ is infinite because each term is greater than the corresponding term of the harmonic series. Here the comparison test was used effectively.

The series that provoked the greatest discussion and controversy, which occurs when l and m in (11) are 1, is

(12)
$$1 - 1 + 1 - 1 + \cdots.$$

It seemed clear that by writing the series as

(13)
$$(1 - 1) + (1 - 1) + (1 - 1) + \cdots$$

the sum should be 0. It also seemed clear that by writing the series as

$$1 - (1 - 1) - (1 - 1) - (1 - 1) - \cdots$$

the sum should be 1. However, if one denotes the sum of (12) by S, then $S = 1 - S$, so that $S = 1/2$; and this is in fact Bernoulli's result in (11). Guido Grandi (1671–1742), a professor of mathematics at the University of Pisa, in his little book *Quadratura Circuli et Hyperbolae* (The Quadrature of Circles and Hyperbolas, 1703), obtained the third result by another method. He set $x = 1$ in the expansion

(14)
$$\frac{1}{1 + x} = 1 - x + x^2 - x^3 + \cdots$$

and obtained

$$\frac{1}{2} = 1 - 1 + 1 - 1 + \cdots.$$

Grandi therefore maintained that 1/2 was the sum of the series (12). He also argued that since the sum of (12) in the form (13) was 0, he had proved that the world could be created out of nothing.

In a letter to Christian Wolf (1678–1754), published in the *Acta*,[9] Leibniz also treated the series (12). He agreed with Grandi's result but thought it should be possible to obtain it without resorting to his argument. Instead, Leibniz argued that if one takes the first term, the sum of the first two, the sum of the first three, and so forth, one obtains 1, 0, 1, 0, 1, $\cdots$. Thus 1 and 0 are equally probable; one should therefore take the arithmetic mean, which is also the most probable value, as the sum. This solution was accepted by James and John Bernoulli, Daniel Bernoulli, and, as we shall see, Lagrange. Leibniz conceded that his argument was more metaphysical than mathematical but went on to say that there was more metaphysical truth in mathematics than was generally recognized. However, he was probably much more influenced by Grandi's argument than he himself realized. For when, in later correspondence, Wolf wished to conclude that

$$1 - 2 + 4 - 8 + 16 \cdots = \frac{1}{3}$$

$$1 - 3 + 9 - 27 + 81 \cdots = \frac{1}{4}$$

by using an extension of Leibniz's own probability argument, Leibniz objected. He pointed out that series that have sums have decreasing terms, and (12) is at least a limit of series with decreasing terms, as is evident from (14) by letting x approach 1 from below.

Really extensive work on series began about 1730 with Euler, who aroused tremendous interest in the subject. But there was much confusion in his thinking. To obtain the sum of

$$1 - 1 + 1 - 1 + 1 - \cdots$$

Euler argued that since

(15) $$\frac{1}{1 - x} = 1 + x + x^2 + x^3,$$

then when $x = -1$,

(16) $$\frac{1}{2} = 1 - 1 + 1 - 1 + \cdots$$

so that the sum is 1/2.

Also, when $x = -2$, (15) shows that

(17) $$\frac{1}{3} = 1 - 2 + 2^2 - 2^3 + \cdots;$$

9. *Acta Erud. Supplementum*, 5, 1713, 264–70 = *Math. Schriften*, 5, 382–87.

hence the sum of the right-hand series is 1/3. As a third example, since

(18) $$\frac{1}{(1+x)^2} = (1+x)^{-2} = 1 - 2x + 3x^2 - 4x^3 + \cdots,$$

then for $x = 1$ we have

$$\frac{1}{4} = 1 - 2 + 3 - 4 + \cdots.$$

Again, since

$$\frac{1-x}{(1+x)^2} = (1-x)(1+x)^{-2} = 1 - 3x + 5x^2 - 7x^3 + \cdots,$$

then for $x = 1$ we have

(19) $$0 = 1 - 3 + 5 - 7 + \cdots.$$

There are numerous examples of such arguments in his work.

One sees from (18), for $x = -1$, that

(20) $$\infty = 1 + 2 + 3 + 4 + 5 + \cdots.$$

This Euler accepted. Moreover one sees from (15), for $x = 2$, that

(21) $$-1 = 1 + 2 + 4 + 8 + \cdots.$$

Since the right-hand side of (21) should exceed the right-hand side of (20), the sum $1 + 2 + 4 + 8 + \cdots$ should exceed ∞. According to (21), it yields -1. Euler concluded that ∞ must be a sort of limit between the positive and negative numbers and in this respect resembles 0.

Apropos of (19), Nicholas Bernoulli (1687–1759) said, in a letter to Euler of 1743, that the sum of this series $1 - 3 + 5 - 7 + \cdots$ is $-\infty(-1)^\infty$. Euler's result of 0 he called an unsolvable contradiction. Bernoulli also noted that from (15) one gets, for $x = 2$,

$$-1 = 1 + 2 + 4 + 8 + \cdots;$$

and from

(22) $$\frac{1}{1 - x - x^2} = 1 + x + 2x^2 + 3x^3 + \cdots,$$

for $x = 1$ one gets

$$-1 = 1 + 1 + 2 + 3 + \cdots.$$

The fact that two different series give -1 is also an unsolvable contradiction, for otherwise one could equate the two series.

In one paper Euler did point out that series can be used only for values of x for which they converge. Nevertheless, in the very same paper[10] he concluded that

(23) $$\cdots + \frac{1}{x^2} + \frac{1}{x} + 1 + x + x^2 + x^3 + \cdots = 0.$$

His argument was that

$$\frac{x}{1-x} = x + x^2 + x^3 + \cdots$$

and

$$\frac{x}{x-1} = \frac{1}{1 - \frac{1}{x}} = 1 + \frac{1}{x} + \frac{1}{x^2} + \frac{1}{x^3} + \cdots.$$

But the two left sides add up to 0, while the two right sides add up to the original series.

In an earlier paper,[11] Euler started with the series

(24) $$y = \sin x = x - \frac{x^3}{3!} + \frac{x^5}{5!} - \cdots$$

or

(25) $$1 - \frac{x}{y} + \frac{x^3}{3!y} - \frac{x^5}{5!y} + \cdots = 0.$$

By using algebraic considerations applied to (25) as a *polynomial of infinite degree*, and by using the theorem on the relation between roots and coefficients of an algebraic equation, Euler proved that[12]

$$\frac{1}{1^2} + \frac{1}{3^2} + \frac{1}{5^2} + \cdots = \frac{\pi^2}{8}$$

$$\frac{1}{1^3} - \frac{1}{3^3} + \frac{1}{5^3} - \cdots = \frac{\pi^3}{32}$$

$$\frac{1}{1^4} + \frac{1}{3^4} + \frac{1}{5^4} + \cdots = \frac{\pi^4}{96}$$

$$\frac{1}{1^5} - \frac{1}{3^5} + \frac{1}{5^5} - \cdots = \frac{5\pi^5}{1536}$$

$$\frac{1}{1^6} + \frac{1}{3^6} + \frac{1}{5^6} + \cdots = \frac{\pi^6}{960}$$

. . . .

10. *Comm. Acad. Sci. Petrop.*, 11, 1739, 116–27, pub. 1750 = *Opera*, (1), 14, 350–63.
11. *Comm. Acad. Sci. Petrop.*, 7, 1734/35, 123–34, pub. 1740 = *Opera*, (1), 14, 73–86.
12. He used the symbol p for π until 1739; π had been introduced by William Jones in 1706.

In the same paper, he first gave the product expansion

(26) $$\sin s = \left(1 - \frac{s^2}{\pi^2}\right)\left(1 - \frac{s^2}{4\pi^2}\right)\left(1 - \frac{s^2}{9\pi^2}\right)\cdots.$$

His argument was simply that $\sin s$ has the zeros $\pm\pi$, $\pm 2\pi$, $\cdots$ (he discards the root 0), and so like every polynomial must have a linear factor corresponding to each of its roots. (In 1743[13] and in his *Introductio*,[14] he gave another derivation to meet criticism.) He treated the right side of (26) as a polynomial, which he set equal to zero, and again by using the relation between the roots and the coefficients he deduced that

$$\frac{1}{1^2} + \frac{1}{2^2} + \frac{1}{3^2} + \frac{1}{4^2} + \cdots = \frac{\pi^2}{6}$$

$$\frac{1}{1^4} + \frac{1}{2^4} + \frac{1}{3^4} + \frac{1}{4^4} + \cdots = \frac{\pi^4}{90},$$

and similar sums for higher even powers in the denominator.

In a later paper[15] Euler obtained one of his finest triumphs,

$$\sum_{\nu=1}^{\infty} \frac{1}{\nu^{2n}} = (-1)^{n-1} \frac{(2\pi)^{2n}}{2(2n)!} B_{2n},$$

where the B_{2n} are the Bernoulli numbers (see below). The connection with the Bernoulli numbers was actually established by Euler a little later in his *Institutiones* of 1755.[16] He also gave in the 1740 paper the sum $\sum_{\nu=1}^{\infty}(1/\nu^n)$ for the first few odd values of n but got no general expression for all odd n.

Euler also worked on harmonic series, that is, series such that the reciprocals of the terms are in arithmetic progression. In particular he showed[17] how one can sum a finite number of terms of the ordinary harmonic series by using the logarithm function. He starts with

(27) $$\log\left(1 + \frac{1}{x}\right) = \frac{1}{x} - \frac{1}{2x^2} + \frac{1}{3x^3} - \frac{1}{4x^4} + \cdots.$$

Then

$$\frac{1}{x} = \log\left(\frac{x+1}{x}\right) + \frac{1}{2x^2} - \frac{1}{3x^3} + \frac{1}{4x^4} - \cdots.$$

13. *Opera*, (1), 14, 138–55.
14. *Opera*, (1), 8, 168.
15. *Comm. Acad. Sci. Petrop.*, 12, 1740, 53–96, pub. 1750 = *Opera*, (1), 14, 407–62.
16. Part II, Chap. 5, ¶124 = *Opera*, (1), 10, 327.
17. *Comm. Acad. Sci. Petrop.*, 7, 1734/35, 150–61, pub. 1740 = *Opera*, (1), 14, 87–100.

Now let $x = 1, 2, 3, \ldots, n$. These substitutions give

$$\frac{1}{1} = \log 2 + \frac{1}{2} - \frac{1}{3} + \frac{1}{4} - \frac{1}{5} + \cdots$$

$$\frac{1}{2} = \log \frac{3}{2} + \frac{1}{2 \cdot 4} - \frac{1}{3 \cdot 8} + \frac{1}{4 \cdot 16} - \frac{1}{5 \cdot 32} + \cdots$$

$$\frac{1}{3} = \log \frac{4}{3} + \frac{1}{2 \cdot 9} - \frac{1}{3 \cdot 27} + \frac{1}{4 \cdot 81} - \frac{1}{5 \cdot 243} + \cdots$$

$$\cdots\cdots\cdots\cdots\cdots\cdots\cdots\cdots\cdots\cdots\cdots\cdots\cdots\cdots\cdots$$

$$\frac{1}{n} = \log \frac{n+1}{n} + \frac{1}{2n^2} - \frac{1}{3n^3} + \frac{1}{4n^4} - \frac{1}{5n^5} + \cdots.$$

By adding and noting that each log term is a difference of two logarithms, one gets

$$\frac{1}{1} + \frac{1}{2} + \frac{1}{3} + \cdots + \frac{1}{n} = \log(n+1) + \frac{1}{2}\left(1 + \frac{1}{4} + \frac{1}{9} + \cdots + \frac{1}{n^2}\right)$$

$$- \frac{1}{3}\left(1 + \frac{1}{8} + \frac{1}{27} + \cdots + \frac{1}{n^3}\right) + \frac{1}{4}\left(1 + \frac{1}{16} + \frac{1}{81} + \cdots + \frac{1}{n^4}\right) - \cdots,$$

or

$$(28) \qquad\qquad 1 + \frac{1}{2} + \frac{1}{3} + \cdots + \frac{1}{n} = \log(n+1) + C,$$

where C represents the sum of the infinite set of finite arithmetic sums. The value of C was calculated approximately by Euler (it depends upon n, but for large n the value of n does not affect the result much) and he obtained 0.577218. This C is now known as Euler's constant and is denoted by γ. A more accurate representation of γ is obtained nowadays as follows. Subtract $\log n$ from both sides of (28). Now $\log(n + 1) - \log n = \log(1 + 1/n)$ and this approaches 0 as $n \to \infty$. Hence

$$(29) \qquad\qquad \gamma = \lim_{n \to \infty}\left(1 + \frac{1}{2} + \frac{1}{3} + \cdots + \frac{1}{n} - \log n\right).$$

Incidentally, no simpler form than (29) has been found for Euler's constant, whereas we do have various expressions for π and e. Moreover, we do not know today whether γ is rational or irrational.

In his "De Seriebus Divergentibus"[18] Euler investigated the divergent series

18. *Novi Comm. Acad. Sci. Petrop.*, 5, 1754/5, 205–37, pub. 1760 = *Opera*, (1), 14, 585–617.

(30) $$y = x - (1!)x^2 + (2!)x^3 - (3!)x^4 + \cdots .$$

Formally this series satisfies the differential equation

(31) $$x^2 y' + y = x.$$

But this differential equation has the integrating factor $x^2 e^{-1/x}$, so that

(32) $$y = e^{1/x} \int_0^x \frac{e^{-1/t}}{t} \, dt$$

is a solution that can be shown by L'Hospital's rule to vanish with x. Euler considered the series (30) to be the series expansion of the function in (32) and (32) as the sum of the series (30). In fact he lets $x = 1$ and obtains

$$1 - 1 + 2! - 3! + 4! - \cdots = e \int_0^1 \frac{e^{-1/t}}{t} \, dt.$$

The remarkable fact about the series (30) is that it can be used to obtain good numerical values for the function (32) because, given a value of x, if we neglect all terms beyond a certain one, the absolute value of the remainder can be shown to be smaller than the absolute value of the first of the neglected terms. Hence the series can be used to obtain good numerical approximations to the integral. Euler was using divergent series to advantage. The full significance of what these divergent series accomplished was not appreciated for another 150 years. (See Chap. 47.)

Another famous result of Euler's in the area of series should be noted. In *Ars Conjectandi* James Bernoulli, treating the subject of probability, introduced the now widely used Bernoulli numbers. He sought a formula for the sums of the positive integral powers of the integers and gave the following formula without demonstration:

(33) $$\sum_{k=1}^{n} k^c = \frac{1}{c+1} n^{c+1} + \frac{1}{2} n^c + \frac{c}{2} B_2 n^{c-1} + \frac{c(c-1)(c-2)}{2 \cdot 3 \cdot 4} B_4 n^{c-3}$$
$$+ \frac{c(c-1)(c-2)(c-3)(c-4)}{2 \cdot 3 \cdot 4 \cdot 5 \cdot 6} B_6 n^{c-5} + \cdots .$$

This series terminates at the last positive power of n. The $B_2, B_4, B_6, \ldots$ are the Bernoulli numbers

(34) $$B_2 = \frac{1}{6}, \; B_4 = -\frac{1}{30}, \; B_6 = \frac{1}{42}, \; B_8 = -\frac{1}{30}, \; B_{10} = \frac{5}{66}, \; \cdots .$$

Bernoulli also gave the recurrence relation, which permits one to calculate these coefficients.

Euler's result, the Euler-Maclaurin summation formula, is a generalization.[19] Let $f(x)$ be a real-valued function of the real variable x. Then (in modern notation) the formula reads

$$(35) \quad \sum_{i=0}^{n} f(i) = \int_0^n f(x)\,dx - \frac{1}{2}[f(n) - f(0)] + \frac{B_2}{2!}[f'(n) - f'(0)]$$

$$+ \frac{B_4}{4!}[f'''(n) - f'''(0)] + \cdots + \frac{B_{2k}}{(2k)!}[f^{(2k-1)}(n) - f^{(2k-1)}(0)] + R_k$$

where

$$(36) \qquad\qquad R_k = \int_0^n f^{(2k+1)}(x) P_{2k+1}(x)\,dx.$$

Here n and k are positive integers. $P_{2k+1}(x)$ is the $(2k+1)$th Bernoulli polynomial (which also appears in Bernoulli's *Ars Conjectandi*), which is given by

$$(37) \qquad P_k(x) = \frac{x^k}{k!} + \frac{B_1}{1!}\frac{x^{k-1}}{(k-1)!} + \frac{B_2}{2!}\frac{x^{k-2}}{(k-2)!} + \cdots + \frac{B_k}{k!},$$

wherein $B_1 = -1/2$ and $B_{2k+1} = 0$ for $k = 1, 2, \cdots$. The series

$$(38) \qquad\qquad \sum_{k=1}^{\infty} \frac{B_{2k}}{(2k)!}[f^{(2k-1)}(n) - f^{(2k-1)}(0)]$$

is divergent for almost all $f(x)$ that occur in applications. Nevertheless, the remainder R_k is less than the first term neglected and so the series in (35) gives a useful approximation to

$$\sum_{i=0}^{n} f(i).$$

The Bernoulli numbers B_i are often defined today by a relation given later by Euler,[20] namely,

$$(39) \qquad\qquad t(e^t - 1)^{-1} = \sum_{i=0}^{\infty} B_i \frac{t^i}{i!}.$$

Independently of Euler, Maclaurin[21] arrived at the same summation formula (35) but by a method a little surer and closer to that which we use today. The remainder was first added and seriously treated by Poisson.[22]

Euler also introduced[23] a transformation of series, still known and

19. *Comm. Acad. Sci. Petrop.*, 6, 1732/3, 68–97, pub. 1738 = *Opera*, (1), 14, 42–72; and *Comm. Acad. Sci. Petrop.*, 8, 1736, 147–58, pub. 1741 = *Opera*, (1), 14, 124–37.
20. *Opera*, (1), 14, 407–62.
21. *Treatise of Fluxions*, 1742, p. 672.
22. *Mém. de l'Acad. des Sci.*, *Inst. France*, 6, 1823, 571–602, pub. 1827.
23. *Inst. Cal. Diff.*, 1755, p. 281.

used. Given a series $\sum_{n=0}^{\infty} b_n$, he wrote it as $\sum_{n=0}^{\infty} (-1)^n a_n$. Then by a number of formal algebraic steps he showed that

$$(40) \qquad \sum_{n=0}^{\infty} (-1)^n a_n = \sum_{n=0}^{\infty} (-1)^n \frac{\Delta^n a_0}{2^{n+1}},$$

wherein the Δ^n denotes the nth finite difference (sec. 3). The advantage of this transformation, in modern terms, is to convert a convergent series into a more rapidly converging one. However, for Euler, who did not usually distinguish convergent and divergent series, the transformation could also transform divergent series into convergent ones. If one applies (40) to

$$(41) \qquad 1 - 1 + 1 - 1 + \cdots,$$

then the right side of (40) yields 1/2. Likewise for the series

$$(42) \qquad 1 - 2 + 2^2 - 2^3 + 2^4 \cdots$$

(40) gives

$$(43) \qquad \sum_{n=0}^{\infty} (-1)^n 2^n = \frac{1}{2}(1) + \frac{1}{4}(-1) + \frac{1}{8}(1) - \frac{1}{16}(-1) \cdots = \frac{1}{3}.$$

These results are, of course, the same as those Euler got above (see [16] and [17]) by taking the sum of the series to be the value of the function from which the series is derived.

The spirit of Euler's methods should be clear. He is the great manipulator and pointed the way to thousands of results later established rigorously.

One other famous series must be mentioned. In his *Methodus Differentialis*[24] James Stirling gave the series we now write as

$$(44) \quad \log n! = \left(n + \frac{1}{2}\right) \log n - n + \log \sqrt{2\pi} + \frac{B_2}{1 \cdot 2} \frac{1}{n} + \frac{B_4}{3 \cdot 4} \frac{1}{n^3} + \cdots$$

$$+ \frac{B_{2k}}{(2k-1)(2k)} \frac{1}{n^{2k-1}} + \cdots,$$

which is equivalent to

$$(45) \quad n! = \left(\frac{n}{e}\right)^n \sqrt{2\pi n} \exp\left[\frac{B_2}{1 \cdot 2} \frac{1}{n} + \cdots + \frac{B_{2k}}{(2k-1)2k} \frac{1}{n^{2k-1}} + \cdots\right].$$

Stirling gave the first five coefficients and a recurrence formula for determining the succeeding ones. Though the series for $\log n!$ is divergent, Stirling calculated $\log_{10}(1000!)$, which is 2567 plus a decimal, to ten decimal places by using only a few terms of his series. De Moivre in 1730 (*Miscellanea*

24. 1730, p. 135.

Analytica) gave a similar formula. For large n, $n! \sim (n/e)^n \sqrt{2\pi n}$; though given by de Moivre, it is known as Stirling's approximation.

5. *Trigonometric Series*

The eighteenth-century mathematicians also worked extensively with trigonometric series, especially in their astronomical theory. The usefulness of such series in astronomy is evident from the fact that they are periodic functions and astronomical phenomena are largely periodic. This work was the beginning of a vast subject whose full significance was not appreciated in the eighteenth century. The problem that launched the use of trigonometric series was interpolation, particularly to determine the positions of the planets between those obtained by observation. The same series were introduced in the early work on partial differential equations (see Chap. 22) but curiously the two lines of thought were kept separate even though the same men worked on both problems.

By a trigonometric series is meant any series of the form

$$(46) \qquad \frac{1}{2} a_0 + \sum_{n=1}^{\infty} (a_n \cos nx + b_n \sin nx)$$

with a_n and b_n constant. If such a series represents a function $f(x)$, then

$$(47) \qquad a_n = \frac{1}{\pi} \int_0^{2\pi} f(x) \cos nx \, dx, \qquad b_n = \frac{1}{\pi} \int_0^{2\pi} f(x) \sin nx \, dx$$

for $n = 0, 1, 2, \ldots$. The attainment of these formulas for the coefficients was one of the chief results of the theory, though we shall say nothing at present about the conditions under which these are necessarily the values of a_n and b_n.

As early as 1729 Euler had undertaken the problem of interpolation; that is, given a function $f(x)$ whose values for $x = n$, n positive and integral, are prescribed, to find $f(x)$ for other values of x. In 1747 he applied the method he had obtained to a function arising in the theory of planetary perturbations and secured a trigonometric series representation of the function. In 1753[25] he published the method he had found in 1729.

First he tackled the problem when the given conditions are $f(n) = 1$ for each n and sought a periodic solution that is 1 for integral x. His reasoning is interesting because it illustrates the analysis of the period. He lets $f(x) = y$, and by Taylor's theorem writes

$$(48) \qquad f(x + 1) = y + y' + \frac{1}{2} y'' + \frac{1}{6} y''' + \cdots.$$

25. *Novi Comm. Acad. Sci. Petrop.*, 3, 1750/51, 36–85, pub. 1753 = *Opera*, (1), 14, 463–515.

Since $f(x + 1)$ is to equal $f(x)$, y must satisfy the linear differential equation of infinite order

(49)
$$y' + \frac{1}{2}y'' + \frac{1}{6}y''' + \cdots = 0.$$

He now applied his method of solving linear ordinary differential equations of finite order published in 1743 (see Chap. 21). That is, he set up the auxiliary equation

(50)
$$z + \frac{1}{2}z^2 + \frac{1}{6}z^3 + \cdots = 0.$$

This equation, in view of the series for e, is

$$e^z - 1 = 0.$$

Next he determines the roots of this last equation. He starts with the equation

$$\left(1 + \frac{z}{n}\right)^n = 1,$$

which is a polynomial of the nth degree. According to a theorem which Cotes (1722) and Euler independently in his *Introductio*[26] had proven, this polynomial has the linear factor z and the quadratic factors

$$\left(1 + \frac{z}{n}\right)^2 - 2\left(1 + \frac{z}{n}\right)\cos\frac{2k\pi}{n} + 1, \qquad k = 1, 2, \ldots, < \frac{n}{2}.$$

By virtue of the trigonometric identity for $\sin z$ in terms of $\cos 2z$ these factors are the same as

$$4\left(1 + \frac{z}{n}\right)\sin^2\frac{k\pi}{n} + \frac{z^2}{n^2}.$$

The roots of (50) are not affected if we divide each factor by $4\sin^2 k\pi/n$ (for the respective k), and so the quadratic factors are

$$1 + \frac{z}{n} + \frac{z^2}{4n^2 \sin^2\dfrac{k\pi}{n}}.$$

For $n = \infty$, the term z/n is 0. The quantity $\sin k\pi/n$ is replaced by $k\pi/n$, and so the factors become

$$1 + \frac{z^2}{4k^2\pi^2}.$$

26. Vol. 1, Chap. 14.

To such a factor in the auxiliary equation (50) there correspond the roots $z = \pm i2k\pi$, and hence the integral

$$\alpha_k \sin 2k\pi x + A_k \cos 2k\pi x$$

of (49). The linear factor z mentioned above gives rise to a constant integral. Since $f(0) = 1$ is an initial condition, Euler finally obtains

$$y = 1 + \sum_{k=1}^{\infty} \{\alpha_k \sin 2k\pi x + A_k(\cos 2k\pi x - 1)\}.$$

The coefficients α_k and A_k are still subject to the condition that $f(n) = 1$ for each n.

This paper also contains a result which is formally identical with what came to be called the Fourier expansion of an arbitrary function, as well as the determination of the coefficients by integrals. Specifically Euler showed that the general solution of the functional equation

$$f(x) = f(x - 1) + X(x)$$

is

$$f(x) = \int_0^x X(\xi) \, d\xi + 2 \sum_{n=1}^{\infty} \cos 2n\pi x \int_0^x X(\xi) \cos 2n\pi\xi \, d\xi$$

$$+ 2 \sum_{n=1}^{\infty} \sin 2n\pi x \int_0^x X(\xi) \sin 2n\pi\xi \, d\xi.$$

Here we have a function expressed as a trigonometric series in the year 1750–51. Euler maintained that his was the most general solution of the interpolation problem. If so, it surely included the representation of polynomials by trigonometric series. But, as we shall see in Chapter 22, Euler denied this in the arguments on the vibrating string and related problems.

In 1754 d'Alembert[27] considered the problem of the expansion of the reciprocal of the distance between two planets in a series of cosines of the multiples of the angle between the rays from the origin to the planets, and here too the definite integral expressions for the coefficients in Fourier series can be found.

In another work Euler obtains trigonometric series representations of functions in a totally different fashion.[28] He starts with the geometric series

$$\sum_{n=0}^{\infty} a^n(\cos x + i \sin x)^n, \qquad i = \sqrt{-1}$$

27. *Recherches sur différens points importans du système du monde*, 1754, Vol. II, p. 66.
28. *Novi Comm. Acad. Sci. Petrop.*, 5, 1754/5, 164–204, pub. 1760 = *Opera*, (1), 14, 542–84; see also *Opera*, (1), 15, 435–97, for another method.

and by summing it obtains

$$\frac{1}{1 - a(\cos x + i \sin x)}.$$

He then uses standard formulas to replace powers of $\cos x$ and $\sin x$ by $\cos nx$ and $\sin nx$ (which amounts to de Moivre's theorem) and obtains

$$\frac{1}{1 - a(\cos x + i \sin x)} = \sum_{n=0}^{\infty} a^n(\cos nx + i \sin nx).$$

By multiplying numerator and denominator on the left by the complex conjugate of the denominator, separating the $n = 0$ term on the right and putting it on the left side, dividing through by a, and separating real and imaginary parts, he obtains

$$\frac{a \cos x - a^2}{1 - 2a \cos x + a^2} = \sum_{n=1}^{\infty} a^n \cos nx$$

$$\frac{a \sin x}{1 - 2a \cos x + a^2} = \sum_{n=1}^{\infty} a^n \sin nx.$$

So far his results are not surprising. He now lets $a = \pm 1$ and obtains, for example,

(51) $$\frac{1}{2} = 1 \pm \cos x + \cos 2x \pm \cos 3x + \cos 4x \pm \cdots.$$

(Actually the series are divergent.) He then integrates and obtains

(52) $$\frac{\pi - x}{2} = \sin x + \frac{1}{2} \sin 2x + \frac{1}{3} \sin 3x + \cdots,$$

(which holds for $0 < x < \pi$ and equals 0 for $x = 0$ and π) and

(53) $$\frac{x}{2} = \sin x - \frac{1}{2} \sin 2x + \frac{1}{3} \sin 3x - \frac{1}{4} \sin 4x + \cdots,$$

(which converges in $-\pi < x < \pi$). An integration of the latter and evaluation at $x = 0$ to determine the constant of integration gives

(54) $$\frac{x^2}{4} - \frac{\pi^2}{4} = -\cos x + \frac{1}{4} \cos 2x - \frac{1}{9} \cos 3x + \frac{1}{16} \cos 4x - \cdots.$$

Euler believed that the latter two series [which are convergent in $(-\pi < x < \pi)$] represent the respective functions for all values of x. Moreover, by successively differentiating (51), Euler deduced that

$$\sin x \pm 2 \sin 2x + 3 \sin 3x \pm \cdots = 0$$
$$\cos x \pm 4 \cos 2x + 9 \cos 3x \pm \cdots = 0$$

and other such equations. Daniel Bernoulli, who had also given expansions such as (52), (53), and (54), recognized that the series represent the functions only for certain ranges of x values.

In 1757, while studying perturbations caused by the sun, Clairaut[29] took a far bolder step. He says he will represent *any* function in the form

$$(55) \qquad f(x) = A_0 + 2 \sum_{n=1}^{\infty} A_n \cos nx.$$

He regards the problem as one of interpolation and so uses the function values at the x-values

$$\frac{2\pi}{k}, \qquad \frac{4\pi}{k}, \qquad \frac{6\pi}{k}, \ldots$$

and after some manipulations obtains

$$A_0 = \frac{1}{k} \sum_{\mu} f\left(\frac{2\mu\pi}{k}\right)$$

$$A_n = \frac{1}{k} \sum_{\mu} f\left(\frac{2\mu\pi}{k}\right) \cos \frac{2\mu n\pi}{k}.$$

By letting k become infinite, Clairaut arrives at

$$(56) \qquad A_n = \frac{1}{2\pi} \int_0^{2\pi} f(x) \cos nx \, dx,$$

which is the correct formula for the A_n.

Lagrange, in his research on the propagation of sound,[30] obtained the series (51) and defended the fact that the sum is 1/2. Yet neither Euler nor Lagrange commented on the remarkable fact that they had expressed non-periodic functions in the form of trigonometric series. However, somewhat later they did observe this fact in another connection. D'Alembert had often given the example of $x^{2/3}$ as a function that could not be expanded in a trigonometric series. Lagrange showed him in a letter[31] of August 15, 1768 that $x^{2/3}$ can indeed be expressed in the form

$$x^{2/3} = a + b \cos 2x + c \cos 4x + \cdots.$$

D'Alembert objected and gave counterarguments, such as that the derivatives of the two sides are not equal for $x = 0$. Also, by Lagrange's method one could express $\sin x$ as a cosine series; yet $\sin x$ is an odd function,

29. *Hist. de l'Acad. des Sci.*, Paris, 1754, 545 ff., pub. 1759.
30. *Misc. Taur.*, 1, 1759 = *Œuvres*, 1, 110.
31. *Lagrange, Œuvres*, 13, 116.

whereas the right side would be an even one. The problem was not resolved in the eighteenth century.

In 1777,[32] Euler, working on a problem in astronomy, actually obtained the coefficients of a trigonometric series by using the orthogonality of the trigonometric functions, the method we use today. That is, from

(57)
$$f(x) = \frac{a_0}{2} + \sum_{k=1}^{\infty} a_k \cos \frac{k\pi x}{l}$$

he deduced that

$$a_k = \frac{2}{l} \int_0^l f(s) \cos \frac{k\pi s}{l} \, ds.$$

He had first obtained it, in the immediately preceding paper, in a somewhat complicated fashion, then realized that he could obtain it directly by multiplying both sides of (57) by $\cos(\nu\pi x/l)$, integrating term by term, and applying the relations

$$\int_0^l \cos \frac{\nu\pi x}{l} \cos \frac{k\pi x}{l} \, dx = \begin{cases} 0 & \text{if } \nu \neq k \\ l/2 & \text{if } \nu = k \neq 0 \\ l & \text{if } \nu = k = 0 \end{cases}.$$

Throughout all of the above work on trigonometric series ran the paradox that, although all sorts of functions were being represented by trigonometric series, Euler, d'Alembert, and Lagrange never abandoned the position that *arbitrary* functions could not be represented by such series. The paradox is partially explained by the fact that the trigonometric series were assumed to hold where other evidence, in some cases physical, seemed to assure this fact. They then felt free to assume the series and deduce the formulas for the coefficients. This issue of whether any function can be represented by a trigonometric series became central.

6. *Continued Fractions*

We have already noted (Chap. 13, sec. 2) the use of continued fractions to obtain approximations to irrational numbers. Euler took up this subject. In his first paper on it,[33] entitled "De Fractionibus Continuis," he derived a number of interesting results, such as that every rational number can be expressed as a finite continued fraction. He then gave the expansions

$$e - 1 = 1 + \frac{1}{1+} \frac{1}{2+} \frac{1}{1+} \frac{1}{1+} \frac{1}{4+} \frac{1}{1+} \frac{1}{1+} \frac{1}{6+} \cdots,$$

32. *Nova Acta Acad. Sci. Petrop.*, 11, 1793, 114–32, pub. 1798 = *Opera*, (1), 16, Part 1, 333–55.

33. *Comm. Acad. Sci. Petrop.*, 9, 1737, 98–137, pub. 1744 = *Opera*, (1), 14, 187–215.

which had already appeared in a paper by Cotes in the *Philosophical Trans-actions* of 1714, and

$$\frac{e+1}{e-1} = 2 + \frac{1}{6+}\ \frac{1}{10+}\ \frac{1}{14+}\ \cdots.$$

He showed substantially that e and e^2 are irrational.

The foundations of a theory of continued fractions were laid by Euler in his *Introductio* (Chap. 18). There he showed how to go from a series to a continued fraction representation of the series, and conversely.

Euler's work on continued fractions was used by Johann Heinrich Lambert (1728–77), a colleague of Euler and Lagrange at the Berlin Academy of Sciences, to prove[34] that if x is a rational number (not 0), then e^x and tan x cannot be rational. He thereby proved not only that e^x for positive integral x is irrational, but that all rational numbers have irrational natural (base e) logarithms. From the result on tan x, it follows, since $\tan(\pi/4) = 1$, that neither $\pi/4$ nor π can be rational. Lambert actually proved the convergence of the continued fraction expansion for tan x.

Lagrange[35] used continued fractions to find approximations to the irrational roots of equations, and, in another paper in the same journal,[36] he got approximate solutions of differential equations in the form of con-tinued fractions. In the 1768 paper, Lagrange proved the converse of a theorem that Euler had proved in his 1744 paper. The converse states that a real root of a quadratic equation is a periodic continued fraction.

7. *The Problem of Convergence and Divergence*

We are aware today that the eighteenth-century work on series was largely formal, and that the question of convergence and divergence was certainly not taken too seriously; neither, however, was it entirely ignored.

Newton,[37] Leibniz, Euler, and even Lagrange regarded series as an extension of the algebra of polynomials and hardly realized that they were introducing new problems by extending sums to an infinite number of terms. Consequently, they were not quite prepared to face the problems that infinite series thrust upon them; but the apparent difficulties that did arise caused them at least occasionally to bring up these questions. What is especially interesting is that the correct resolution of the paradoxes and other difficulties was often voiced and just as often ignored.

Even some seventeenth-century men had observed the distinction between convergence and divergence. In 1668 Lord Brouncker, treating

34. *Hist. de l'Acad. de Berlin*, 1761, 265–322, pub. 1768 = *Opera*, 2, 112–59.
35. *Nouv. Mém. de l'Acad. de Berlin*, 23, 1767, 311–52, pub. 1769 = *Œuvres*, 2, 539–78, and 24, 1768, 111–80, pub. 1770 = *Œuvres*, 2, 581–652.
36. 1776 = *Œuvres*, 4, 301–34.
37. See the quotation from Newton in Chap. 17, sec. 3.

the relation between $\log x$ and the area under $y = 1/x$, demonstrated the convergence of the series for $\log 2$ and $\log 5/4$ by comparison with a geometric series. Newton and James Gregory, who made much use of numerical values of series to calculate logarithmic and other function tables and to evaluate integrals, were aware that the sums of series can be finite or infinite. The terms "convergent" and "divergent" were actually used by James Gregory in 1668, but he did not develop the ideas. Newton recognized the need to consider convergence but did no more than affirm that power series converge for small values of the variable at least as well as the geometric series. He also remarked that some series can be infinite for some values of x and so be useless.

Leibniz, too, felt some concern about convergence and noted in a letter of October 25, 1713, to John Bernoulli what is now a theorem, that a series whose terms alternate in sign and decrease in absolute value monotonically to zero converges.[38]

Maclaurin, in his *Treatise of Fluxions* (1742), used series as a regular method for integration. He says, "When a fluent cannot be represented accurately in algebraic terms, it is then to be expressed by a converging series." That the terms of a convergent series must continually decrease and become less than any quantity howsoever small that can be assigned he also recognized. "In that case a few terms at the beginning of the series will be nearly equal to the value of the whole." In the *Treatise* Maclaurin gave the integral test (independently discovered by Cauchy) for the convergence of an infinite series: $\sum_n \phi(n)$ converges if and only if $\int_a^\infty \phi(x)$ is finite, provided that $\phi(x)$ is finite and of the same sign for $a \leq x \leq \infty$. Maclaurin gave it in geometrical form.

Some ideas about convergence were also expressed by Nicholas Bernoulli (1687–1759) in letters to Leibniz of 1712 and 1713. In a letter of April 7, 1713,[39] Bernoulli says the series

$$(1 + x)^n = 1 + nx + \frac{n(n-1)}{2} x^2 + \cdots$$

has no sum when x is negative and numerically greater than 1 if n is fractional and has an even denominator. That is, the (arithmetical) divergence of a series is not the only reason for a series not to have a sum. Thus for $x > 1$ both series

$$(1 - x)^{-1/3} = 1 + \frac{1}{3} x + \frac{1 \cdot 4}{3 \cdot 6} x^2 + \frac{1 \cdot 4 \cdot 7}{3 \cdot 6 \cdot 9} x^3 + \cdots$$

$$(1 - x)^{-1/2} = 1 + \frac{1}{2} x + \frac{1 \cdot 3}{2 \cdot 4} x^2 + \frac{1 \cdot 3 \cdot 5}{2 \cdot 4 \cdot 6} x^3 + \cdots$$

38. *Math. Schriften*, 3, 922–23. Leibniz also gave an incorrect proof in a letter to John of January 10, 1714 = *Math. Schriften*, 3, 926.
39. Leibniz: *Math. Schriften*, 3, 980–84.

are divergent, but the first series has a possible value and the second an imaginary value. One cannot distinguish the two by examining the series because the *remainders are missing*. However, Nicholas did not set up a clear concept of convergence. In a reply of June 28, 1713, Leibniz[40] uses the term "advergent" for series that converge (roughly in our sense) and agrees that non-advergent series may be impossible or infinitely large.

There is no doubt that Euler saw some of the difficulties with divergent series, and in particular the difficulty in using them for computations, but he certainly was unclear about the concepts of convergence and divergence. He did recognize that the terms must become infinitely small for convergence. The letters described below tell us, indirectly, something of his views.

Nicholas Bernoulli (1687–1759), in correspondence with Euler during 1742–43, had challenged some of Euler's ideas and work. He pointed out that Euler's use in his paper of 1734/35 (see sec. 4) of

$$\sin s = s - \frac{s^3}{3!} + \frac{s^5}{5!} + \cdots = \left(1 - \frac{s^2}{\pi^2}\right)\left(1 - \frac{s^2}{4\pi^2}\right)\left(1 - \frac{s^2}{9\pi^2}\right)\cdots$$

does yield

$$\frac{\pi^2}{6} = 1 + \frac{1}{2^2} + \frac{1}{3^2} + \frac{1}{4^2} + \cdots,$$

but a proof of the convergence of the basic series in s is missing. In a letter of April 6, 1743,[41] he says he cannot imagine that Euler can believe a divergent series gives the exact value of some quantity or function. He points out that the remainder is lacking. Thus $1/(1 - x)$ cannot equal $1 + x + x^2 + \cdots$ because the remainder, namely, $x^{\infty+1}/(1 - x)$, is missing.

In another letter of 1743, Bernoulli says Euler must distinguish between a finite sum and a sum of an infinite number of terms. There is no last term in the latter case. Hence one cannot use for infinite polynomials (as Euler did) the relation between the roots and coefficients of a polynomial of finite degree. For polynomials with an infinite number of terms one cannot speak of the sum of the roots.

Euler's answers to these letters of Bernoulli are not known. In writing to Goldbach on August 7, 1745,[42] Euler refers to Bernoulli's argument that divergent series such as

$$+1 - 2 + 6 - 24 + 120 - 720 - \cdots$$

have no sum but says that these series have a definite *value*. He notes that we should not use the term "sum" because this refers to actual addition. He then states the general principle which explains what he means by a

40. *Math. Schriften*, 3, 986.
41. Fuss: *Correspondance*, 2, 701 ff.
42. Fuss: *Correspondance*, 1, 324.

definite value. He points out that the divergent series come from finite algebraic expressions and then says that the value of the series is *the value of the algebraic expression from which the series comes.* In the paper of 1754/55 (sec. 4), he adds, "Whenever an infinite series is obtained as the development of some closed expression, it may be used in mathematical operations as the equivalent of that expression, even for values of the variable for which the series diverges." He repeats the first principle in his *Institutiones* of 1755:

> Let us say, therefore, that the sum of any infinite series is the finite expression, by the expansion of which the series is generated. In this sense the sum of the infinite series $1 - x + x^2 - x^3 + \cdots$ will be $1/(1 + x)$, because the series arises from the expansion of the fraction, whatever number is put in place of x. If this is agreed, the new definition of the word sum coincides with the ordinary meaning when a series converges; and since divergent series have no sum in the proper sense of the word, no inconvenience can arise from this terminology. Finally, by means of this definition, we can preserve the utility of divergent series and defend their use from all objections.[43]

It is fairly certain that Euler meant to limit the doctrine to power series.

In writing to Nicholas Bernoulli in 1743, Euler did say that he had had grave doubts as to the use of divergent series but that he had never been led into error by using his definition of sum.[44] To this Bernoulli replied that the same series might arise from the expansion of two different functions and, if so, the sum would not be unique.[45] Euler then wrote to Goldbach (in the letter of August 7, 1745): "Bernoulli gives no examples and I do not believe it possible that the same series could come from two truly different algebraic expressions. Hence it follows unquestionably that any series, divergent or convergent, has a definite sum or value."

There is an interesting sequel to this argument. Euler rested on his contention that the sum of series such as

(58) $$1 - 1 + 1 - 1 + 1 \cdots$$

could be the value of the function from which the series comes. Thus the above series comes from $1/(1 + x)$ when $x = 1$, and so has the value $1/2$. However, Jean-Charles (François) Callet (1744–99), in an unpublished memorandum submitted to Lagrange (Lagrange approved it for publication in the *Mémoires* of the Academy of Sciences of Paris but it was never published), pointed out some forty years later that

(59) $$\frac{1 + x + \cdots + x^{m-1}}{1 + x + \cdots + x^{n-1}} = \frac{1 - x^m}{1 - x^n}$$

$$= 1 - x^m + x^n - x^{n+m} + x^{2n} - \cdots.$$

43. Paragraphs 108–11.
44. *Opera Posthuma*, 1, 536.
45. April 6, 1743; Fuss: *Correspondance*, 2, 701 ff.

Hence for $x = 1$ (and $m < n$), since the left side is m/n, the sum of the right side must also be m/n, where m and n are at our disposal.

Lagrange[46] considered Callet's objection and argued that it was incorrect. He used Leibniz's probability argument thus: Suppose $m = 3$ and $n = 5$. Then the *full* series on the right side of (59) is

$$1 + 0 + 0 - x^3 + 0 + x^5 + 0 + 0 - x^8 + 0 + x^{10} + 0 - \cdots.$$

Now if one takes for $x = 1$, the sum of the first term, the first two, the first three, ..., then in each five of these partial sums three are equal to 1 and two are equal to 0. Hence the most probable value (mean value) is 3/5; and this is the value of the series in (59) for $m = 3$ and $n = 5$. Incidentally, Poisson, without mentioning Lagrange, repeats Lagrange's argument.[47]

Euler did say that great care should be exercised in the summation of divergent series. He also made a distinction between divergent series and semiconvergent series, such as (58), that oscillate in value as more and more terms are added but do not become infinite. Certainly he recognized the distinction between convergent series and divergent ones. In one case (1747), where he used infinite series to calculate the attraction that the earth, as an oblate spheroid, exerts on a particle at the pole, he says the series "converges vehemently."

Lagrange, too, showed some awareness of the distinction between convergence and divergence. In his earlier writings he was indeed lax on this matter. In one paper[48] he says that a series will represent a number if it converges to its extremity, that is, if its nth term approaches 0. Later, toward the end of the eighteenth century, when he worked with Taylor's series, he gave what we call Taylor's theorem,[49] namely,

$$f(x + h) = f(x) + f'(x)h + f''(x)\frac{h^2}{2!} + \cdots + f^{(n)}(x)\frac{h^n}{n!} + R_n$$

where

$$R_n = f^{(n+1)}(x + \theta h)\frac{h^{n+1}}{(n+1)!}$$

and θ is between 0 and 1 in value. This expression for R_n is still known as Lagrange's form of the remainder. Lagrange said that the Taylor (infinite)

46. *Mém. de l'Acad. des Sci., Inst. France*, 3, 1796, 1–11, pub. 1799; this article does not appear in the *Œuvres*.
47. *Jour. de l'Ecole Poly.*, 12, 1823, 404–509. If one insists on using the full *power* series, then Lagrange's argument makes more sense. It can be rigorized by applying Frobenius's definition of summability (Chap. 47, sec. 4).
48. *Hist. de l'Acad. de Berlin*, 24, 1770 = *Œuvres*, 3, 5–73, p. 61 in particular.
49. *Théorie des fonctions*, 2nd ed., 1813, Chap. 6 = *Œuvres*, 9, 69–85. The mean value theorem of the differential calculus, $f(b) - f(a) = f'(c)(b - a)$, is due to Lagrange (1797). Later it was used to derive Taylor's theorem as in modern books.

serics should not be used without consideration of the remainder. However, he did not investigate the idea of convergence or the relation of the value of the remainder to the convergence of the infinite series. He thought that one need consider only a finite number of terms of the series, enough to make the remainder small. Convergence was considered later by Cauchy, who stressed Taylor's theorem as primary, as well as the fact that to obtain a convergent series the remainder must approach 0.

D'Alembert, too, distinguished convergent from divergent series. In his article "Série" in the *Encyclopédie* he says, "When the progression or series approaches some finite quantity more and more, and, consequently, the terms of the series, or quantities of which it is composed, go on diminishing, one calls it a convergent series, and if one continues to infinity, it will finally become equal to this quantity. Thus $1/2 + 1/4 + 1/8 + 1/16 + \cdots$ form a series which always approaches 1 and which will become equal to it finally when the series is continued to infinity." In 1768 d'Alembert expressed doubts about the use of nonconvergent series. He said, "As for me, I avow that all the reasonings based on series that are not convergent . . . appear to me very suspect, even when the results are in accord with truths arrived at in other ways."[50] In view of the effective uses of series by John Bernoulli and Euler, the doubts such as d'Alembert expressed went unheeded in the eighteenth century. In this same volume d'Alembert gave a test for the absolute convergence of the series $u_1 + u_2 + u_3 + \cdots$, namely, if for all n greater than some fixed value r, the ratio $|u_{n+1}/u_n| < \rho$ where ρ is independent of n and less than 1, the series converges absolutely.[51]

Edward Waring (1734–98), Lucasian professor of mathematics at Cambridge University, held advanced views on convergence. He taught that

$$1 + \frac{1}{2^n} + \frac{1}{3^n} + \frac{1}{4^n} + \cdots$$

converges when $n > 1$ and diverges when $n < 1$. He also gave (1776) the well-known test for convergence and divergence, now known as the ratio test and attributed to Cauchy. The ratio of the $(n + 1)$st to the nth term is formed, and if the limit as $n \to \infty$ is less than 1, the series converges; if greater than 1, the series diverges. No conclusion may be drawn when the limit is 1.

Though Lacroix said several nonsensical things about series in the 1797 edition of his influential *Traité du calcul différentiel et du calcul intégral*, he was more cautious in his second edition. Speaking of

$$\frac{a}{a - x} = 1 + \frac{x}{a} + \frac{x^2}{a^2} + \frac{x^3}{a^3} + \cdots,$$

50. *Opuscules mathématiques*, 5, 1768, 183.
51. Pages 171–82.

he says that one should speak of the series as a *development* of the function because the series does not always have the *value* of the function to which it belongs.[52] The series, he says, gives the value of the function only for $|x| < |a|$. He continues with a thought already expressed by Euler, that the infinite series is nevertheless tied in with the function for all x. In any analytical work involving the series we would be right to conclude that we are dealing with the function. Thus if we discover some property of the series, we may be sure this property holds for the function. To perceive the truth of this assertion, it is sufficient to observe that the series verifies the equation that characterizes the function. For example, for $y = a/(a - x)$ we have

$$a - (a - x)y = 0.$$

But if one substitutes the series for y in this last equation, he will see that the series also satisfies it. One knows, Lacroix continues, that it would be the same for any other example; and he points to the great number presented in the text.

It is fair to say that in the eighteenth-century work on infinite series the formal view dominated. On the whole, the mathematicians even resented any limitations, such as the need to think about convergence. Their work produced useful results, and they were satisfied with this pragmatic sanction. They did exceed the bounds of what they could justify, but they were at least prudent in their use of divergent series. As we shall see, the insistence on restricting the use of series to convergent ones won out during most of the nineteenth century. But the eighteenth-century men were ultimately vindicated; two vital ideas that they glimpsed in infinite series were later to gain acceptance. The first was that divergent series can be useful for numerical approximations of functions; the second, that a series may represent a function in analytical operations, even though the series is divergent.

Bibliography

Bernoulli, James: *Ars Conjectandi*, 1713, reprinted by Culture et Civilisation, 1968.
————: *Opera*, 2 vols., 1744, reprinted by Birkhaüser, 1968.
Bernoulli, John: *Opera Omnia*, 4 vols., 1742, reprinted by Georg Olms, 1968.
Burkhardt, H.: "Trigonometrische Reihen und Integrale bis etwa 1850," *Encyk. der math. Wiss.*, B. G. Teubner, 1914–15, 2, Part 1, pp. 825–1354.
————: "Entwicklungen nach oscillirenden Functionen," *Jahres. der Deut. Math.-Verein.*, Vol. 10, 1908, pp. 1–1804.
————: "Über den Gebrauch divergenter Reihen in der Zeit 1750–1860," *Math. Ann.*, 70, 1911, 189–206.

52. 1810–19, 3 vols.; Vol. 1, p. 4.

Cantor, Moritz: *Vorlesungen über Geschichte der Mathematik*, B. G. Teubner, 1898, Vol. 3, Chaps. 85, 86, 97, 109, 110.

Dehn, M., and E. D. Hellinger: "Certain Mathematical Achievements of James Gregory," *Amer. Math. Monthly*, 50, 1943, 149–63.

Euler, Leonhard: *Opera Omnia*, (1), Vols. 10, 14, and 16 (2 parts), B. G. Teubner and Orell Füssli, 1913, 1924, 1933, and 1935.

Fuss, Paul Heinrich von: *Correspondance mathématique et physique de quelques célèbres géomètres du XVIIIème siècle*, 2 vols., 1843, Johnson Reprint Corp., 1967.

Hofmann, Joseph E.: "Über Jakob Bernoullis Beiträge zur Infinitesimalmathematik," *L'Enseignement Mathématique*, (2), 2, 1956, 61–171; also published separately by Institut de Mathématiques, Genève, 1957.

Montucla, J. F.: *Histoire des mathématiques*, A. Blanchard (reprint), 1960, Vol. 3, pp. 206–43.

Reiff, R. A.: *Geschichte der unendlichen Reihen*, H. Lauppsche Buchhandlung, 1889; Martin Sändig (reprint), 1969.

Schneider, Ivo: "Der Mathematiker Abraham de Moivre (1667–1754)," *Archive for History of Exact Sciences*, 5, 1968, 177–317.

Smith, David Eugene: *A Source Book in Mathematics*, Dover (reprint), 1959, Vol. 1, pp. 85–90, 95–98.

Struik, D. J.: *A Source Book in Mathematics, 1200–1800*, Harvard University Press, 1969, pp. 111–15, 316–24, 328–33, 338–41, 369–74.

Turnbull, H. W.: *James Gregory Tercentenary Memorial Volume*, Royal Society of Edinburgh, 1939.

————: *The Correspondence of Issac Newton*, Cambridge University Press, 1959, Vol. 1.

21

Ordinary Differential Equations in the Eighteenth Century

> A traveler who refuses to pass over a bridge until he has personally tested the soundness of every part of it is not likely to go far; something must be risked, even in mathematics.
>
> HORACE LAMB

1. *Motivations*

The mathematicians sought to use the calculus to solve more and more physical problems and soon found themselves obliged to handle a new class of problems. They wrought more than they had consciously sought. The simpler problems led to quadratures that could be evaluated in terms of the elementary functions. Somewhat more difficult ones led to quadratures which could not be so expressed, as was the case for elliptic integrals (Chap. 19, sec. 4). Both of these types fall within the purview of the calculus. However, solution of the still more complicated problems demanded specialized techniques; thus the subject of differential equations arose.

Several classes of physical problems motivated the investigations in differential equations. Problems in the area now generally known as the theory of elasticity were one class. A body is elastic if it deforms under the action of a force and recovers its original shape when the force is removed. The most practical problems are concerned with the shapes assumed by beams, vertical and horizontal, when loads are applied. These problems, treated empirically by the builders of the great medieval cathedrals, were approached mathematically in the seventeenth century by men such as Galileo, Edme Mariotte (1620?–84), Robert Hooke (1635–1703), and Wren. The behavior of beams is one of the two sciences Galileo treats in the *Dialogues Concerning Two New Sciences.* Hooke's investigation of springs led to his discovery of the law that states that the restoring force of a spring that is stretched or contracted is proportional to the stretch or contraction. The men of the eighteenth century, armed with more mathematics, began their work in elasticity by tackling such problems as the shape assumed by an inelastic but flexible rope suspended from two fixed points, the shape of an inelastic but flexible cord or chain suspended from one fixed point and set into vibration, the shape assumed by an elastic vibrating string held fixed at its ends, the shape of a rod when fixed at its ends and subject to a load, and the shape when the rod is set into vibration.

The pendulum continued to interest the mathematicians. The exact differential equation for the circular pendulum, $d^2\theta/dt^2 + g/l \sin \theta = 0$, defied treatment, and even the approximate one obtained by replacing $\sin \theta$ by θ had yet to be treated analytically. Moreover, the period of a circular pendulum is not strictly independent of the amplitude of the motion, and the search was undertaken for a curve along which the bob of a pendulum must swing for the period to be strictly independent of the amplitude. Huygens had solved this geometrically by introducing the cycloid; but the analytical solution was yet to be fashioned.

The pendulum was closely connected with two other major investigations of the eighteenth century, the shape of the earth and the verification of the inverse square law of gravitational attraction. The approximate period of a pendulum, $T = 2\pi\sqrt{l/g}$, was used to measure the force of gravity at various points on the surface of the earth because the period depends on the acceleration g determined by that force. By measuring successive lengths along a meridian, each length corresponding to a change of one degree in latitude, one can, with the aid of some theory and the values of g, determine the shape of the earth. In fact by using the observed variation in the period at various places on the earth's surface, Newton deduced that the earth bulges at the equator.

After Newton had, by his theoretical argument, concluded that the equatorial radius was 1/230 longer than the polar radius (this value is 30 percent too large), the European scientists were eager to confirm it. One method would be to measure the length of a degree of latitude near the equator and near a pole. If the earth were flattened, one degree of latitude would be slightly longer at the poles than at the equator.

Jacques Cassini (1677–1756) and members of his family made such measurements and in 1720 gave an opposite result. They found that the pole-to-pole diameter was 1/95 longer than the equatorial diameter. To settle the question the French Academy of Sciences sent out two expeditions in the 1730s, one to Lapland under the mathematician Pierre L. M. de Maupertuis and the other to Peru. Maupertuis's party included a fellow mathematician Alexis-Claude Clairaut. Their measurements confirmed that the earth was flattened at the poles; Voltaire hailed Maupertuis as the "flattener of the poles and the Cassinis." Actually Maupertuis's value was 1/178, which was less accurate than Newton's. The question of the shape of the earth remained a major subject and for a long time it was open as to whether the shape was an oblate spheroid, a prolate spheroid, a general ellipsoid, or some other figure of revolution.

The related problem, verifying the law of gravitation, could be handled if the shape of the earth were known. Given the shape, one could determine the centripetal force needed to keep an object on or near the surface of the rotating earth. Then, knowing the acceleration g due to the force of gravity at the surface, one could test whether the full force of gravity, which supplies the centripetal acceleration and g, is indeed the inverse square law. Clairaut, one of the men who questioned the law, believed at one time that it should be of the form $F = A/r^2 + B/r^3$. The two problems of the law of attraction and the shape of the earth are further intertwined, because when the earth is treated as a rotating fluid in equilibrium, the conditions for equilibrium involve the attraction of the particles of the fluid on each other.

The physical field of interest that dominated the century was astronomy. Newton had solved what is called the two-body problem, that is, the motion of a single planet under the gravitational attraction of the sun, wherein each body is idealized to have a point mass. He had also made some steps toward treating the major three-body problem, the behavior of the moon under the attraction of the earth and sun (Chap. 17, sec. 3). However, this was just the beginning of the efforts to study the motions of the planets and their satellites under the gravitational attraction of the sun and the mutual attraction of all the other bodies. Even Newton's work in the *Principia*, though constituting in effect the solution of differential equations, had to be translated into analytical form, which was done gradually during the eighteenth century. It was begun, incidentally, by Pierre Varignon, a fine French mathematician and physicist, who sought to free dynamics from the encumbrance of geometry. Newton did solve some differential equations in analytical form, for example in his *Method of Fluxions* of 1671 (Chap. 17, sec. 3); and in his *Tractatus* of 1676, he observed that the solution of $d^n y/dx^n = f(x)$ is arbitrary to the extent of an $(n - 1)$st degree polynomial in x. In the third edition of the *Principia*, Proposition 34, Scholium, he confines himself to a statement of which shapes of surfaces of revolution offer least resistance to motion in a fluid; but in a letter to David Gregory of 1694 he explains how he got his results and in the explanation uses differential equations.

Among problems of astronomy, the motion of the moon received the greatest attention, because the common method of determining longitude of ships at sea (Chap. 16, sec. 4), as well as other methods recommended in the seventeenth century, depended on knowing at all times the direction of the moon from a standard position (which from late in the century was Greenwich, England). It was necessary to know this direction of the moon to within 15 seconds of angle to determine the time at Greenwich to within 1 minute; even such an error could lead to an error of 30 kilometers in the determination of a ship's position. But with the tables of the moon's position available in Newton's time such accuracy was far from attainable. Another reason for the interest in the theory of the motion of the moon is that it could be used to predict eclipses, which in turn were a check on the entire astronomical theory.

The subject of ordinary differential equations arose in the problems just sketched. As mathematics developed, the subject of partial differential equations led to further work in ordinary differential equations. So did the branches now known as differential geometry and the calculus of variations. In this chapter we shall consider the problems leading directly to the basic early work in ordinary differential equations, that is, equations involving derivatives with respect to only one independent variable.

2. First Order Ordinary Differential Equations

The earliest work in differential equations, like the late seventeenth- and early eighteenth-century work in the calculus, was first disclosed in letters from one mathematician to another, many of which are no longer available, or in publications that often repeated results established or claimed in letters. The announcement of a result by one man frequently provoked the claim by another that he had done precisely the same thing earlier, which, in view of the bitter rivalries that existed, might or might not have been true. Some proofs were merely sketched, and it is not clear that the authors had the complete story; likewise, purported general methods of solution were merely illustrated by particular examples. For these reasons, even if we ignore the entire question of rigor, it is difficult to credit the results obtained to the right man.

In the *Acta Eruditorum* of 1693[1] Huygens speaks explicitly of differential equations, and Leibniz in another article in the same journal and year[2] says differential equations are functions of pieces of the characteristic triangle. The point we usually learn first about ordinary differential equations, that they can arise by eliminating the arbitrary constants from a given function and its derivatives, was not made until about 1740 and is due to Alexis Fontaine des Bertins: (1705–1771).

James Bernoulli was among the earliest to use the calculus in solving analytically problems of ordinary differential equations. In May of 1690[3] he published his solution of the problem of the isochrone, though an analytic solution had already been given by Leibniz. This problem is to find a curve along which a pendulum takes the same time to make a complete oscillation whether it swings through a wide or small arc. The differential equation, in Bernoulli's symbols, was

$$dy\sqrt{b^2y - a^3} = dx\sqrt{a^3}.$$

1. *Œuvres*, 10, 512–14.
2. *Math. Schriften*, 5, 306.
3. *Acta Erud.*, 1690, 217–19 = *Opera*, 1, 421–24.

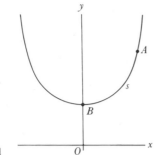

Figure 21.1

Bernoulli concluded from the equality of the differentials that the integrals (the word is used for the first time) must be equal and gave

$$\frac{2b^2y - 2a^3}{3b^2} \sqrt{b^2y - a^3} = x\sqrt{a^3}$$

as the solution. The curve is, of course, the cycloid.

In the same paper of 1690 James Bernoulli posed the problem of finding the curve assumed by a flexible inextensible cord hung freely from two fixed points, the curve Leibniz called the catenary. The problem had been considered as far back as the fifteenth century by Leonardo da Vinci. Galileo thought the curve was a parabola. Huygens affirmed that this was not correct and showed, largely by physical reasoning, that if the total load of cord and any weights suspended from it is uniform per horizontal foot, the curve is a parabola. For the true catenary the weight per foot *along the cable* is uniform.

In the *Acta* for June of 1691, Leibniz, Huygens, and John Bernoulli published independent solutions. Huygens's was geometrical and unclear. John Bernoulli[4] gave a solution by the method of the calculus. The full explanation is in his calculus text of 1691. It is the one now given in calculus and mechanics texts and is based on the equation

(1)
$$\frac{dy}{dx} = \frac{s}{c}$$
.

where s is arc length from B to some arbitrary point A (Fig. 21.1) and c depends upon the weight per unit length of the cord. This differential equation leads to what we now write as $y = c \cosh(x/c)$. Leibniz too obtained this result by calculus methods.

John Bernoulli was immensely proud that he had been able to solve the catenary problem and that his brother James, who had proposed it,

4. *Acta Erud.*, 1691, 274–76 = *Opera*, 1, 48–51.

had not. In a letter of September 29, 1718, to Pierre Rémond de Montmort (1678–1719) he boasts,[5]

> The efforts of my brother were without success; for my part, I was more fortunate, for I found the skill (I say it without boasting, why should I conceal the truth?) to solve it in full and to reduce it to the rectification of the parabola. It is true that it cost me study that robbed me of rest for an entire night. It was much for those days and for the slight age and practice I then had, but the next morning, filled with joy, I ran to my brother, who was still struggling miserably with this Gordian knot without getting anywhere, always thinking like Galileo that the catenary was a parabola. Stop! Stop! I say to him, don't torture yourself any more to try to prove the identity of the catenary with the parabola, since it is entirely false. The parabola indeed serves in the construction of the catenary, but the two curves are so different that one is algebraic, the other is transcendental. . . . But then you astonish me by concluding that my brother found a method of solving this problem. . . . I ask you, do you really think, if my brother had solved the problem in question, he would have been so obliging to me as not to appear among the solvers, just so as to cede me the glory of appearing alone on the stage in the quality of the first solver, along with Messrs. Huygens and Leibniz?

During the years 1691 and 1692 James and John also solved the problem of the shape assumed by a flexible inelastic hanging cord of variable density, an elastic cord of constant thickness, and a cord acted on at each point by a force directed to a fixed center. John also solved the converse problem: Given the equation of the curve assumed by an inelastic hanging cord, to find the law of variation of density of the cord with arc length. John's solutions are the ones often found in mechanics texts. James published in the *Acta* of 1691 the proof that of all shapes that a given cord hung from two fixed points can take, the catenary has the lowest center of gravity.

In the *Acta* of 1691 James Bernoulli derived the equation for the tractrix, the curve (Fig. 21.2) for which the ratio of PT to OT is constant for any point P on the curve. James first derived

$$\frac{dy}{ds} = \frac{y}{a},$$

where s is arc length. From this equation he deduced that

(2)
$$\int y \, dx = \int dy \sqrt{a^2 - y^2}$$

and

(3)
$$\int y^2 \, dx = -\frac{1}{3} \sqrt{(a^2 - y^2)^3},$$

5. Johann Bernoulli, *Der Briefwechsel von Johann Bernoulli*, Birkhäuser Verlag, 1955, 97–98.

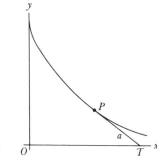

Figure 21.2

which he left as characteristic integrals for the curve. (Equation [2] can be integrated to yield $x + \sqrt{a^2 - y^2} = a \log \left[(a + \sqrt{a^2 - y^2})/y \right]$.)

Leibniz hit upon the technique of separating variables in ordinary differential equations and communicated it in a letter to Huygens of 1691. Thus he solved an equation of the form $y(dx/dy) = f(x)g(y)$, by writing $dx/f(x) = g(y) \, dy/y$, and then was able to integrate both sides. He did not formulate the general method. He also reduced (1691) the homogeneous differential equations of first order, $y' = f(y/x)$, to quadratures. He let $y = vx$ and substituted in the equation. The equation is then separable. Both these ideas, separation of variables and solution of homogeneous equations, were explained more fully by John Bernoulli in the *Acta Eruditorum* of 1694. Then Leibniz, in 1694, showed how to reduce the linear first order ordinary differential equation $y' + P(x)y = Q(x)$ to quadratures. His method utilized a change in the dependent variable. In general, Leibniz solved only first order ordinary differential equations.

James Bernoulli then proposed in the *Acta* of 1695[6] the problem of solving what is now called Bernoulli's equation:

$$(4) \qquad \frac{dy}{dx} = P(x)y + Q(x)y^n.$$

Leibniz in 1696[7] showed it can be reduced to a linear equation (first degree in y and y') by the change of variable $z = y^{1-n}$. John Bernoulli gave another method. In the *Acta* of 1696 James solved it essentially by separation of variables.

In 1694 Leibniz and John Bernoulli introduced the problem of finding the curve or family of curves that cut a given family of curves at a given angle. John Bernoulli called the cutting curves trajectories and pointed out, on the basis of Huygens's work on light, that this problem is important in finding the paths of rays of light traveling in a non-uniform medium because

6. Page 553.
7. *Acta Erud.*, 1696, 145.

these rays cut what are called the wave fronts of light orthogonally. The problem did not become public until 1697, when John posed it as a challenge to James, who solved a few special cases. John obtained the differential equation of the orthogonal trajectories of a particular family of curves and solved it in 1698.[8] Leibniz found the orthogonal trajectories of a family of curves thus: Consider $y^2 = 2bx$, where b is the parameter of the family (a term he introduced). From this equation $y \, dy/dx = b$. Leibniz then lets $b = -y \, dx/dy$, substitutes in $y^2 = 2bx$, and obtains $y^2 = -2xy \, dx/dy$ as the differential equation of the trajectories. The solution is $a^2 - x^2 = y^2/2$. Though he solved just special cases, he conceived the general problem and method.

The subject of orthogonal trajectories remained dormant until 1715, when Leibniz, aiming primarily at Newton, challenged the English mathematicians to discover the general method of finding the orthogonal trajectories of a given family of curves. Newton, tired from a day at the mint, solved the problem before going to sleep and the solution was published in the *Philosophical Transactions* of 1716.[9] Newton also showed how to find the curves cutting a given family at a constant angle or at an angle varying with each curve of the given family according to a given law. The method is not too different from the modern one, though Newton used ordinary differential equations of the second order.

Further work on this problem was done by Nicholas Bernoulli (1695–1726), in 1716. Jacob Hermann (1678–1733), a student of James Bernoulli, gave in the *Acta* of 1717 the rule that if $F(x, y, c) = 0$ is the given family of curves, then $y' = -F_x/F_y$, where F_x and F_y are the partial derivatives of F, and the orthogonal trajectory has F_y/F_x as its slope.[10] Hence, Hermann said, the ordinary differential equation of the orthogonal trajectories of $F(x, y, c) = 0$ is

$$(5) \qquad\qquad F_y \, dx = F_x \, dy.$$

He solved (5) for c, substituted this value in the original equation $F(x, y, c) = 0$, and solved the resulting differential equation. This method is really Leibniz's, but more explicitly stated. It is more customary today to find the true differential equation satisfied by $F = 0$; this equation does not contain parameter c. In it we replace y' by $-1/y'$ and so obtain the differential equation of the orthogonal trajectories.

John Bernoulli posed other trajectory problems for the English, his particular bête noire being Newton. Since the English and the Continentals were already at odds, the challenges were marked by bitterness and hostility.

8. *Opera*, 1, 266.
9. *Phil. Trans.*, 29, 1716, 399–400.
10. *Acta Erud.*, 1717, 349 ff. Also in John Bernoulli, *Opera*, 2, 275–79.

John Bernoulli then solved the problem of finding the motion of a projectile in a medium whose resistance is proportional to any power of the velocity. The differential equation here is

$$(6) \qquad m\frac{dv}{dt} - kv^n = mg.$$

Exact first order equations, that is, equations $M(x, y)\, dx + N(x, y) \cdot dy = 0$, for which $M\, dx + N\, dy$ is an exact differential of some function $z = f(x, y)$, were also recognized. Clairaut, who is famous for his work on the shape of the earth, had given the condition $\partial M/\partial y = \partial N/\partial x$ that the equation be exact in his papers of 1739 and 1740 (Chap. 19, sec. 6). The condition was also given independently by Euler in a paper written in 1734–35.[11] If the equation is exact then, as Clairaut and Euler pointed out, it can be integrated.

When a first order equation is not exact it is often possible to multiply the equation by a quantity, called an integrating factor, that makes it exact. Though integrating factors had been used in special problems of first order ordinary differential equations, it was Euler who realized (in the 1734/35 paper) that this concept furnishes a method; he set up classes of equations in which integrating factors will work. He also proved that if two integrating factors of any first order ordinary differential equations are known then their ratio is a solution of the equation. Clairaut independently introduced the idea of an integrating factor in his 1739 paper and in the 1740 paper added to the theory. All the elementary methods of solving first order equations were known by 1740.

3. Singular Solutions

Singular solutions are not obtainable from the general solution by giving a definite value to the constant of integration; that is, they are not particular solutions. This was observed by Brook Taylor in his *Methodus Incrementorum*[12] while solving a particular first order second degree equation. Leibniz in 1694 had already noted that an envelope of a family of solutions is also a solution. Singular solutions were more fully explored by Clairaut and Euler.

Clairaut's work of 1734[13] dealt with the equation that now bears his name,

$$(7) \qquad y = xy' + f(y').$$

Let y' be denoted by p. Then

$$(8) \qquad y = xp + f(p).$$

11. *Comm. Acad. Sci. Petrop.*, 7, 1734/35, 174–93, pub. 1740 = *Opera*, (1), 22, 36–56.
12. 1715, p. 26.
13. *Hist. de l'Acad. des Sci., Paris*, 1734, 196–215.

By differentiating with respect to x Clairaut obtained

$$p = p + \{x + f'(p)\}\frac{dp}{dx}.$$

Then

(9) $$\frac{dp}{dx} = 0 \quad \text{and} \quad x + f'(p) = 0.$$

The equation $dp/dx = 0$ leads to $y' = c$, and then from the original equation we have

(10) $$y = cx + f(c).$$

This is the general solution and is a family of straight lines. The second factor, $x + f'(p) = 0$, may be used together with the original equation to eliminate p; this yields a new solution, which is the singular solution. To see that it is the envelope of the general solution we take (10) and differentiate with respect to c. Then

(11) $$x + f'(c) = 0.$$

The envelope is the curve that results from eliminating c between (10) and (11). But these two are exactly the same as the two equations that yield the singular solution. The fact that the singular solution is an envelope was not yet appreciated, but Clairaut was explicit that the singular solution was not included in the general solution.

Clairaut and Euler had given a method of finding the singular solution by working from the differential equation itself, that is, by eliminating y' from $f(x, y, y') = 0$ and $\partial f/\partial y' = 0$. This fact and the fact that singular solutions are not contained in the general solution puzzled Euler. In his *Institutiones* of 1768[14] he gave a criterion for distinguishing the singular solution from a particular integral, which could be used when the general solution was not known. D'Alembert[15] sharpened this criterion. Then Laplace[16] extended the notion of singular solutions (he called them particular integrals) to equations of higher order and to differential equations in three variables.

Lagrange[17] made a systematic study of singular solutions and their connection with the general solution. He gave the general method of obtaining the singular solution from the general solution by elimination of the constant in a clear and elegant way that surpasses Laplace's contribution. Given the general solution $V(x, y, \alpha) = 0$, Lagrange's method was to find $dy/d\alpha$,

14. Vol. 1, pp. 393 ff.
15. *Hist. de l'Acad. des Sci., Paris*, 1769, 85 ff., pub. 1772.
16. *Hist. de l'Acad. des Sci., Paris*, 1772, Part 1, 344 ff., pub. 1775 = *Œuvres* 8, 325–66.
17. *Nouv. Mém. de l'Acad. de Berlin*, 1774, pub. 1776 = *Œuvres*, 4, 5–108.

set it equal to 0, and eliminate α from this equation and $V = 0$. The same procedure can be used with $dx/d\alpha = 0$. He also gave further information on Clairaut's and Euler's method of obtaining the singular solution from the differential equation. Finally, Lagrange gave the geometrical interpretation of the singular solution as the envelope of the family of integral curves. There are a number of special difficulties in the theory of singular solutions which he did not recognize. For example, he did not realize that other singular curves, which may appear in the equation obtained by eliminating y' from $f(x, y, y') = 0$ and $\partial f/\partial y' = 0$ are not singular solutions, or that a singular solution may contain a branch that is a particular solution. The full theory of singular solutions was developed in the nineteenth century and given its present form by Cayley and Darboux in 1872.

4. Second Order Equations and the Riccati Equations

Second order ordinary differential equations arose in physical problems as early as 1691. James Bernoulli tackled the problem of the shape of a sail under the pressure of the wind, the *velaria* problem, and was led to a second order equation $d^2x/ds^2 = (dy/ds)^3$ where s is arc length. John Bernoulli treated this problem in his calculus text of 1691 and established that it is the same mathematically as the catenary problem. Second order equations appear next in the attack on the problem of determining the shape of a vibrating elastic string fixed at both ends—for example, the violin string. In tackling this problem Taylor was pursuing an old theme. The whole subject of mathematics and musical sounds, begun by the Pythagoreans, was continued by men of the medieval period and brought to the fore in the seventeenth century. Benedetti, Beeckman, Mersenne, Descartes, Huygens, and Galileo are prominent in this work, though no new mathematical results are worth noting here. The fact that a string can vibrate in many modes, that is, halves, thirds, etc., and that the tone produced by a string vibrating in k parts is the kth harmonic or $(k - 1)$st overtone (the fundamental is the first harmonic) was well known in England by 1700, largely through the experimental work of Joseph Sauveur (1653–1716).

Brook Taylor[18] derived the fundamental frequency of a stretched vibrating string. He solved the equation $a^2\ddot{x} = \dot{s}y\ddot{y}$, where $\dot{s}$ equals $\sqrt{\dot{x}^2 + \dot{y}^2}$ and the differentiation is with respect to time, and gave $y = A \sin{(x/a)}$ as the form of the string at any time. Here $a = l/\pi$, where l is the length of the string. Taylor's result for the fundamental frequency (in modern notation) is

$$v = \frac{1}{2l} \sqrt{\frac{T}{\sigma}},$$

18. *Phil. Trans.*, 28, 1713, 26–32, pub. 1714; also in *Phil. Trans. Abridged*, 6, 1809, 7–12, 14–17.

where T is the tension in the string, $\sigma = m/g$, m is the mass per unit length, and g is the acceleration of gravity.

In his effort to treat the vibrating string, John Bernoulli, in a letter of 1727 to his son Daniel and in a paper,[19] considered the weightless elastic string loaded with n equal and equally spaced masses. He derived the fundamental frequency of the system when there are $1, 2, \ldots, 6$ masses. (There are other frequencies of oscillation of the system of masses.) John recognized that the force on each mass is $-K$ times its displacement, and solved $d^2x/dt^2 = -Kx$, thus integrating the equation of simple harmonic motion by analytic methods. He then passed to the continuous string which, like Taylor, he proved must have the shape of a sine curve (at any instant) and calculated the fundamental frequency. Here he solved $d^2y/dx^2 = -ky$. Neither Taylor nor John Bernoulli treated the higher modes of elastic vibrating bodies.

In 1728 Euler began to consider second order equations. His interest in these was aroused partly by his work in mechanics. He had worked, for example, on pendulum motion in resisting media, which leads to a second order differential equation. For the king of Prussia he worked on the effect of the resistance of air on projectiles. Here he took over the work of the Englishman Benjamin Robins, improved it, and wrote a German version (1745). This was translated into French and English and used by the artillery.

He also considered[20] a class of second order equations that he reduced to first order by a change of variables. For example, he considered the equation

$$(12) \qquad\qquad ax^m \, dx^p = y^n \, dy^{p-2} \, d^2y$$

or in derivative form

$$(13) \qquad\qquad \left(\frac{dy}{dx}\right)^{p-2} \frac{d^2y}{dx^2} = \frac{ax^m}{y^n}.$$

Euler introduced the new variables t and v by means of the equations

$$(14) \qquad\qquad y = e^v t(v), \qquad x = e^{\alpha v}$$

wherein α is a constant to be determined. The equations (14) may be regarded as parametric equations for x and y in terms of v, so that one can now calculate dy/dx and d^2y/dx^2 and by substitution in (13) obtain a second order equation in t as a function v. Euler then fixes α so as to eliminate the exponential factor, and v no longer appears explicitly. A further transformation, namely, $z = dv/dt$, reduces the second order equation to first order.

The details of this method are not worth pursuing because they apply to just one class of second order equations, but historically this piece of work

19. *Comm. Acad. Sci. Petrop.*, 3, 1728, 13–28, pub. 1732 = *Opera*, 3, 198–210.
20. *Comm. Acad. Sci. Petrop.*, 3, 1727, 124–37, pub. 1732 = *Opera*, (1), 22, 1–14.

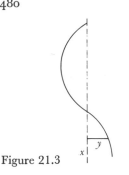

Figure 21.3

is significant because it does initiate the systematic study of second order equations and because Euler here introduces the exponential function, which, as we shall see, plays an important role in solving second and higher order equations.

Before leaving St. Petersburg in 1733, Daniel Bernoulli completed a paper, "Theorems on Oscillations of Bodies Connected by a Flexible Thread and of a Vertically Suspended Chain."[21] He starts with the hanging chain suspended from the upper end, weightless but loaded with equally spaced weights. He finds, when the chain is set into vibration, that the system has different modes of (small) oscillation about a vertical line through the point of suspension. Each of these modes has its own characteristic frequency.[22] Then, for an oscillating, uniformly heavy hanging chain of length l, he gives that the displacement y at a distance x from the bottom (Fig. 21.3) satisfies the equation

$$(15) \qquad \alpha \frac{d}{dx}\left(x\frac{dy}{dx}\right) + y = 0,$$

and the solution is an infinite series, which (in modern notation) can be expressed as

$$(16) \qquad y = AJ_0(2\sqrt{x/\alpha}),$$

where J_0 is the zero-th order Bessel function (of the first kind).[23] Moreover, α is such that

$$(17) \qquad J_0(2\sqrt{l/\alpha}) = 0,$$

21. *Comm. Acad. Sci. Petrop.*, 6, 1732/33, 108–22, pub. 1738.
22. In the case of n masses each mass has its own motion, which is a sum of n sinusoidal terms, each having one of the characteristic frequencies. The entire system has n different principal modes, each with one of the characteristic frequencies. Which of these are present depends on the initial conditions.
23. $J_n(x) = \left(\frac{x}{2}\right)^n \sum\limits_{k=0}^{\infty} \frac{(-1)^k(x/2)^{2k}}{k!\,(k+n)!}$ for n positive integral or 0.

where l is the length of the chain. He asserts that (17) has infinitely many roots, which become smaller and smaller and approach 0, and he gives the largest value for α. For each α there is a mode of oscillation and a characteristic frequency.

He now says, "Nor would it be difficult to derive from this theory a theory of musical strings agreeing with those given by Taylor and by my father . . . Experiment shows that in musical strings there are intersections [nodes] similar to those for vibrating chains." Actually, Bernoulli here goes beyond Taylor and his father in recognizing the higher modes or harmonics of a vibrating string.

His paper on the hanging chain also treats the oscillating chain of non-uniform thickness; here he introduces the differential equation

$$(18) \qquad \alpha \frac{d}{dx}\left(g(x)\frac{dy}{dx}\right) + y\frac{dg(x)}{dx} = 0,$$

where $g(x)$ is the distribution of the weight along the chain. For $g(x) = x^2/l^2$ he gives a series solution that can be expressed in modern notation as

$$(19) \qquad y = 2A\left(\frac{2x}{\alpha}\right)^{-1/2} J_1(2\sqrt{2x/\alpha})$$

with

$$J_1(2\sqrt{2l/\alpha}) = 0.$$

J_1 is the Bessel function of first order and first kind.

What is missing in Daniel Bernoulli's solutions is, first, any reference to the displacement as a function of the *time*, so that his work remains mathematically in the realm of ordinary differential equations; and second, any suggestion that the simple modes (the harmonics), which he explicitly recognizes as real motions, may be superposed to form more complicated ones.

After having entered upon the subject of musical sounds with a book, *Tentamen Novae Theoriae Musicae ex Certissimis Harmoniae Principiis Dilucide Expositae* (An Investigation into a New, Clearly Presented Theory of Music Based on Incontestable Principles of Harmony), written by 1731 and published in 1739,[24] Euler followed up Daniel Bernoulli's work in a paper, "On the Oscillations of a Flexible Thread Loaded with Arbitrarily Many Weights."[25] Euler's results are much the same as Bernoulli's, except that Euler's mathematics is clearer. For one form of the continuous chain, that is, the special case in which the weight is proportional to x^n, Euler has to solve

$$\frac{x}{n+1}\frac{d^2y}{dx^2} + \frac{dy}{dx} + \frac{y}{\alpha} = 0.$$

24. *Opera*, (3), 1, 197–427.
25. *Comm. Acad. Sci. Petrop.*, 8, 1736, 30–47, pub. 1741 = *Opera*, (2), 10, 35–49.

He derives the series solution, which in modern notation is [26]

$$y = Aq^{-n/2}I_n(2\sqrt{q}), \qquad q = -\frac{(n+1)x}{\alpha}.$$

The n here is general so that Euler is introducing Bessel functions of arbitrary real index. He also gives the integral solution

$$y = A\frac{\int_0^1 (1-t^2)^{(2n-1)/2}\cosh\left(2t\sqrt{\frac{(n+1)x}{\alpha}}\right)\,dt}{\int_0^1 (1-\tau^2)^{(2n-1)/2}\,d\tau}.$$

This is perhaps the earliest case of a solution of a second order differential equation expressed as an integral.

In a paper of 1739 [27] Euler took up the differential equations of the harmonic oscillator $\ddot{x} + kx = 0$ and the forced oscillation of the harmonic oscillator

(20) $$M\ddot{x} + Kx = F\sin\omega_a t.$$

He obtained the solutions by quadratures and discovered (really rediscovered, since others had found it earlier) the phenomenon of resonance; that is, if ω is the natural frequency $\sqrt{K/M}$ of the oscillator, which obtains when $F = 0$, then when ω_a/ω approaches 1 the forced oscillations have larger and larger amplitudes and become infinite.

In an attempt to set up a model for the transmission of sound in air, Euler considers in his paper "On the Propagation of Pulses Through an Elastic Medium" [28] n masses M connected by like (weightless) springs and lying in a horizontal line PQ. The motion considered is longitudinal, that is, along PQ. For the kth mass he obtains

$$M\ddot{x}_k = K(x_{k+1} - 2x_k + x_{k-1}), \qquad k = 1, 2, \ldots, n$$

where K is the spring constant and x_k is the displacement of the kth mass. He obtains the correct characteristic frequencies for each mass and the general solution

(21) $$x_k = \sum_{r=1}^{n} A_r \sin\frac{rk\pi}{n+1}\cos\left(2\sqrt{K/M}t\,\frac{\sin r\cdot\pi/2}{n+1}\right),$$

26. For general ν (including complex values)

$$I_\nu(z) = \sum_{n=0}^{\infty} \frac{(z/2)^{\nu+2n}}{n!\,\Gamma(\nu+n+1)}.$$

The $I_\nu(z)$ are called modified Bessel functions.
27. *Comm. Acad. Sci. Petrop.*, 11, 1739, 128–49, pub. 1750 = *Opera*, (2), 10, 78–97.
28. *Novi Comm. Acad. Sci. Petrop.*, 1, 1747/48, 67–105, pub. 1750 = *Opera*, (2), 10, 98–131.

wherein $k = 1, 2, \ldots, n$. Thus he not only obtains the individual modes for each mass but the general motion of that mass as a sum of simple harmonic modes. The particular modes that may appear depend on the initial conditions, that is, on how the masses are set into motion. All these results are interpretable in terms of the transverse motion (perpendicular to PQ) of the loaded string.

Some of the equations already treated, for example, the Bernoulli equation, are nonlinear; that is, as an equation in the variables y, y', and y'' (if present), terms of the second or higher degree occur. Among such equations of first order, a few are of special interest because they are intimately involved with linear second order equations. In the early history of ordinary differential equations the nonlinear Riccati equation

$$(22) \qquad \frac{dy}{dx} = a_0(x) + a_1(x)y + a_2(x)y^2$$

commanded a great deal of attention.

The Riccati equation acquired importance when it was introduced by Jacopo Francesco, Count Riccati of Venice (1676–1754), who worked in acoustics, to help solve second order ordinary differential equations. He considered curves whose radii of curvature depend only on the ordinates and was led [29] to

$$x^m \frac{d^2x}{dp^2} = \frac{d^2y}{dp^2} + \left(\frac{dy}{dp}\right)^2$$

(Riccati wrote $x^m\, d^2x = d^2y + (dy)^2$), wherein we must understand that x and y depend upon p. By changes of variables Riccati obtained

$$x^m \frac{dq}{dx} = \frac{du}{dx} + \frac{u^2}{q},$$

which is of first order. He assumed next that q is a power of x, for example, x^n, and arrived at the form

$$(23) \qquad \frac{du}{dx} + \frac{u^2}{x^n} - nx^{m+n-1}.$$

He then showed how to solve (23) for special values of n by the method of separation of variables for ordinary differential equations. Later, several of the Bernoullis determined other values of n for which solution of (23) by separation of variables was possible.

Riccati's work is significant not only because he treated second order differential equations but because he had the idea of reducing second order equations to first order. This idea of reducing the order of an ordinary

29. *Acta Erud.*, 1724, 66–73.

differential equation by one device or another will be seen to be a major method in the treatment of higher order ordinary differential equations.

Euler in 1760[30] considered the Riccati equation

$$(24) \qquad \frac{dz}{dx} + z^2 = ax^n$$

and showed that if one knows a particular integral v, then the transformation

$$z = v + u^{-1}$$

produces a linear equation. Moreover, if one knows two particular integrals, one can reduce the problem of solving the original differential equation to quadratures.

D'Alembert[31] was the first to consider the general form (22) of the Riccati equation and to use the term "Riccati equation" for this form. He started with

$$(25) \qquad \frac{d^2S}{dx^2} = \frac{-\lambda^2 x \pi^2 S}{2aLe}$$

and let

$$(26) \qquad S = \exp\left[\int p\,dx\right], \qquad p = f(x),$$

from which he obtained the form (22) for an equation in p as a function of x.

5. Higher Order Equations

In December of 1734 Daniel Bernoulli wrote to Euler, who was in St. Petersburg, that he had solved the problem of the transverse displacement y of an elastic bar (a steel or wooden one-dimensional body) fixed at one end in a wall and free at the other. Bernoulli obtained the differential equation

$$(27) \qquad K^4 \frac{d^4y}{dx^4} = y,$$

where K is a constant, x is the distance from the free end of the bar, and y is the vertical displacement at that point from the unbent position of the bar. Euler, in a reply written before June 1735, said he too had discovered this equation and was unable to integrate it except by using series, and that he did obtain four separate series. These series represented circular and exponential functions, but Euler did not realize it at this time.

30. *Novi Comm. Acad. Sci. Petrop.*, 8, 1760/61, 3–63, pub. 1763 = *Opera*, (1), 22, 334–94, and 9, 1762/63, 154–69, pub. 1764 = *Opera*, (1), 22, 403–20.
31. *Hist. de l'Acad. de Berlin*, 19, 1763, 242 ff., pub. 1770.

Four years later, in a letter to John Bernoulli (September 15, 1739), Euler indicated that his solution can be represented as

(28) $$y = A\left[\left(\cos\frac{x}{K} + \cosh\frac{x}{K}\right) - \frac{1}{b}\left(\sin\frac{x}{K} + \sinh\frac{x}{K}\right)\right]$$

where b is determined by the condition that $y = 0$ when $x = l$ so that

$$b = \frac{\sin\dfrac{l}{K} + \sinh\dfrac{l}{K}}{\cos\dfrac{l}{K} + \cosh\dfrac{l}{K}}.$$

The problems of elasticity led Euler to consider the mathematical problem of solving general linear equations with constant coefficients; and in the letter of September 15, 1739, he wrote John Bernoulli that he had succeeded. Bernoulli wrote back that he had already considered such equations in 1700, even with variable coefficients. Actually he had considered only a special third order equation and had shown how to reduce it to one of second order.

In his publication of this work[32] Euler considers the equation

(29) $$0 = Ay + B\frac{dy}{dx} + C\frac{d^2y}{dx^2} + D\frac{d^3y}{dx^3} + \cdots + L\frac{d^ny}{dx^n},$$

where the coefficients are constants. The equation is called homogeneous because the term independent of y and its derivatives is 0. He points out that the general solution must contain n arbitrary constants and that the solution will be a sum of n particular solutions, each multiplied by an arbitrary constant. Then he makes the substitution

$$y = \exp\left[\int r\, dx\right],$$

r constant, and obtains the equation in r,

$$A + Br + Cr^2 + \cdots + Lr^n = 0,$$

which is called the characteristic or indicial or auxiliary equation. When q is a simple real root of this equation, then

$$a \exp\left[\int q\, dx\right]$$

is a solution of the original differential equation. When the characteristic equation has a multiple root q, Euler lets $y = e^{qx}u(x)$ and substitutes in the differential equation. He finds that

(30) $$y = e^{qx}(\alpha + \beta x + \gamma x^2 + \cdots + \kappa x^{k-1})$$

32. *Misc. Berolin.*, 7, 1743, 193–242 = *Opera*, (1), 22, 108–49.

is a solution involving k arbitrary constants if the root q appears k times in the characteristic equation. He also treats the cases of conjugate complex roots and multiple complex roots. Thus Euler disposes completely of the homogeneous linear equation with constant coefficients.

Somewhat later[33] he treated the nonhomogeneous nth-order linear ordinary differential equation. His method was to multiply the differential by $e^{\alpha x} dx$, integrate both sides, and proceed to determine α so as to reduce the equation to one of lower order. Thus to treat

$$(31) \qquad C\frac{d^2y}{dx^2} + B\frac{dy}{dx} + Ay = X(x),$$

he multiplies through by $e^{\alpha x} dx$ and obtains

$$\int\left[e^{\alpha x}C\frac{d^2y}{dx} + e^{\alpha x}B\frac{dy}{dx} + e^{\alpha x}Ay\right] dx = \int e^{\alpha x}X \, dx.$$

But for proper A', B', and α the left side must be

$$e^{\alpha x}\left(A'y + B'\frac{dy}{dx}\right).$$

By differentiating this quantity and comparing with the original equation he finds that

$$(32) \qquad B' = C, \qquad A' = B - \alpha C, \qquad A' = \frac{A}{\alpha},$$

whence, from the second two equations,

$$(33) \qquad A - B\alpha + C\alpha^2 = 0.$$

Thus α, A', and B' are found and the original equation is reduced to

$$(34) \qquad A'y + B'\frac{dy}{dx} = e^{-\alpha x}\int e^{\alpha x}X \, dx.$$

An integrating factor for this equation is $e^{\beta x} dx$ where $\beta = A'/B'$, so that from (32) he has $\alpha\beta = A/C$ and $\alpha + \beta = B/C$ and therefore, by (33), α and β are the two roots of $A - B\alpha + C\alpha^2 = 0$.

The method applies to linear nth-order ordinary differential equations with constant coefficients. The order is reduced step by step as in the above example. Euler also took care of the cases of equal roots of the equation in α and of complex roots.

Following the work on linear ordinary differential equations with constant coefficients, Lagrange took the step to non-constant coefficients.[34]

33. *Novi Comm. Acad. Sci. Petrop.*, 3, 1750/51, 3–35, pub. 1753 = *Opera*, (1), 22, 181–213.
34. *Misc. Taur.*, 3, 1762/65, 179–86 = *Œuvres*, 1, 471–78.

This led, as we shall see, to the concept of the adjoint equation. Lagrange starts with

$$(35) \qquad Ly + M\frac{dy}{dt} + N\frac{d^2y}{dt^2} + \cdots = T$$

where L, M, N, and T are functions of t. For simplicity we shall stick to second order equations. Lagrange multiplies by $z\,dt$ where $z(t)$ is as yet undetermined, and integrates by parts, thus:

$$\int Mzy'\,dt = Mzy - \int (Mz)'y\,dt$$

$$\int Nzy''\,dt = Nzy' - (Nz)'y + \int (Nz)''y\,dt.$$

Then the original differential equation becomes

$$y[Mz - (Nz)'] + y'(Nz) + \int [Lz - (Mz)' + (Nz)'']y\,dt = \int Tz\,dt.$$

The bracket under the integral sign may be treated as an ordinary differential equation in z and set equal to 0. If it can be solved for $z(t)$, there remains an ordinary differential equation for y of lower order than the original. The new differential equation in z is called the adjoint of the original equation, a term introduced by Lazarus Fuchs in 1873. Lagrange used no special term for it.

To treat the equation for z (the adjoint equation), Lagrange proceeds in the same way to reduce the order. He multiplies by $w(t)\,dt$, proceeds as above, and arrives at an equation in w that will reduce the order of the equation in z. The equation in w turns out to be the original equation (35) except that the right side is 0. Hence Lagrange discovered the theorem that the adjoint of the adjoint of the original nonhomogeneous ordinary differential equation is the original but homogeneous equation. Euler did essentially the same thing in 1778. He had seen Lagrange's work but apparently forgot about it.

In further work on homogeneous linear ordinary differential equations with variable coefficients, Lagrange[35] extended to these equations some of the results Euler had obtained for linear ordinary differential equations with constant coefficients. Lagrange found that the general solution of the homogeneous equation is a sum of independent particular solutions, each multiplied by an arbitrary constant, and that knowing m particular integrals of the nth-order homogeneous equation, one can reduce the order by m.

35. *Misc. Taur.*, 3, 1762/65, 190–99 = *Œuvres*, 1, 481–90.

6. *The Method of Series*

We have already had occasion to note that some differential equations were solved by means of infinite series. The importance of this method even today warrants a few specific remarks on the subject. Series solutions have been used so widely since 1700 that we must restrict ourselves to a few examples.

We know that Newton used series to integrate somewhat complicated functions even where only quadrature was involved. He also used them to solve first order equations. Thus, to integrate

$$(36) \qquad \dot{y} = 2 + 3x - 2y + x^2 + x^2 y,$$

Newton assumes

$$(37) \qquad y = A_0 + A_1 x + A_2 x^2 + \cdots.$$

Then

$$(38) \qquad \dot{y} = A_1 + 2A_2 x + 3A_3 x^2 + \cdots.$$

Substitution of (37) and (38) in (36) and equating coefficients of like powers of x yields

$$A_1 = 2 - 2A_0, \qquad 2A_2 = 3 - 2A_1, \qquad 3A_3 = 1 + A_0 - 2A_2 + \cdots.$$

Thus we determine the A_i except for A_0. The fact that A_0 is undetermined and that therefore there is an infinite number of solutions was noted, but the significance of an arbitrary constant was not fully appreciated until about 1750. Leibniz solved some elementary differential equations by the use of infinite series [36] and also used the above method of undetermined coefficients.

Euler put the method of series in the fore from about 1750 on, to solve differential equations that could not be integrated in closed form. Though he worked with specific differential equations, and the details of what he did are often complicated, his method is what we use today. He assumes a solution of the form

$$y = x^\lambda (A + Bx + Cx^2 + \cdots),$$

substitutes for y and its derivatives in the differential equation, and determines λ and the coefficients $A, B, C, \ldots$ from the condition that each power of x in the resulting series must have a zero coefficient. Thus the ordinary differential equation which arose from his work on the oscillating membrane [37] (see Chap. 22, sec. 3), namely,

$$\frac{d^2 u}{dr^2} + \frac{1}{r}\frac{du}{dr} + \left(\alpha^2 - \frac{\beta^2}{r^2}\right) = 0,$$

36. *Acta Erud.*, 1693 = *Math. Schriften*, 5, 285–88.
37. *Novi Comm. Acad. Sci. Petrop.*, 10, 1764, 243–60, pub. 1766 = *Opera*, (2), 10, 344–59.

now called the Bessel equation, Euler solved by an infinite series. He gives
the solution

$$u(r) = r^\beta \left\{ 1 - \frac{1}{1 \cdot (\beta + 1)} \left(\frac{\alpha r}{2}\right)^2 + \frac{1}{1 \cdot 2 (\beta + 1)(\beta + 2)} \left(\frac{\alpha r}{2}\right)^4 \right.$$

$$\left. - \frac{1}{1 \cdot 2 \cdot 3 (\beta + 1)(\beta + 2)(\beta + 4)} \left(\frac{\alpha r}{2}\right)^6 + \cdots \right\},$$

which is, except for a factor depending only on β, what we now write as
$J_\beta(r)$. In further work on these functions he showed that for values of β
that are odd halves of an integer, the series reduce to elementary functions.
He noted further that $u(r)$, for real β, has an infinite number of zeros and
gave an integral representation for $u(r)$. Finally, for $\beta = 0$ and $\beta = 1$ he
gave the second linearly independent series solution of the differential
equation.

Euler in the *Institutiones Calculi Integralis*[38] treated the hypergeometric
differential equation

(39) $$x(1 - x) \frac{d^2 y}{dx^2} + [c - (a + b + 1)x] \frac{dy}{dx} - aby = 0$$

and gave the series solution

(40) $$y = 1 + \frac{a \cdot b}{1 \cdot c} x + \frac{a(a + 1)b(b + 1)}{1 \cdot 2 \cdot c(c + 1)} x^2$$

$$+ \frac{a(a + 1)(a + 2)b(b + 1)(b + 2)}{1 \cdot 2 \cdot 3 \cdot c(c + 1)(c + 2)} x^3 + \cdots.$$

The above form of (39) and of the solution (40) he gave again in his main
paper on this subject, written in 1778.[39] He had written other papers on
what he called the hypergeometric series, but there the term referred to
another series originally introduced by Wallis. The term "hypergeometric,"
to describe the differential equation (39) and the series (40), is due to
Johann Friedrich Pfaff (1765–1825), Gauss's friend and teacher. The series
(40) for y is now denoted by $F(a, b, c; z)$. In this notation Euler gave the
famous relations

$$F(-n, b, c; z) = (1 - z)^{c+n-b} F(c + n, c - b, c; z)$$

(41) $$F(-n, b, c; z) = \frac{n!}{c(c + 1) \cdots (c + n - 1)} \cdot$$

$$\int_0^1 t^{-n-1} (1 - t)^{c+n-1} (1 - tz)^{-b} \, dt.$$

38. Vol. 2, 1769, Chaps. 8–11.
39. *Nova Acta Acad. Sci. Petrop.*, 12, 1794, 58–70, pub. 1801 = *Opera*, (1), 16_2, 41–55.

7. Systems of Differential Equations

The differential equations involved thus far in the study of elasticity were rather simple, because the mathematicians were using crude physical principles and were still struggling to grasp better ones. In the area of astronomy, however, the physical principles, chiefly Newton's laws of motion and the law of gravitation, were clear and the mathematical problems far deeper. The basic mathematical problem in studying the motion of two or more bodies, each moving under the gravitational attraction of the others, is that of solving a system of ordinary differential equations, though the problem often reduces to solving a single equation.

Beyond isolated occurrences, the work on systems dealt primarily with problems of astronomy. The basis for writing the differential equations is Newton's second law of motion, $f = ma$, where f is the force of attraction. This is a vector law, which means that each component of f produces an acceleration in the direction of the component. Euler in a paper of 1750[40] gave the analytical form of Newton's second law as

$$(42) \qquad f_x = m\frac{d^2x}{dt^2}, \qquad f_y = m\frac{d^2y}{dt^2}, \qquad f_z = m\frac{d^2z}{dt^2}.$$

Here he assumes fixed rectangular axes. He also points out that for point bodies, that is, bodies that can be treated as though their masses were concentrated at one point, m is the total mass, and for distributed masses m is dM.

We shall consider briefly the formulation of the differential equations. Let us suppose that a fixed body of mass M is at the origin and that a moving body of mass m is at (x, y, z). Then the components of the gravitational force in the axial directions (Fig. 21.4) are

$$f_x = -\frac{GMmx}{r^3}, \qquad f_y = -\frac{GMmy}{r^3}, \qquad f_z = -\frac{GMmz}{r^3},$$

where G is the gravitational constant and $r = \sqrt{(x^2 + y^2 + z^2)}$. One can readily show that the moving body stays in one plane so that the system of equations (42) reduces to

$$(43) \qquad \frac{d^2x}{dt^2} = -\frac{kx}{r^3}, \qquad \frac{d^2y}{dt^2} = -\frac{ky}{r^3}$$

with $k = GM$. In polar coordinates these equations become

$$(44) \qquad \frac{d^2r}{dt^2} - r\left(\frac{d\theta}{dt}\right)^2 = -\frac{k}{r^2}$$

$$r\frac{d^2\theta}{dt^2} + 2\frac{dr}{dt}\frac{d\theta}{dt} = 0.$$

40. *Hist. de l'Acad. de Berlin,* 6, 1750, 185-217, pub. 1752 = *Opera,* (2), 5, 81-108.

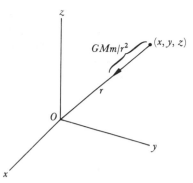

Figure 21.4

In this case of one body moving under the attracting force of another (fixed) body, the two differential equations can be combined into one involving x and y or r and θ because, for example, the second polar equation can be integrated to yield $r^2\, d\theta/dt = C$, and the value of $d\theta/dt$ can be substituted in the first equation. It turns out that the moving body describes a conic section with the position of the first body as a focus.

If the two bodies move, each subject to the attraction of the other, then the differential equations are slightly different. Let m_1 and m_2 be the masses of two spherical bodies with spherically symmetric mass and with $m_1 + m_2 = M$. Choose a fixed coordinate system (usually the center is taken at the center of mass of the two bodies) and let (x_1, y_1, z_1) be the coordinates of one body and (x_2, y_2, z_2) the coordinates of the other; let r be the distance $\sqrt{(x_1 - x_2)^2 + (y_1 - y_2)^2 + (z_1 - z_2)^2}$. Then the system of equations that describes their motion is

$$m_1 \frac{d^2 x_1}{dt^2} = -km_1 m_2 \frac{(x_1 - x_2)}{r^3}, \qquad m_1 \frac{d^2 y_1}{dt^2} = -km_1 m_2 \frac{(y_1 - y_2)}{r^3}$$

$$m_1 \frac{d^2 z_1}{dt^2} = -km_1 m_2 \frac{(z_1 - z_2)}{r^3}, \qquad m_2 \frac{d^2 x_2}{dt^2} = -km_1 m_2 \frac{(x_2 - x_1)}{r^3}$$

$$m_2 \frac{d^2 y_2}{dt^2} = -km_1 m_2 \frac{(y_2 - y_1)}{r^3}, \qquad m_2 \frac{d^2 z_2}{dt^2} = -km_1 m_2 \frac{(z_2 - z_1)}{r^3}.$$

This is a system of six second order equations whose solution calls for twelve integrals, each with an arbitrary constant of integration. These are determined by the three coordinates of initial position and three components of initial velocity of each body. The equations can be solved, and the solution shows that each body moves in a conic section with respect to the common center of mass of the two bodies.

Actually this problem of the motion of two spheres under the mutual attractive force of gravitation was solved geometrically by Newton in the

Principia (Book I, Section 11). However, the analytical work was not taken up for some time. In mechanics the French followed Descartes's system until Voltaire, after visiting London in 1727, returned to champion Newton's. Even Cambridge, Newton's own university, continued to teach natural philosophy from the text of Jacques Rohault (1620–75), a Cartesian. In addition, the most eminent mathematicians of the late seventeenth century —Huygens, Leibniz, and John Bernoulli—were opposed to the concept of gravitation and hence to its application. Analytical methods of treating planetary motion were undertaken by Daniel Bernoulli, who received a prize from the French Academy of Sciences for a paper of 1734 on the two-body problem. It was handled completely by Euler in his book *Theoria Motuum Planetarum et Cometarum.*[41]

If we have n bodies, each spherical and with spherically symmetric mass distribution (density a function of radius), they will attract each other as though their masses were at their centers. Let $m_1, m_2, \ldots, m_n$ represent the masses and (x_i, y_i, z_i) the (variable) coordinates of the ith mass with respect to a fixed system of axes; let r_{ij} be the distance from m_i to m_j. Then the x-components of the forces acting on m_1 are

$$-\frac{k}{r_{12}^3} m_1 m_2 (x_1 - x_2), \quad -\frac{k}{r_{13}^3} m_1 m_3 (x_1 - x_3), \ldots, \quad -\frac{k}{r_{1n}^3} m_1 m_n (x_1 - x_n),$$

with similar expressions for the y- and z-components of the force. Each body has such components of force acting on it.

The differential equations of the motion of the ith body are then

$$m_i(d^2 x_i/dt^2) = -km_i \sum_{j=1}^{n} m_j[(x_i - x_j)/r_{ij}^3]$$

(45)
$$m_i(d^2 y_i/dt^2) = -km_i \sum_{j=1}^{n} m_j[(y_i - y_j)/r_{ij}^3]$$

$$m_i(d^2 z_i/dt^2) = -km_i \sum_{j=1}^{n} m_j[(z_i - z_j)/r_{ij}^3]$$

with $j \neq i$ and $i = 1, 2, \ldots, n$. There are $3n$ equations, each of second order. The origin can be chosen to be the center of mass of the n bodies or it can be chosen in one of the bodies, for example, the sun. There are $6n$ integrals, of which 10 can be found somewhat readily; these are the only ones known in the general problem.

The problem of n bodies, indeed even of three bodies, cannot be solved exactly. Hence the investigations of this problem have taken two general

41. 1744 = *Opera*, (2), 28, 105–251.

directions. The first is a search for whatever general theorems one can deduce, which may at least shed some light on the motions. The second is a search for approximate solutions that may be useful for a period of time subsequent to some instant at which data may be available; this is known as the method of perturbations.

The first type of investigation produced some theorems on the motion of the center of mass of n bodies, which were given by Newton in his *Principia*. For example, the center of mass of the n bodies moves with uniform speed in a straight line. The ten integrals mentioned above, which are consequences of what are called conservation laws of motion, also constitute theorems of the first type. These integrals were known to Euler. There are also some exact results for special cases of the problem of three bodies, and these are due to one of the masters of celestial mechanics, Joseph-Louis Lagrange.

Lagrange (1736–1813) was of French and Italian extraction. As a boy he was unimpressed with mathematics, but while still at school he read an essay by Halley on the merits of Newton's calculus and became excited about the subject. At the age of nineteen he became a professor of mathematics at the Royal Artillery School of Turin, the city of his birth. He soon contributed so much to mathematics that even at an early age he was recognized as one of the period's greatest mathematicians. Though Lagrange worked in many branches of mathematics—the theory of numbers, the theory of algebraic equations, the calculus, differential equations and the calculus of variations—and in many branches of physics, his chief interest was the application of the law of gravitation to planetary motion. He said, in 1775, "The arithmetical researches are those which have cost me most trouble and are perhaps the least valuable." Archimedes was Lagrange's idol.

Lagrange's most famous work, his *Mécanique analytique* (1788; second edition, 1811–15; posthumous edition, 1853), extended, formalized, and crowned Newton's work on mechanics. Lagrange had once complained that Newton was a most fortunate man, since there was but one universe and Newton had already discovered its mathematical laws. However, Lagrange had the honor of making apparent to the world the perfection of the Newtonian theory. Though the *Mécanique* is a classic of science and is significant also for the theory and use of ordinary differential equations, Lagrange had trouble finding a publisher.

The particular exact solutions obtained in the three-body problem were given by Lagrange in a prize paper of 1772, *Essai sur le problème des trois corps*.[42] One of these solutions states that it is possible to set these bodies in motion so that their orbits are similar ellipses all described in the same time and with the center of mass of the three bodies as a common focus.

42. *Hist. de l'Acad. des Sci., Paris*, 9, 1772 = *Œuvres*, 6, 229–331.

Another solution assumes the three bodies are started from the three summits of an equilateral triangle. They will then move as though attached to the triangle which itself rotates about the center of mass of the bodies. The third solution assumes the three bodies are projected into motion from positions on a straight line. For appropriate initial conditions they will continue to be fixed on that line while the line rotates in a plane about the center of mass of the bodies. These three cases had no physical reality for Lagrange, but the equilateral triangle case was found in 1906 to apply to the sun, Jupiter, and an asteroid named Achilles.

The second type of problem involving n bodies deals, as already noted, with approximate solutions or the theory of perturbations. Two spherical bodies acted upon by their mutual gravitational attraction move along conic sections. This motion is said to be unperturbed. Any departure from such motions, whether in position or velocity, however caused, is perturbed motion. If there are two spheres but there is resistance from the medium in which they move, or if the two bodies are no longer spherical but, say, oblate spheroids, or if more than two bodies are involved, then the orbits of the bodies are no longer conic sections. Before the use of telescopes the perturbations were not striking. In the eighteenth century the calculation of perturbations became a major mathematical problem, and Clairaut, d'Alembert, Euler, Lagrange, and Laplace all made contributions. Laplace's work in this area was the most outstanding.

Pierre-Simon de Laplace (1749–1827) was born to reasonably well-to-do parents in the town of Beaumont, Normandy. It seemed likely that he would become a priest but at the University of Caen, which he entered at the age of sixteen, he took to mathematics. He spent five years at Caen and while there wrote a paper on the calculus of finite differences. After finishing his studies, Laplace went to Paris with letters of recommendation to d'Alembert, who ignored him. Then Laplace wrote d'Alembert a letter containing an exposition of the general principles of mechanics; this time d'Alembert took notice, sent for Laplace, and got him the position of professor of mathematics at the Ecole Militaire in Paris.

Even as a youth Laplace published prolifically. A statement made in the Paris Academy of Sciences shortly after his election in 1773 pointed out that no one so young had presented so many papers on such diverse and difficult subjects. In 1783 he replaced Bezout as an examiner in artillery and examined Napoleon. During the Revolution he was made a member of the Commission on Weights and Measures but was later expelled, along with Lavoisier and others, for not being a good republican. Laplace retired to Melun, a small city near Paris, where he worked on his celebrated and popular *Exposition du système du monde* (1796). After the Revolution he became a professor at the Ecole Normale, where Lagrange by this time was also teaching, and served on a number of government committees. Then he became successively Minister of the Interior, a member of the Senate, and chancellor of that body. Though honored by Napoleon with the title of count, Laplace voted against him in 1814 and rallied to Louis XVIII, who named Laplace marquis and peer of France.

During these years of political activity he continued to work in science. Between 1799 and 1825 there appeared the five volumes of his *Mécanique céleste*. In this work Laplace presented "complete" analytic solutions of the mechanical problems posed by the solar system. He took as little from observational data as possible. The *Mécanique céleste* embodies discoveries and results of Newton, Clairaut, d'Alembert, Euler, Lagrange, and Laplace himself. The masterpiece was so complete that his immediate successors could add little. Perhaps its only defect is that Laplace frequently neglected to acknowledge the sources of his results and left the impression that they were all his own.

In 1812 he published his *Théorie analytique des probabilités*. The introduction to the second edition (1814) consists of a popular essay known as the *Essai philosophique sur les probabilités*. It contains the famous passage to the effect that the future of the world is completely determined by the past and that one possessed of the mathematical knowledge of the state of the world at any given instant could predict the future.

Laplace made many important discoveries in mathematical physics, some of which we shall take up in later chapters. Indeed, he was interested in anything that helped to interpret nature. He worked on hydrodynamics, the wave propagation of sound, and the tides. In the field of chemistry, his work on the liquid state of matter is classic. His studies of the tension in the surface layer of water, which accounts for the rise of liquids inside a capillary tube, and of the cohesive forces in liquids, are fundamental. Laplace and Lavoisier designed an ice calorimeter (1784) to measure heat and measured the specific heat of numerous substances; heat to them was still a special kind of matter. Most of Laplace's life was, however, devoted to celestial mechanics. He died in 1827. His last words were reported to be, "What we know is very slight; what we don't know is immense"—though De Morgan states that they were "Man follows only phantoms."

Laplace is often linked with Lagrange, but they are not similar in their personal qualities or their work. Laplace's vanity kept him from giving sufficient credit to the works of those whom he considered rivals; in fact, he used many ideas of Lagrange without acknowledgment. When Laplace is mentioned in connection with Lagrange it is always to commend the personal qualities of the latter. Lagrange is the mathematician, very careful in his writings, very clear, and very elegant. Laplace created a number of new mathematical methods that were subsequently expanded into branches of mathematics, but he never cared for mathematics except as it helped him to study nature. When he encountered a mathematical problem in his physical research he solved it almost casually and merely stated, "It is easy to see that . . ." without bothering to show how he had worked out the result. He confessed, however, that it was not easy to reconstruct his own work. Nathaniel Bowditch (1773–1838), the American mathematician and astronomer, who translated four of the five volumes of the *Mécanique céleste* and added explanations, said that whenever he came across the phrase "It is easy to see that . . ." he knew he had hours of hard work ahead to fill the gaps. Actually, Laplace was impatient of the mathematics and wanted to get on with the applications. Pure mathematics did not interest him and his contributions to it were by-products of his great work in natural philosophy. Mathematics was a means, not an end, a tool one ought to perfect in order to solve the problems of science.

Laplace's work, so far as it is relevant to the subject of this chapter, deals with approximate solutions of problems of planetary motion. The possibility of obtaining useful solutions by any approximate method rests on the following factors. The solar system is dominated by the sun, which contains 99.87 percent of the entire matter in the system. This means that the orbits of the planets are nearly elliptical, because the perturbing forces of the planets on each other are small. Nevertheless, Jupiter has 70 percent of the planetary mass. Also, the earth's moon is quite close to the earth and so each is affected by the other. Hence perturbations do have to be considered.

The three-body problem, especially sun, earth, and moon, was studied most in the eighteenth century, partly because it is the next step after the two-body problem and partly because an accurate knowledge of the moon's motion was needed for navigation. In the case of the sun, earth, and moon, some advantage can be taken of the fact that the sun is far from the other two bodies and can be treated as exercising only a small influence on the relative motion of the earth and moon. In the case of the sun and two planets, one planet is usually regarded as perturbing the motion of the other around the sun. If one of the planets is small, its gravitational effect on the other planet can be neglected, but the effect of the larger one on the small one must be taken into account. These special cases of the three-body problem are called the restricted three-body problem.

The theory of perturbations for the three-body problem was first applied to the motion of the moon; this was done geometrically by Newton in the third book of the *Principia*. Euler and Clairaut tried to obtain exact solutions for the general three-body problem, complained about the difficulties, and of course resorted to approximate methods. Here Clairaut made the first real progress (1747) by using series solutions of the differential equations. He then had occasion to apply his result to the motion of Halley's comet, which had been observed in 1531, 1607, and 1682. It was expected to be at the perihelion of its path around the earth in 1759. Clairaut computed the perturbations due to the attraction of Jupiter and Saturn and predicted in a paper read to the Paris Academy on November 14, 1758, that the perihelion would occur on April 13, 1759. He remarked that the exact time was uncertain to the extent of a month because the masses of Jupiter and Saturn were not known precisely and because there were slight perturbations caused by other planets. The comet reached its perihelion on March 13.

To compute perturbations the method called variation of the elements or variation of parameters—or variation of the constants of integration—was created and is the most effective one. We are obliged to confine ourselves to its mathematical principles; we shall therefore examine it without taking into account the full physical background.

The mathematical method of variation of parameters for the problem of three bodies goes back to Newton's *Principia*. After treating the motion of the moon about the earth and obtaining the elliptical orbit, Newton took account of the effects of the sun on the moon's orbit by considering variations in the latter. The method was used in isolated instances to solve nonhomogeneous equations by John Bernoulli in the *Acta Eruditorum* of 1697[43] and by Euler in 1739, in treating the second order equation $y'' + k^2 y = X(x)$. It was first used to treat perturbations of planetary motions by Euler in his paper of 1748,[44] which treated the mutual perturbations of Jupiter and Saturn and won a prize from the French Academy. Laplace wrote many papers on the method.[45] It was fully developed by Lagrange in two papers.[46]

The method of variation of parameters for a single ordinary differential equation was applied by Lagrange to the nth-order equation

$$Py + Qy' + Ry'' + \cdots + Vy^{(n)} = X,$$

where X, P, Q, R, ... are functions of x. For simplicity we shall suppose that we have a second order equation.

In the case $X = 0$, Lagrange knew that the general solution is

(46) $$y = ap(x) + bq(x)$$

where a and b are integration constants and p and q are particular integrals of the homogeneous equation. Now, says Lagrange, let us regard a and b as functions of x. Then

(47) $$\frac{dy}{dx} = ap' + bq' + pa' + qb'.$$

43. Page 113.
44. *Opera*, (2), 25, 45–157.
45. See, for example, *Hist. de l'Acad. des Sci.*, Paris, 1772, Part 1, 651 ff., pub. 1775 = *Œuvres*, 8, 361–66, and *Hist. de l'Acad. des Sci.*, Paris, 1777, 373 ff., pub. 1780 = *Œuvres*, 9, 357–80.
46. *Nouv. Mém. de l'Acad. de Berlin*, 5, 1774, 201 ff., and 6, 1775, 190 ff. = *Œuvres*, 4, 5–108 and 151–251.

Lagrange now sets

(48) $$pa' + qb' = 0;$$

that is, the part of y' that results from the variability of a and b he sets equal to 0. From (47) we have, in view of (48),

(49) $$\frac{d^2y}{dx^2} = ap'' + bq'' + p'a' + q'b'.$$

If the equation were of higher order than the second, Lagrange would now set $p'a' + q'b' = 0$ and find d^3y/dx^3. Since, in our case, the equation is of second order, he would keep all the terms in (49).

He now substitutes the expressions for y, dy/dx and d^2y/dx^2 given by (46), (47), and (49) in the original equation. Since (46) is a solution of the homogeneous equation for constant a and b and (48) throws out some of the terms that result from the variability of a and b, there remains after the substitution

(50) $$p'a' + q'b' = \frac{X}{R}.$$

This equation and equation (48) give two algebraic equations in the two unknown functions a' and b'. These two simultaneous equations can be solved for a' and b' in terms of the known functions p, q, p', q', X, and R. Then a and b can each be obtained by an integration or are at least reduced to a quadrature. With these values of a and b, (46) gives a solution of the original nonhomogeneous equation. This solution together with the solution of the homogeneous equation is the complete solution of the nonhomogeneous equation.

The method of variation of parameters was treated in more general fashion by Lagrange[47] and he showed its applicability to many problems of physics. In the paper of 1808 he applied it to a system of three second order equations. The technique is of course more complicated, but again the basic idea is to treat the six constants of integration of the solution of the corresponding homogeneous system as variables and to determine them so that the expression will satisfy the nonhomogeneous system.

During the period in which they were developing the method of variation of parameters and afterward, Lagrange and Laplace wrote a number of key papers on basic problems of the solar system. In his crowning work, the *Mécanique céleste*, Laplace summarized the scope of their results:

> We have given, in the first part of this work, the general principles of the equilibrium and motion of bodies. The application of these principles

47. *Mém. de l'Acad. des Sci., Inst. France*, 1808, 267 ff. = *Œuvres*, 6, 713–68.

to the motions of the heavenly bodies had conducted us, by geometrical [analytical] reasoning, without any hypothesis, to the law of universal attraction, the action of gravity and the motion of projectiles being particular cases of this law. We have then taken into consideration a system of bodies subjected to this great law of nature and have obtained, by a singular analysis, the general expressions of their motions, of their figures, and of the oscillations of the fluids which cover them. From these expressions we have deduced all the known phenomena of the flow and ebb of the tide; the variations of the degrees and the force of gravity at the surface of the earth; the precession of the equinoxes; the libration of the moon; and the figure and rotation of Saturn's rings. We have also pointed out the reason that these rings remain permanently in the plane of the equator of Saturn. Moreover, we have deduced, from the same theory of gravity, the principal equations of the motions of the planets, particularly those of Jupiter and Saturn, whose great inequalities have a period of above 900 years.[48]

Laplace concluded that nature ordered the celestial machine "for an eternal duration, upon the same principles which prevail so admirably upon the earth, for the preservation of individuals and for the perpetuity of the species."

As the mathematical methods for solving differential equations were improved and as new physical facts about the planets were acquired, efforts were made throughout the nineteenth and twentieth centuries to obtain better results on the various subjects Laplace mentions, in particular on the n-body problem and the stability of the solar system.

8. *Summary*

As we have seen, the attempt to solve physical problems, which at first involved no more than quadratures, led gradually to the realization that a new branch of mathematics was being created, namely, ordinary differential equations. By the middle of the eighteenth century the subject of differential equations became an independent discipline and the solution of such equations an end in itself.

The nature of what was regarded and sought as a solution gradually changed. At first, mathematicians looked for solutions in terms of elementary functions; soon they were content to express an answer as a quadrature that might not be effected. When the major attempts to find solutions in terms of elementary functions and quadratures failed, mathematicians became content to seek solutions in infinite series.

The problem of solution in closed form was not forgotten, but instead of attempting to solve in that manner the particular differential equations

48. Preface to Vol. 3.

that arose from physical problems, the mathematicians sought differential equations that permit solutions in terms of a finite number of elementary functions. A great number of differential equations integrable in this manner were found. D'Alembert (1767) worked on this problem and included elliptic integrals among acceptable answers. A typical approach to this problem, made by Euler (1769), among others, was to start with a differential equation whose integration in closed form could be effected and then derive other differential equations from the known one. Another approach was to look for conditions under which the series solution might contain only a finite number of terms.

An interesting but unfruitful piece of work by Marie-Jean-Antoine-Nicolas Caritat de Condorcet (1743–94) in *Du calcul intégral* (1765) was his attempt to bring order and method out of the many separate methods and tricks used to solve differential equations. He listed such operations as differentiation, elimination, and substitution and sought to reduce all methods to these canonical operations. The work led nowhere. In line with this plan Euler showed that where separation of variables is possible a multiplying (integrating) factor will work, but not conversely. Also he showed that separation of variables will not do for higher order differential equations. As for substitutions, he found no general principles for obtaining them.[49] Finding substitutions is as difficult as solving the differential equations directly. However, transformation can reduce the order of a differential equation. Euler used this idea to solve the nth-order nonhomogeneous linear ordinary differential equation, and even in the case of the homogeneous equation he thought of each exp $[\int p \, dx]$ as giving, for the proper value of p, a first order factor of the ordinary differential equation. Reducing the order was also Riccati's plan. A number of other methods, including Lagrange's undetermined multipliers, were devised. At first this method was believed to be general but it did not prove to be so.

The search for general methods of integrating ordinary differential equations ended about 1775. Much new work was yet to be done with ordinary differential equations, particularly those resulting from the solution of partial differential equations. But no major new methods beyond those already surveyed here were discovered for a hundred years or so, until operator methods and the Laplace transform were introduced at the end of the nineteenth century. In fact interest in general methods of solution receded, because methods of one form or another adequate to those types required in applications were obtained. Broad, comprehensive principles for the solution of ordinary differential equations are still lacking. On the whole the subject has continued to be a series of separate techniques for the various types.

49. *Institutiones Calculi Integralis*, 1, 290.

Bibliography

Bernoulli, James: *Opera*, 2 vols., 1744, reprinted by Birkhaüser, 1968.

Bernoulli, John: *Opera Omnia*, 4 vols., 1742, reprinted by Georg Olms, 1968.

Berry, Arthur: *A Short History of Astronomy*, Dover (reprint), 1961, Chaps. 9–11.

Cantor, Moritz: *Vorlesungen über Geschichte der Mathematik*, B. G. Teubner, 1898 and 1924, Vol. 3, Chaps. 100 and 118, Vol. 4, Sec. 27.

Delambre, J. B. J.: *Histoire de l'astronomie moderne*, 2 vols., 1821, Johnson Reprint Corp., 1966.

Euler, Leonhard: *Opera Omnia*, Orell Füssli, Series 1, Vols. 22 and 23, 1936 and 1938; Series 2, Vols. 10 and 11, Part 1, 1947 and 1957.

Hofmann, J. E.: "Über Jakob Bernoullis Beiträge zur Infinitesimal-mathematik," *L'Enseignement Mathématique*, (2), 2, 61–171. Published separately by Institut de Mathématiques, Geneva, 1957.

Lagrange, Joseph-Louis: *Œuvres de Lagrange*, Gauthier-Villars, 1868–1873, relevant papers in Vols. 2, 3, 4, and 6.

————: *Mécanique analytique*, 1788; 4th ed., Gauthier-Villars, 1889. The fourth edition is an unchanged reproduction of the third edition of 1853.

Lalande, J. de: *Traité d'astronomie*, 3 vols., 1792, Johnson Reprint Corp., 1964.

Laplace, Pierre-Simon: *Œuvres complètes*, Gauthier-Villars, 1891–1904, relevant papers in Vols. 8, 11 and 13.

————: *Traité de mécanique céleste*, 5 vols., 1799–1825. Also in *Œuvres complètes*, Vols. 1–5, Gauthier-Villars, 1878–82. English trans. of Vols. 1–4 by Nathaniel Bowditch, 1829–39, Chelsea (reprint), 1966.

————: *Exposition du système du monde*, 1st ed., 1796, 6th ed. in *Œuvres complètes*, Gauthier-Villars, 1884, Vol. 6.

Montucla, J. F.: *Histoire des mathématiques*, 1802, Albert Blanchard (reprint), 1960, Vol. 3, 163–200; Vol. 4, 1–125.

Todhunter, I.: *A History of the Mathematical Theories of Attraction and the Figure of the Earth*, 1873, Dover (reprint), 1962.

Truesdell, Clifford E.: *Introduction to Leonhardi Euleri Opera Omnia, Vol. X et XI Seriei Secundae*, in Euler, *Opera Omnia*, (2), 11, Part 2, Orell Füssli, 1960.

22

Partial Differential Equations
in the Eighteenth Century

Mathematical Analysis is as extensive as nature herself.
JOSEPH FOURIER

1. *Introduction*

As in the case of ordinary differential equations, the mathematicians did not consciously create the subject of partial differential equations. They continued to explore the same physical problems that had led to the former subject; and as they secured a better grasp of the physical principles underlying the phenomena, they formulated mathematical statements that are now comprised in partial differential equations. Thus, whereas the displacement of a vibrating string had been studied separately as a function of time and as a function of the distance of a point on the string from one end, the study of the displacement as a function of both variables and the attempt to comprehend all the possible motions led to a partial differential equation. The natural continuation of this study, namely, the investigation of the sounds created by the string as they propagate in air, introduced additional partial differential equations. After studying these sounds the mathematicians took up the sounds given off by horns of all shapes, organ pipes, bells, drums, and other instruments.

Air is one type of fluid, as the term is used in physics, and happens to be compressible. Liquids are (virtually) incompressible fluids. The laws of motion of such fluids and, in particular, the waves that can propagate in both became a broad field of investigation that now constitutes the subject of hydrodynamics. This field, too, gave rise to partial differential equations.

Throughout the eighteenth century, mathematicians continued to work on the problem of the gravitational attraction exerted by bodies of various shapes, notably the ellipsoid. While basically this is a problem of triple integration, it was converted by Laplace into a problem of partial differential equations in a manner we shall examine shortly.

2. *The Wave Equation*

Though specific partial differential equations appear as early as 1734 in the work of Euler[1] and in 1743 in d'Alembert's *Traité de dynamique*, nothing worth noting was done with them. The first real success with partial differential equations came in renewed attacks on the vibrating-string problem, typified by the violin string. The approximation that the vibrations are small was imposed to make the partial differential equation tractable. Jean Le Rond d'Alembert (1717–83), in his papers of 1746[2] entitled "Researches on the Curve Formed by a Stretched String Set into Vibrations," says he proposes to show that infinitely many curves other than the sine curve are modes of vibration.

We may recall from the preceding chapter that in the first approaches to the vibrating string, it was regarded as a "string of beads." That is, the string was considered to contain n discrete equal and equally spaced weights joined to each other by pieces of weightless, flexible, and elastic thread. To treat the continuous string, the number of weights was allowed to become infinite while the size and mass of each was decreased, so that the total mass of the increasing number of individual "beads" approached the mass of the continuous string. There were mathematical difficulties in passing to the limit, but these subtleties were ignored.

The case of a discrete number of masses had been treated by John Bernoulli in 1727 (Chap. 21, sec. 4). If the string is of length l and lies along $0 \leq x \leq l$, and if x_k is the abscissa of the kth mass, $k = 1, 2, \cdots, n$ (the nth mass at $x = l$ is motionless), then

$$x_k = k\frac{l}{n}, \qquad k = 1, 2, \cdots, n.$$

By analyzing the force on the kth mass, Bernoulli had shown that if y_k is the displacement of the kth mass, then

$$\frac{d^2y_k}{dt^2} = \left(\frac{na}{l}\right)^2 (y_{k+1} - 2y_k + y_{k-1}), \qquad k = 1, 2, \cdots, n-1,$$

where $a^2 = lT/M$, T is the tension in the string (which is taken to be constant as the string vibrates), and M the total mass. D'Alembert replaced y_k by $y(t, x)$ and l/n by Δx. Then

$$\frac{\partial^2 y(t, x)}{\partial t^2} = a^2 \left[\frac{y(t, x + \Delta x) - 2y(t, x) + y(t, x - \Delta x)}{(\Delta x)^2}\right].$$

1. *Comm. Acad. Sci. Petrop.*, 7, 1734/35, 184–200, pub. 1740 = *Opera*, (1), 22, 57–75.
2. *Hist. de l'Acad. de Berlin*, 3, 1747, 214–19 and 220–49, pub. 1749.

He now observed that as n becomes infinite so that Δx approaches 0, the bracketed expression becomes $\partial^2 y/\partial x^2$. Hence

$$(1) \qquad \frac{\partial^2 y(t, x)}{\partial t^2} = a^2 \frac{\partial^2 y(t, x)}{\partial x^2},$$

where a^2 is now T/σ, σ being the mass per unit length. Thus what is now called the wave equation in one spatial dimension appears for the first time.

Since the string is fixed at the endpoints $x = 0$ and $x = l$, the solution must satisfy the boundary conditions

$$(2) \qquad y(t, 0) = 0, \qquad y(t, l) = 0.$$

At $t = 0$ the string is displaced into some shape $y = f(x)$ and then released, which means that each particle starts with zero initial velocity. These initial conditions are expressed mathematically as

$$(3) \qquad y(0, x) = f(x), \qquad \left.\frac{\partial y(t, x)}{\partial t}\right|_{t=0} = 0$$

and they must also be satisfied by the solution.

This problem was solved by d'Alembert in so clever a manner that it is often reproduced in modern texts. We shall not take space for all the details. He proved first that

$$(4) \qquad y(t, x) = \frac{1}{2} \phi(at + x) + \frac{1}{2} \psi(at - x),$$

where ϕ and ψ are as yet unknown functions.

Thus far d'Alembert had deduced that *every* solution of the partial differential equation (1) is the sum of a function of $(at + x)$ and a function of $(at - x)$. The converse is easy to show by direct substitution of (4) into (1). Of course d'Alembert had yet to satisfy the boundary and initial conditions. The condition $y(t, 0) = 0$ applied to (4) gives, for all t,

$$(5) \qquad \frac{1}{2} \phi(at) + \frac{1}{2} \psi(at) = 0.$$

Since for any x, $ax + t = at'$ for some value of t', we may say that for any x and t

$$(6) \qquad \phi(x + at) = -\psi(x + at).$$

Then the condition $y(t, l) = 0$, becomes, in view of (4) and (6),

$$(7) \qquad \frac{1}{2} \phi(at + l) = \frac{1}{2} \phi(at - l);$$

and since this is an identity in t, it shows that ϕ must be periodic in $at + x$ with period $2l$.

The condition

(8)
$$\frac{\partial y(t, x)}{\partial t}\bigg|_{t=0} = 0$$

yields, from (4) and the fact that $\phi = -\psi$,

(9)
$$\phi'(x) = \phi'(-x).$$

On integration this becomes

(10)
$$\phi(x) = -\phi(-x),$$

and thus ϕ is an odd function of x. If we now use the fact that $\phi = -\psi$ in (4), form $y(0, x)$ and use (10), we have that

(11)
$$y(0, x) = \phi(x),$$

and since the initial condition is $y(0, x) = f(x)$ we have

(12)
$$\phi(x) = f(x) \quad \text{for} \quad 0 \le x \le l.$$

To sum up,

(13)
$$y(t, x) = \frac{1}{2}\phi(at + x) - \frac{1}{2}\phi(at - x),$$

where ϕ is subject to the above conditions of periodicity and oddness. Moreover, if the initial state is $y(0, x) = f(x)$, then (12) must hold between 0 and l. Thus there would be just one solution for a given $f(x)$. Now d'Alembert regarded functions as analytic expressions formed by the processes of algebra and the calculus. Hence if two such functions agree in one interval of x-values, they must agree for every value of x. Since $\phi(x) = f(x)$ in $0 \le x \le l$ and ϕ had to be odd and periodic, then $f(x)$ had to meet the same conditions. Finally, since $y(t, x)$ had to satisfy the differential equation, it had to be twice differentiable. But $y(0, x) = f(x)$, and so $f(x)$ had to be twice differentiable.

Within a few months of seeing d'Alembert's 1746 papers, Euler wrote his own paper, "On the Vibration of Strings," which was presented on May 16, 1748.[3] Though in method of solution he followed d'Alembert, Euler by this time had a totally different idea as to what functions could be admitted as initial curves and therefore as solutions of partial differential equations. Even before the debate on the vibrating-string problem, in fact in a work of 1734, he allowed functions formed from parts of different well-known curves and even formed by drawing curves freehand. Thus the curve (Fig. 22.1) formed by an arc of a parabola in the interval (a, c) and by an arc of a third degree curve in the interval (c, b) constituted one curve or one function under

3. *Nova Acta Erud.*, 1749, 512–27 = *Opera*, (2), 10, 50–62; also in French by Euler, *Hist. de l'Acad. de Berlin*, 4, 1748, 69–85 = *Opera*, (2), 10, 63–77.

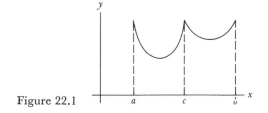

Figure 22.1

this concept. Euler called such curves discontinuous, though in modern terminology they are continuous with discontinuous derivatives. In his text, the *Introductio* of 1748, he stuck to the notion that was standard in the eighteenth century, that a function must be given by a single analytical expression. However, the physics of the vibrating-string problem seems to have been his compelling reason for bringing his new concept of function to the fore. He accepted any function defined by a formula $\phi(x)$ in $-l \leq x \leq l$, and regarded $\phi(x + 2l) = \phi(x)$ to be the definition of the curve outside $(-l, l)$. In a later paper[4] he goes further; he says that

$$(14) \qquad\qquad y = \phi(ct + x) + \psi(ct - x),$$

with arbitrary ϕ and ψ, is a solution of

$$(15) \qquad\qquad \frac{1}{c^2}\frac{\partial^2 y}{\partial t^2} = \frac{\partial^2 y}{\partial x^2}.$$

This follows by substitution in the differential equation. But the initial curve is equally satisfactory, whether it is expressed by some equation, or whether it is traced in any fashion not expressible by an equation. Of the initial curve, only the part in $0 \leq x < l$ is relevant. The continuation of this part is not to be taken into consideration. The different parts of this curve are thus not joined to each other by any law of continuity (single analytic expression); it is only by the description that they are joined together. For this reason it may be impossible to comprise the entire curve in one equation, except when by chance the curve is some sine function.

In 1755 Euler gave as a new definition of function, "If some quantities depend on others in such a way as to undergo variation when the latter are varied, then the former are called functions of the latter." And in another paper[5] he says that the parts of a "discontinuous" function do not belong to one another and are not determined by one single equation for the whole extent of the function. Moreover, given the initial shape in $0 \leq x \leq l$, one repeats it in reverse order in $-l \leq x \leq 0$ (so as to make it odd) and conceives the continual repetition of this curve in each interval of length $2l$ to infinity.

4. *Hist. de l'Acad. de Berlin*, 9, 1753, 196–222, pub. 1755 = *Opera*, (2), 10, 232–54.
5. *Novi Comm. Acad. Sci. Petrop.*, 11, 1765, 67–102, pub. 1767 = *Opera*, (1), 23, 74–91.

Then, if this curve $[y = f(x)]$ is used to represent the initial function, after the time t the ordinate that will answer to the abscissa x of the string in vibration will be (cf. [13] and [12])

(16) $$y = \frac{1}{2}f(x + ct) + \frac{1}{2}f(x - ct).$$

In his basic 1749 paper Euler points out that all possible motions of the vibrating string are periodic in time whatever the shape of the string; that is, the period is (usually) the period of what we now call the fundamental. He also realized that individual modes whose periods are one half, one third, and so on of the basic (fundamental) period can occur as the vibrating figure. He gives such special solutions as

(17) $$y(t, x) = \sum A_n \sin \frac{n\pi x}{l} \cos \frac{n\pi ct}{l}$$

when the initial shape is

(18) $$y(0, x) = \sum A_n \sin \frac{n\pi x}{l},$$

but does not say whether the summation covers a finite or infinite number of terms. Nevertheless he has the idea of superposition of modes. Thus Euler's main point of disagreement with d'Alembert is that he would admit all kinds of initial curves, and therefore non-analytic solutions, whereas d'Alembert accepted only analytic initial curves and solutions.

In introducing his "discontinuous" functions, Euler appreciated that he had taken a big step forward. He wrote to d'Alembert on December 20, 1763, that "considering such functions as are subject to no law of continuity [analyticity] opens to us a wholly new range of analysis."[6]

The solution of the vibrating-string problem was given in entirely different form by Daniel Bernoulli; this work stirred up another ground for controversy about the allowable solutions. Daniel Bernoulli (1700–82), the son of John Bernoulli, was a professor of mathematics at St. Petersburg from 1725 to 1733 and then, successively, professor of medicine, metaphysics, and natural philosophy at Basle. His chief work was in hydrodynamics and elasticity. In the former area, he won a prize for a paper on the flow of the tides; he also contemplated the application of the theory of the flow of liquids to the flow of blood in human blood vessels. He was a skilled experimentalist and through experimental work discovered the law of attraction of static electric charges before 1760. This law is usually credited to Charles Coulomb. Bernoulli's *Hydrodynamica* (1738), which contains studies that appeared in a number of papers, is the first major text in its field. It has a chapter on the

6. *Opera* (2), 11, sec. 1, 2.

mechanical theory of heat (as opposed to heat as a substance) and gives many results on the theory of gases.

In his paper of 1732/33 cited in the preceding chapter, Bernoulli had expressly stated that the vibrating string could have higher modes of oscillations. In a later paper[7] on composite oscillation of weights on a loaded vertical flexible string, he made the following remark:

> Similarly, a taut musical string can produce its isochronous tremblings in many ways and even according to theory infinitely many, ... and moreover in each mode it emits a higher or lower note. The first and most natural mode occurs when the string in its oscillations produces a single arch; then it makes the slowest oscillations and gives out the deepest of all its possible tones, fundamental to all the rest. The next mode demands that the string produce two arches on the opposite sides [of the string's rest position] and then the oscillations are twice as fast, and now it gives out the octave of the fundamental sound.

Then he describes the higher modes. He does not give the mathematics but it seems evident that he had it.

In a paper on the vibrations of a bar and the sounds given off by the vibrating bar,[8] Bernoulli not only gives the separate modes in which the bar can vibrate but says distinctly that both sounds (the fundamental and a higher harmonic) can exist together. This is the first statement of the co-existence of small harmonic oscillations. Bernoulli based it on his physical understanding of how the bar and the sounds can act but gave no mathematical evidence that the sum of two modes is a solution.

When he read d'Alembert's first paper of 1746 and Euler's paper of 1749 on the vibrating string, he hastened to publish the ideas he had had for many years.[9] After indulging in sarcasm about the abstractness of d'Alembert's and Euler's works, he reasserts that many modes of a vibrating string can exist simultaneously (the string then responds to the sum or superposition of all the modes) and claims that this is all that Euler and d'Alembert have shown. Then comes a major point. He insists that *all* possible initial curves are representable as

$$(19) \qquad f(x) = \sum_{n=1}^{\infty} a_n \sin \frac{n\pi x}{l}$$

because there are enough constants a_n to make the series fit any curve. Hence, he asserts, *all* subsequent motions would be

$$(20) \qquad y(t, x) = \sum_{n=1}^{\infty} a_n \sin \frac{n\pi x}{l} \cos \frac{n\pi ct}{l}.$$

7. *Comm. Acad. Sci. Petrop.*, 12, 1740, 97–108, pub. 1750.
8. *Comm. Acad. Sci. Petrop.*, 13, 1741/43, 167–96, pub. 1751.
9. *Hist. de l'Acad. de Berlin*, 9, 1753, 147–72 and 173–95, pub. 1755.

Thus every motion corresponding to any initial curve is no more than a sum of sinusoidal periodic modes, and the combination has the frequency of the fundamental. However, he gives no mathematical arguments to back up his contentions; he relies on the physics. In this paper of 1753 Bernoulli states:

> My conclusion is that all sonorous bodies include an infinity of sounds with a corresponding infinity of regular vibrations. . . . But it is not of this multitude of sounds that Messrs. d'Alembert and Euler claim to speak. . . . Each kind [each fundamental mode generated by some initial curve] multiplies an infinite number of times to accord to each interval an infinite number of curves, such that each point starts, and achieves at the same instant, these vibrations while, following the theory of Mr. Taylor, each interval between two nodes should assume the form of the companion of the cycloid [sine function] extremely elongated.
>
> We then remark that the chord AB cannot make vibrations only conforming to the first figure [fundamental] or second [second harmonic] or third and so forth to infinity but that it can make a combination of these vibrations in all possible combinations, and that all new curves given by d'Alembert and Euler are only combinations of the Taylor vibrations.

In this last remark Bernoulli ascribes knowledge to Taylor that Taylor never displayed. This apart, however, Bernoulli's contentions were enormously significant.

Euler objected at once to Bernoulli's last assertion. In fact Euler's 1753 paper presented to the Berlin Academy (already referred to above) was in part a reply to Bernoulli's two papers. Euler emphasizes the importance of the wave equation as the starting point for the treatment of the vibrating string. He praises Bernoulli's recognition that many modes can exist simultaneously so that the string can emit many harmonics in one motion, but denies, as did d'Alembert, that all possible motions can be expressed by (20). He admits that an initial curve such as

(21) $$f(x) = \frac{c \sin(ax/l)}{1 - a \cos(ax/l)}, \qquad |a| < 1,$$

can be expressed by a series such as (19). Bernoulli would be borne out if every function could be represented by an infinite trigonometric series, but this Euler regards as impossible. A sum of sine functions is, Euler says, an odd periodic function. But in his solution (see [16]),

(22) $$y(t, x) = \frac{1}{2}f(x + ct) + \frac{1}{2}f(x - ct),$$

f is arbitrary (discontinuous, in Euler's sense) and so cannot be expressed as a sum of sine functions. In fact, he says, f could be a combination of arcs spread out over the infinite x domain and be odd and periodic; yet because it is discontinuous (in Euler's sense), it cannot be expressed as a sum of sine curves. His own solution, he affirms, is not limited in any respect. The initial curve, in fact, need not be expressible by an equation (single analytic expression).

Euler also pointed, in this instance rightly, to the Maclaurin series and said this could not represent any arbitrary function; hence neither could an infinite sine series. All he would grant was that Bernoulli's trigonometric series represented special solutions; and indeed he (Euler) had obtained such solutions in his 1749 paper (see [17] and [18]).

D'Alembert, in his article "Fondamental" in Volume 7 (1757) of the *Encyclopédie*, also attacked Bernoulli. He did not believe that all odd and periodic functions could be represented by a series such as (19), because the series is twice differentiable but all odd and periodic functions need not be so. However, even when the initial curve is sufficiently differentiable—and d'Alembert did require that it be twice differentiable in his 1746 paper— it need not be representable in Bernoulli's form. On the same ground d'Alembert objected to Euler's discontinuous curves. Actually d'Alembert's requirement that the initial curve $y = f(x)$ must be twice differentiable was correct, because a solution derived from an $f(x)$ that does not have a second derivative at some value or values of x must satisfy special conditions at such singular points.

Bernoulli did not retreat from his position. In a letter of 1758[10] he repeats that he had in the a_n an infinite number of coefficients at his disposal, and so by choosing them properly could make the series in (19) agree with any function $f(x)$ at an infinite number of points. In any case he insisted that (20) was the most general solution. The argument between d'Alembert, Euler, and Bernoulli continued for a decade with no agreement reached. The essence of the problem was the extent of the class of functions that could be represented by the sine series, or, more generally, by Fourier series.

In 1759, Lagrange, then young and unknown, entered the controversy. In his paper, which dealt with the nature and propagation of sound,[11] he gave some results on that subject and then applied his method to the vibrating string. He proceeded as though he were tackling a new problem but repeated much that Euler and Daniel Bernoulli had done before. Lagrange, too, started with a string loaded with a finite number of equal and equally spaced masses and then passed to the limit of an infinite number of masses. Though he criticized Euler's method as restricting the results to continuous (analytic) curves, Lagrange said he would prove that Euler's

10. *Jour. des Sçavans*, March 1758, 157–66.
11. *Misc. Taur.*, 1_3, 1759, i-x, 1–112 = *Œuvres*, 1, 39–148.

conclusion, that any initial curve can serve, is correct. We shall pass at once to Lagrange's conclusion for the continuous string. He had obtained

$$(23) \quad y(x, t) = \frac{2}{l} \sum_{r=1}^{\infty} \sin \frac{r\pi x}{l} \sum_{q=1}^{\infty} \sin \frac{r\pi x}{l} \, dx \left[Y_q \cos \frac{c\pi rt}{l} + \frac{l}{r\pi c} V_q \sin \frac{c\pi rt}{l} \right].$$

Here Y_q and V_q are the initial displacement and initial velocity of the qth mass. He then replaced Y_q and V_q by $Y(x)$ and $V(x)$ respectively. Lagrange regarded the quantities

$$(24) \qquad \sum_{q=1}^{\infty} \sin \frac{r\pi x}{l} Y(x) \, dx \quad \text{and} \quad \sum_{q=1}^{\infty} \sin \frac{r\pi x}{l} V(x) \, dx$$

as integrals and he took the integration operation outside the summation $\sum_{r=1}^{\infty}$. From these moves there resulted

$$(25) \qquad y(x, t) = \left(\frac{2}{l} \int_0^l Y(x) \sum_{r=1}^{\infty} \sin \frac{r\pi x}{l} \, dx \right) \sin \frac{r\pi x}{l} \cos \frac{r\pi ct}{l}$$

$$+ \left(\frac{2}{\pi c} \int_0^l V(x) \sum_{r=1}^{\infty} \frac{1}{r} \sin \frac{r\pi x}{l} \, dx \right) \sin \frac{r\pi x}{l} \sin \frac{r\pi ct}{l}.$$

The interchange of summation and integration not only introduced divergent series but spoiled whatever chance Lagrange might have had to recognize

$$(26) \qquad \int_0^l Y(x) \sin \frac{r\pi x}{l} \, dx$$

as a Fourier coefficient. After other long, difficult, and dubious steps, Lagrange obtained Euler's and d'Alembert's result

$$(27) \qquad y = \phi(ct + x) + \psi(ct - x).$$

He concluded that the above derivation put the theory of this great geometer [Euler]

> beyond all doubt and established on direct and clear principles which rest in no way on the law of continuity [analyticity] which Mr. d'Alembert requires; this, moreover, is how it can happen that the same formula that has served to support and prove the theory of Mr. Bernoulli on the mixture of isochronous vibrations when the number of bodies is ... finite shows us its insufficiency ... when the number of these bodies becomes infinite. In fact, the change that this formula undergoes in passing from one case to the other is such that the simple motions which made up the absolute motions of the whole system annul each

other for the most part, and those which remain are so disfigured and altered as to become absolutely unrecognizable. It is truly annoying that so ingenious a theory ... is shown false in the principal case, to which all the small reciprocal motions occurring in nature may be related.

All of this is almost total nonsense.

Lagrange's main basis for contending that his solution did not require the initial curve $Y(x)$ and initial velocity $V(x)$ to be restricted is that he did not apply differentiation to them. But if one were to rigorize what he did do, restrictions would have to be made.

Euler and d'Alembert criticized Lagrange's work but actually did not hit at the main failings; they picked on details in his "prodigious calculations," as Euler put it. Lagrange tried to answer these criticisms. The replies and rebuttals on both sides are too extensive to relate here, though many are revealing of the thinking of the times. For example, Lagrange replaced $\sin \pi/m$ for $m = \infty$ by π/m and $\sin v\pi/2m$ by $v\pi/2m$ for $m = \infty$. D'Alembert allowed the first but not the second, because the values of v involved were comparable with m. The objection that a series of the form

$$\cos x + \cos 2x + \cos 3x + \cdots$$

might be divergent was also raised by d'Alembert and answered by Lagrange with the argument, common at that time, that the value of the series is the value of the function from which the series comes.

Though Euler did criticize mathematical details, his overall response to Lagrange's paper, communicated in a letter of October 23, 1759,[12] was to commend Lagrange's mathematical skill and to state that it put the whole discussion beyond all quibbling, and that everyone must now recognize the use of irregular and discontinuous (in Euler's sense) functions in this class of problems.

On October 2, 1759, Euler had written Lagrange: "I am delighted to learn that you approve my solution ... which d'Alembert has tried to undermine by various cavils, and that for the sole reason that he did not get it himself. He has threatened to publish a weighty refutation; whether he really will I do not know. He thinks he can deceive the semi-learned by his eloquence. I doubt whether he is serious, unless perhaps he is thoroughly blinded by self-love."[13]

In 1760/61 Lagrange, seeking to answer criticisms of d'Alembert and Bernoulli that had been communicated by letter, gave a different solution of the vibrating-string problem.[14] This time he starts directly with the wave

12. Lagrange, Œuvres, 14, 164-70.
13. Œuvres, 14, 162-64.
14. Misc. Taur., 2_2, 1760/61, 11-172, pub. 1762 = Œuvres, 1, 151-316.

equation (with $c = 1$), and by multiplying by an unknown function and further steps he reduces the partial differential equation to the solution of two ordinary differential equations. Then by still further steps, not all correct, Lagrange obtains the solution

$$y(t, x) = \frac{1}{2}f(x + t) + \frac{1}{2}f(x - t) - \frac{1}{2}\int_0^{x+t} g \, dx + \frac{1}{2}\int_0^{x-t} g \, dx,$$

wherein $f(x) = y(0, x)$ and $g(x) = \partial y/\partial t$ at $t = 0$ are the given initial data. As Lagrange shows, this agrees with d'Alembert's result. But then, without reference to his own work, he tries to convince his readers that he did not use any law of continuity (analyticity) for the initial curve. It is true that he did not use any direct operation of differentiation on the initial function. But, in this paper also, to justify rigorously his limit procedures, one cannot avoid assumptions about the continuity and differentiability of the initial functions.

The debate raged throughout the 1760s and 1770s. Even Laplace entered the fray in 1779,[15] and sided with d'Alembert. D'Alembert continued it in a series of booklets, entitled *Opuscules*, which began to appear in 1768. He argued against Euler, on the ground that Euler admitted too general initial curves, and against Daniel Bernoulli, on the ground that his (d'Alembert's) solutions could not be represented as a sum of sine curves, so that Bernoulli's solutions were not general enough. The idea that the infinite series of trigonometric functions $\sum_{n=1}^{\infty} a_n \sin nx$ might be made to fit any initial curve because there is an infinity of a_n's to be determined (Daniel Bernoulli had so contended) was rejected by Euler as impossible to execute. He also raised the question of how a trigonometric series could represent the initial curve when only a part of the string is disturbed initially. Euler, d'Alembert, and Lagrange continued throughout to deny that a trigonometric series could represent any analytic function, to say nothing of more arbitrary functions.

Many of the arguments each presented were grossly incorrect; and the results, in the eighteenth century, were inconclusive. One major issue, the representability of an arbitrary function by trigonometric series, was not settled until Fourier took it up. Euler, d'Alembert, and Lagrange were on the threshold of discovering the significance of Fourier series but did not appreciate what lay before them. Judging by the knowledge of the times, all three men and Bernoulli were correct in their main contentions. D'Alembert, following a tradition established since Leibniz's time, insisted that functions must be analytical, so that any problem not solvable in such terms was unsolvable. He was also correct in the argument he gave that $y(t, x)$

15. *Mém. de l'Acad. des Sci.*, Paris, 1779, 207–309, pub. 1782 = *Œuvres*, 10, 1–89.

must be periodic in x. However, he failed to realize that, given any arbitrary function in, say, $0 \leq x \leq l$, this function could be repeated in every interval $[nl, (n + 1)l]$ for integral n and so be periodic. Of course, such a periodic function might not be representable by one (closed) formula. Euler and Lagrange were, at least in their time, justified in believing that not all "discontinuous" functions could be represented by Fourier series, yet equally right in believing (though they did not have proof) that the initial curve can be very general. It need not be analytic, nor need it be periodic. Bernoulli did adopt the correct position on physical grounds but could not back it up with the mathematics.

One of the very curious features of the debate on the trigonometric series representation of functions is that all the men involved knew that non-periodic functions can be represented (in an interval) by trigonometric series. Reference to Chapter 20 (sec. 5) will show that Clairaut, Euler, Daniel Bernoulli, and others had actually produced such representations; many of their papers also had the formulas for the coefficients of the trigonometric series. Practically all of this work was in print by 1759, the year in which Lagrange presented his basic paper on the vibrating string. He could then have inferred that any function has a trigonometric expansion and could have read off the formulas for the coefficients, but failed to do so. Only in 1773, when the heat of the controversy was past, did Daniel Bernoulli notice that the sum of a trigonometric series may represent different algebraic expressions in different intervals. Why did all these results have no influence on the controversy concerning the vibrating string? It may be explained in several ways. Many of the results on the representation of quite general functions by trigonometric series were in papers on astronomy, and Daniel Bernoulli may not have read these and so could not point to them in defense of his position. Euler and d'Alembert, who must have known Clairaut's work of 1757 (Chap. 20, sec. 5), were probably not inclined to study it, since it refuted their own arguments. Also, this astronomical work by Clairaut was soon superseded and forgotten. On the other hand, whereas Euler used trigonometric series, as in his work on interpolation theory, to represent polynomial expressions, he did not accept the general fact that quite arbitrary functions could be so represented; the existence of such a series representation, where he used it, was assured by other means.

Another issue, how a partial differential equation with analytic coefficients (e.g. constants) could have a non-analytic solution, was not really clarified. In the case of ordinary differential equations, if the coefficients are analytical, the solutions must be. However, this is not true for partial differential equations. Though Euler was correct in saying that solutions with corners are admissible (and he did insist on it), determination of the singularities that are admissible in the solution of partial differential equations was still far in the future.

3. *Extensions of the Wave Equation*

While the controversy over the vibrating string was being carried on, the interest in musical instruments prompted further work, not only on vibrations of physical structures but also on hydrodynamical questions which concern the propagation of sound in air. Mathematically, these involve extensions of the wave equation.

In 1762 Euler took up the problem of the vibrating string with variable thickness. He had been stimulated by one of the principal questions of musical aesthetics. Jean-Philippe Rameau (1683–1764) had explained (1726) that the consonance of a musical sound is due to the fact that the component tones of any one sound are harmonics of the fundamental tone; that is, their frequencies are integral multiples of the fundamental frequency. But Euler, in his *Tentamen Novae Theoriae Musicae* (1739)[16] maintained that only in proper musical instruments were the overtones harmonics of the fundamental tone. He therefore undertook to show that the string of variable thickness or nonuniform density $\sigma(x)$ and tension T gives off inharmonic overtones.

The partial differential equation becomes

(28)
$$\frac{1}{c^2}\frac{\partial^2 y}{\partial t^2} = \frac{\partial^2 y}{\partial x^2},$$

with c now a function of x. The first substantial results were obtained by Euler in a paper, "On the Vibratory Motion of Non-Uniformly Thick Strings."[17] The general solution he declares to be beyond the power of analysis. He obtains a solution in the special case where the mass distribution σ is given by

$$\sigma = \frac{\sigma_0}{\left(1 + \dfrac{x}{\alpha}\right)^4},$$

wherein σ_0 and α are constants. Then

$$y = \left(1 + \frac{x}{\alpha}\right)\left[\phi\left(\frac{x}{1 + \dfrac{x}{\alpha}} + c_0 t\right) + \psi\left(\frac{x}{1 + \dfrac{x}{\alpha}} - c_0 t\right)\right],$$

where $c_0 = \sqrt{T/\sigma_0}$. The frequencies of the modes or harmonics are given by

$$\nu_k = \frac{k}{2l}\left(1 + \frac{l}{\alpha}\right)\sqrt{T/\sigma_0}, \qquad k = 1, 2, 3, \cdots.$$

16. *Opera*, (3), 1, 197–427.
17. *Novi Comm. Acad. Sci. Petrop.*, 9, 1762/63, 246–304, pub. 1764 = *Opera*, (2), 10, 293–343.

Thus the ratio of two successive frequencies is the same as for a string of uniform thickness, but the fundamental frequency is no longer inversely proportional to the length.

In this paper of 1762/63 Euler also considered the vibrations of a string composed of two lengths, a and b, of different thicknesses m and n. He derived the equation for the frequencies ω of the modes. These turn out to be solutions of

$$(29) \qquad m \tan \frac{\omega a}{m} + n \tan \frac{\omega b}{m} = 0,$$

and he solves for ω in special cases. The solutions of (29) are called the characteristic values or eigenvalues of the problem. These values are, as we shall see, of prime importance in the theory of partial differential equations. It is almost evident from (29) that the characteristic frequencies are not integral multiples of the fundamental one.

However, Euler took up this question again in another paper on the vibrating string of variable thickness,[18] and starting with (28) he shows that there are functions $c(x)$ for which the frequencies of the higher modes are not integral multiples of the fundamental.

D'Alembert, too, took up the string of variable thickness.[19] Here he used a significant method of solution that he had introduced earlier for the string of constant density. In this earlier attempt at the vibrating-string problem d'Alembert had introduced the idea of separation of variables, which is now a basic method of solution in partial differential equations.[20] To solve

$$\frac{\partial^2 y(t, x)}{\partial t^2} = a^2 \frac{\partial^2 y(t, x)}{\partial x^2}$$

d'Alembert sets

$$y = h(t)\, g(x),$$

substitutes this in the differential equation, and obtains

$$(30) \qquad \frac{1}{a^2} \frac{h''(t)}{h(t)} = \frac{g''(x)}{g(x)}.$$

He then argues, as we do now, that since g''/g does not vary when t does, it must be a constant, and by the like argument applied to h''/h, this expression too must be a constant. The two constants are equal and are denoted by A. Thus he gets the two separate ordinary differential equations

$$(31) \qquad \begin{aligned} h''(t) - a^2 A h(t) &= 0 \\ g''(x) - A g(x) &= 0. \end{aligned}$$

18. *Misc. Taur.*, 3, 1762/65, 25–59, pub. 1766 = *Opera*, (2), 10, 397–425.
19. *Hist. de l'Acad. de Berlin*, 19, 1763, 242 ff., pub. 1770.
20. *Hist. de l'Acad. de Berlin*, 6, 1750, 335–60, pub. 1752.

Since a and A are constants, each of these equations is readily solvable, and d'Alembert gets

$$y(t, x) = h(t)\, g(x) = [Me^{a\sqrt{A}\cdot t} + Ne^{-a\sqrt{A}\cdot t}][Pe^{\sqrt{A}\cdot x} + Qe^{-\sqrt{A}\cdot x}].$$

The end-conditions, $y(t, 0) = 0$ and $y(t, l) = 0$, led d'Alembert to assert that $g(x)$ must be of the form $k \sin Rx$ and that $h(t)$ must be of the same form because $y(t, x)$ must be periodic in t. He left the matter there. Daniel Bernoulli had used the idea of separation of variables in 1732 in his treatment of the vibrations of a chain suspended from one end, but d'Alembert was more explicit, despite the fact that he did not complete the solution.

In his 1763 paper d'Alembert wrote the wave equation as

$$\frac{\partial^2 y}{\partial t^2} = X(x)\, \frac{\partial^2 y}{\partial x^2}$$

and sought solutions of the form

$$u = \zeta(x)\, \cos \lambda \pi t.$$

He obtained for ζ the equation

(32)
$$\frac{d^2 \zeta}{dx^2} = \frac{-\lambda^2 \pi^2 \zeta}{X(x)}.$$

D'Alembert now had to determine ζ so that it was 0 at both ends of the string. By a detailed analysis he showed that there are values of λ for which ζ meets this condition. He did not appreciate in this work that there are infinitely many values of λ. The significance of the investigation is that it is another step in the direction of boundary value or eigenvalue problems for ordinary differential equations.

The transverse oscillation of a *heavy* continuous horizontal cord was taken up by Euler. In the paper, "On the Modifying Effect of Their Own Weight on the Motion of Strings,"[21] Euler obtains the differential equation

$$\frac{1}{c^2} y_{tt} = \frac{g}{c^2} + y_{xx}.$$

For constant c and with fixed ends at $x = 0$ and $x = l$, Euler finds that

$$y = -\frac{(1/2)gx(x - l)}{c^2} + \phi(ct + x) + \psi(ct - x).$$

Thus the results are the same as for the "weightless" string (where the gravitational force is ignored), except that the oscillation takes place about the parabolic figure of equilibrium

$$y = -\frac{(1/2)gx(x - l)}{c^2}.$$

21. *Acta Acad. Sci Petrop.*, 1, 1781, 178–90, pub. 1784 = *Opera*, (2), 11, 324–34, but dating from 1774.

As we shall see in a moment, Euler had introduced all the Bessel functions of the first kind in a paper on the vibrating drum (see also Chap. 21, secs. 4 and 6) and in this 1781 paper he remarks that it is possible to express any motion by a series of Bessel functions (despite the fact that he had argued against Daniel Bernoulli's claim, in the vibrating-string problem, that any function can be represented as a series of trigonometric functions).

Papers on the vibrating string and the hanging chain, of which the above are just samples, were published by many other men up to the end of the century. The authors continued to disagree, correct each other, and make all sorts of errors in doing so, including contradicting what they themselves had previously said and even proven. They made assertions, contentions, and rebuttals on the basis of loose arguments and often just personal predilections and convictions. Their references to papers to prove their contentions did not prove what they claimed. They also resorted to sarcasm, irony, invective, and self-praise. Mingled with these attacks were seeming agreements expressed in order to curry favor, particularly with d'Alembert, who had considerable influence with Frederick II of Prussia and as director of the Berlin Academy of Sciences.

The second order partial differential equation problems described thus far involved only one space variable and time. The eighteenth century did not go much beyond this. In a paper of 1759 [22] Euler took up the vibration of a rectangular drum, thus considering a two-dimensional body. He obtained for the vertical displacement z of the surface of the drum

$$(33) \qquad \frac{1}{c^2}\frac{\partial^2 z}{\partial t^2} = \frac{\partial^2 z}{\partial x^2} + \frac{\partial^2 z}{\partial y^2},$$

wherein x and y represent the coordinates of any point on the drum and c is determined by the mass and tension. Euler tried

$$z = v(x,y)\sin(\omega t + \alpha)$$

and found that

$$0 = \frac{\omega^2 v}{c^2} + \frac{\partial^2 v}{\partial x^2} + \frac{\partial^2 v}{\partial y^2}.$$

This equation has sinusoidal solutions of the form

$$v = \sin\left(\frac{\beta x}{a} + B\right)\sin\left(\frac{\gamma y}{b} + C\right)$$

where

$$\frac{\omega^2}{c^2} = \frac{\beta^2}{a^2} + \frac{\gamma^2}{b^2}.$$

22. *Novi Comm. Acad. Sci. Petrop.*, 10, 1764, 243–60, pub. 1766 = *Opera*, (2), 10, 344–59.

The dimensions of the drum are a and b, so that $0 \leq x \leq a$ and $0 \leq y \leq b$. When the initial velocity is 0, B and C may be taken to be 0. If the boundaries are fixed, then $\beta = m\pi$ and $\gamma = n\pi$ where m and n are integers. Then, since $\omega = 2\pi\nu$ where ν is the frequency per second, he obtains readily that the frequencies are

$$\nu = \frac{1}{2}c\sqrt{\frac{m^2}{a^2} + \frac{n^2}{b^2}}.$$

He then considers a circular drum and transforms (33) to polar coordinates (a highly original step), obtaining

(34) $$\frac{1}{c^2}\frac{\partial^2 z}{\partial t^2} = \frac{\partial^2 z}{\partial r^2} + \frac{1}{r}\frac{\partial z}{\partial r} + \frac{1}{r^2}\frac{\partial^2 z}{\partial \phi^2}.$$

He now tries solutions of the form

(35) $$z = u(r)\sin(\omega t + A)\sin(\beta\phi + B)$$

so that $u(r)$ satisfies

(36) $$u'' + \frac{1}{r}u' + \left(\frac{\omega^2}{c^2} - \frac{\beta^2}{r^2}\right)u = 0.$$

Here Bessel's equation appears in the current form (Cf. Chap. 21, sec. 6). Euler then calculates a power series solution

$$u\left(\frac{\omega}{c}r\right) = r^\beta\left\{1 - \frac{1}{1(\beta + 1)}\left(\frac{\omega}{c}\frac{r}{2}\right)^2 + \frac{1}{1\cdot 2(\beta + 1)(\beta + 2)}\left(\frac{\omega}{c}\frac{r}{2}\right)^4 + \cdots\right\},$$

which we would write now as

$$u\left(\frac{\omega}{c}r\right) = \left(\frac{c}{\omega}\right)^\beta 2^\beta \Gamma(\beta + 1)J_\beta\left(\frac{\omega}{c}r\right).$$

Since the edge $r = a$ must remain fixed,

(37) $$J_\beta\left(\frac{\omega}{c}a\right) = 0.$$

It also follows from (35), since z must be of period 2π in ϕ, that β is an integer. Euler asserts that for a fixed β there are infinitely many roots ω so that infinitely many simple sounds result. However, he did not calculate these roots. He did attempt to find a second solution of (36) but failed to do so. The theory of the vibrating membrane was derived independently by Poisson[23] and is often credited solely to him.

Euler, Lagrange, and others worked on the propagation of sound in air. Euler wrote on the subject of sound frequently from the time he was twenty

23. *Mém. de l'Acad. des Sci.*, Paris, (2), 8, 1829, 357–570.

years old (1727) and established this field as a branch of mathematical physics. His best work on the subject followed his major papers of the 1750s on hydrodynamics. Air is a compressible fluid, and the theory of the propagation of sound is part of fluid mechanics (and of elasticity, because air is also an elastic medium). However, to treat the propagation of sound he made reasonable simplifications of the general hydrodynamical equations.

Three fine and definitive papers were read to the Berlin Academy in 1759. In the first, "On the Propagation of Sound,"[24] Euler considers the propagation of sound in one space dimension. After some approximations, which amount to considering waves of small amplitude, he is led to the one-dimensional wave equation

$$\frac{\partial^2 y}{\partial t^2} = 2gh \frac{\partial^2 y}{\partial x^2},$$

where y is the amplitude of the wave at the point x and at time t, g is the acceleration of gravity and h is a constant relating the pressure and density. This equation, as Euler of course recognized, is the same as that for the vibrating string and he did nothing mathematically new in solving it.

In his second paper[25] Euler gives the two-dimensional equation of propagation in the form

$$(38) \qquad \frac{\partial^2 x}{\partial t^2} = c^2 \frac{\partial^2 x}{\partial X^2} + c^2 \frac{\partial^2 y}{\partial X\, \partial Y}, \qquad \frac{\partial^2 y}{\partial t^2} = c^2 \frac{\partial^2 y}{\partial Y^2} + c^2 \frac{\partial^2 x}{\partial X\, \partial Y},$$

where x and y are the wave amplitudes in the X-direction and Y-direction respectively, or the components of the displacement, and $c = \sqrt{2gh}$. He gives the plane wave solution

$$x = \alpha\phi(\alpha X + \beta Y + c\sqrt{\alpha^2 + \beta^2}\, t), y = \beta\phi(\alpha X + \beta Y + c\sqrt{\alpha^2 + \beta^2}\, t),$$

where ϕ is an arbitrary function and α and β are arbitrary constants. Then letting

$$(39) \qquad v = \frac{\partial x}{\partial X} + \frac{\partial y}{\partial Y}$$

(v is called the divergence of the displacement), he gets the two-dimensional wave equation

$$(40) \qquad \frac{1}{c^2}\frac{\partial^2 v}{\partial t^2} = \frac{\partial^2 v}{\partial X^2} + \frac{\partial^2 v}{\partial Y^2}.$$

He also states the need for the superposition of solutions to obtain the most general solution of the problem in order to meet some initial condition, that is, the value of v or of x and y at $t = 0$.

24. *Mém. de l'Acad. de Berlin*, 15, 1759, 185–209, pub. 1766, = *Opera*, (3), 1, 428–51.
25. *Mém. de l'Acad. de Berlin*, 15, 1759, 210–40, pub. 1766 = *Opera*, (3), 1, 452–83.

Euler then shows how he can get the differential equation whose solutions are called cylindrical waves because the wave spreads out like an expanding cylinder. He lets $Z = \sqrt{X^2 + Y^2}$ and introduces $v = f(Z, t)$ where f is arbitrary. By letting $x = vX$ and $y = vY$, he obtains from (40)

$$\frac{1}{c^2} \frac{\partial^2 v}{\partial t^2} = \frac{3}{Z} \frac{\partial v}{\partial Z} + \frac{\partial^2 v}{\partial Z^2}.$$

He also obtains in this paper in a similar manner the three-dimensional wave equation

(41)
$$\frac{1}{c^2} \frac{\partial^2 v}{\partial t^2} = \frac{\partial^2 v}{\partial X^2} + \frac{\partial^2 v}{\partial Y^2} + \frac{\partial^2 v}{\partial Z^2},$$

where v is again the divergence of the displacement (x, y, z). Euler gives plane wave and spherical wave solutions using the kind of substitution just indicated for cylindrical waves. The basic equation for spherical waves is

$$\frac{1}{c^2} \frac{\partial^2 s}{\partial t^2} = \frac{4}{V} \frac{\partial s}{\partial V} + \frac{\partial^2 s}{\partial V^2}, \qquad V = \sqrt{X^2 + Y^2 + Z^2}.$$

Much of the above work on spherical and cylindrical waves was also done independently by Lagrange at the end of the year 1759. Each communicated his results to the other. Though there are many details in which Lagrange's work differs from Euler's, there are no major mathematical points worthy of being related here.

From the propagation of sound waves in air it was but a step to the study of the sounds given off by musical instruments that employ air motion. This study was initiated by Daniel Bernoulli in 1739. Bernoulli, Euler, and Lagrange wrote numerous papers on the tones given off by an almost incredible variety of such instruments. In a publication of 1762 Daniel Bernoulli showed that at the open end of a cylindrical tube (organ pipe) no condensation of air can take place.[26] At a closed end the air particles must be at rest. He concluded from this that a tube closed at both ends or open at both ends has the same fundamental mode as a tube of half the length but open at one end and closed at the other. He also discovered the theorem that for closed organ pipes the frequencies of the overtones are odd multiples of the frequency of the fundamental. In the same paper Bernoulli took up pipes of other than cylindrical form, in particular the conical pipe, for which he obtained expressions for the individual tones (modes) but recognized that these hold only for infinite cones and not for the truncated one. For the (infinite) conical pipe the overtones proved to be harmonic to the fundamental. Bernoulli confirmed many of his theoretical results by experiments.

26. *Mém. de l'Acad. des Sci., Paris*, 1762, 431–85, pub. 1764.

Euler, too, studied cylindrical pipes and non-cylindrical figures of revolution[27] and considered reflection at open and closed ends. The efforts of these men were directed toward understanding flutes; organ pipes; all sorts of horns of hyperboloidal, conical, and cylindrical shape; trumpets; bugles; and other wind instruments.

On the whole these efforts to solve partial differential equations in three and four variables were limited, mainly because the solutions were expressed in series involving several variables as opposed to simpler trigonometric series in x and t separately (cf. [20]). But the mathematicians knew too little about the functions that appeared in these more complicated series and about methods of determining the coefficients. Such methods were soon developed.

It is worth mentioning that Euler, in considering the sound of a bell and reconsidering some of the problems of the vibrations of rods, was led to fourth order partial differential equations. However, he was unable to do much with them and in fact, for the rest of the century no progress was made with them.

4. Potential Theory

The development of the subject of partial differential equations was furthered by another class of physical investigations. One of the major problems of the eighteenth century was the determination of the amount of gravitational attraction one mass exerts on another, the prime cases being the attraction of the sun on a planet, of the earth on a particle exterior or interior to it, and of the earth on another extended mass. When the two masses are very far apart compared to their sizes, it is possible to treat them as point masses; but in other cases, notably the earth attracting a particle, the extent of the earth must be taken into account. Clearly the shape of the earth must be known if one is to calculate the gravitational attraction its distributed mass exerts on a particle or another distributed mass. Although the precise shape remained a subject for investigation (Chap. 21, sec. 1), it was already clear by 1700 that it must be some form of ellipsoid, perhaps an oblate spheroid (an ellipsoid generated by revolving an ellipse around the minor axis). For the solid oblate spheroid the force of attraction both on an external and on an internal particle cannot be calculated as though the mass were concentrated at the center.

In a prize paper of 1740 on the tides, and in his *Treatise of Fluxions* (1742), Maclaurin proved that, for a fluid of uniform density under constant angular rotation, the oblate spheroid is an equilibrium shape. Then Maclaurin proved synthetically that, given two confocal homogeneous ellipsoids

27. *Novi Comm. Acad. Sci. Petrop.*, 16, 1771, 281-425, pub. 1772 = *Opera*, (2), 13, 262-369.

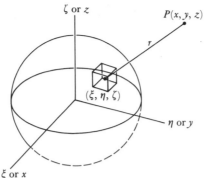

Figure 22.2

of revolution, the attractions of the two bodies on the same particle external to both, provided the particle be on the prolongation of the axis of revolution or in the plane of the equator, will be proportional to the volumes. Some other limited results were also established geometrically in the nineteenth century by James Ivory (1765–1842) and Michel Chasles.

The geometrical approach to the problem of gravitational attraction used by Newton, Maclaurin, and others is good only for special bodies and special locations of the attracted masses. This approach soon gave way to analytical methods, which one finds first in papers by Clairaut before 1743 and especially in his famous book *Théorie de la figure de la terre* (1743), in which he considers both the shape of the earth and gravitational attraction.

Let us first note some facts about the analytical formulation. The gravitational force exerted by an extended body on a unit mass P regarded as a particle is the sum of the forces exerted by all the small masses that make up the body. If $d\xi\, d\eta\, d\zeta$ is a small volume of the body (Fig. 22.2), so small that it may be regarded as a particle centered at the point (ξ, η, ζ), and if P has the coordinates (x, y, z), the attraction exerted by the small mass of density ρ on the unit particle is a vector directed from P to the small mass and, in view of the Newtonian law of gravitation, the components of this vector are (Chap. 21, sec. 7)

$$-k\rho\,\frac{x-\xi}{r^3}\,d\xi\, d\eta\, d\zeta, \qquad -k\rho\,\frac{y-\eta}{r^3}\,d\xi\, d\eta\, d\zeta, \qquad -k\rho\,\frac{z-\zeta}{r^3}\,d\xi\, d\eta\, d\zeta,$$

where k is the constant in the Newtonian law and

$$r = \sqrt{(x-\xi)^2 + (y-\eta)^2 + (z-\zeta)^2}.$$

Of course ρ may be a function of ξ, η, and ζ or, in the case of a homogeneous body, a constant.

The force exerted by the entire body on the unit mass at P has the components

$$f_x = -k \iiint \rho \frac{x - \xi}{r^3} \, d\xi \, d\eta \, d\zeta$$

(42) $$f_y = -k \iiint \rho \frac{y - \eta}{r^3} \, d\xi \, d\eta \, d\zeta$$

$$f_z = -k \iiint \rho \frac{z - \zeta}{r^3} \, d\xi \, d\eta \, d\zeta,$$

wherein the integral is extended over the entire attracting body. These integrals are finite and correct also when P is inside the attracting body.

Instead of treating each component of the force separately, it is possible to introduce one function $V(x, y, z)$ whose partial derivatives with respect to x, y, and z respectively are the three components of the force. This function is

(43) $$V(x, y, z) = \iiint \frac{\rho}{r} \, d\xi \, d\eta \, d\zeta.$$

By differentiating under the integral sign with respect to x, y, and z which are involved in r, one obtains

$$\frac{\partial V}{\partial x} = \frac{1}{k} f_x, \qquad \frac{\partial V}{\partial y} = \frac{1}{k} f_y, \qquad \frac{\partial V}{\partial z} = \frac{1}{k} f_z,$$

and these equations also hold when P is inside the attracting body. The function V is called a potential function. When problems involving the three components f_x, f_y, and f_z can be reduced to the problem of working with V, there is the advantage of working with one function instead of three.

If one knows the distribution of mass inside the body, which means knowing ρ as a function of ξ, η, and ζ, and if one knows the precise shape of the body, one can sometimes calculate V by actually evaluating the integral. However, for most shapes of bodies this triple integral is not integrable in terms of simple functions. Moreover, we do not know the true distribution of mass inside the earth and other bodies. Hence V must be determined in other ways. The principal fact about V is that for points (x, y, z) *outside* the attracting body, it satisfies the partial differential equation

(44) $$\frac{\partial^2 V}{\partial x^2} + \frac{\partial^2 V}{\partial y^2} + \frac{\partial^2 V}{\partial z^2} = 0,$$

in which we note that ρ does not appear. This differential equation is known as the potential equation and as Laplace's equation.

The idea that a force can be derived from a potential function, and even the term "potential function," were used by Daniel Bernoulli in

Hydrodynamica (1738). The potential equation itself appears for the first time in one of Euler's major papers composed in 1752, "Principles of the Motion of Fluids."[28] In dealing with the components u, v, and w of the velocity of any point in a fluid, Euler had shown that $u\,dx + v\,dy + w\,dz$ must be an exact differential. He introduces the function S such that $dS = u\,dx + v\,dy + w\,dz$. Then

$$u = \frac{\partial S}{\partial x}, \qquad v = \frac{\partial S}{\partial y}, \qquad w = \frac{\partial S}{\partial z}.$$

But the motion of incompressible fluids is subject to what is called the law of continuity, namely,

$$(45) \qquad \frac{\partial u}{\partial x} + \frac{\partial v}{\partial y} + \frac{\partial w}{\partial z} = 0,$$

which expresses mathematically the fact that no matter is destroyed or created during the motion. Then it follows that

$$\frac{\partial^2 S}{\partial x^2} + \frac{\partial^2 S}{\partial y^2} + \frac{\partial^2 S}{\partial z^2} = 0.$$

How to solve this equation generally, Euler says, is not known; so he considers just special cases where S is a polynomial in x, y, and z. The function S was later (1868) called by Helmholtz the velocity potential. In a paper published in 1762[29] Lagrange reproduced all of these quantities, which he took over from Euler without acknowledgment, though he did improve the order of the ideas and the expressions.

Before we can investigate the work done to solve the potential equation in behalf of gravitational attraction, we must review some efforts to evaluate this attraction directly by means of the integrals (42) or the equivalents in other coordinate systems.

In a paper written in 1782 but published in 1785, entitled "Recherches sur l'attraction des sphéroïdes,"[30] Legendre, interested in the attraction exerted by solids of revolution, proved the theorem: If the attraction of a solid of revolution is known for every external point on the prolongation of its axis, then it is known for every external point. He first expressed the component of the force of attraction in the direction of the radius vector r by means of

$$(46) \quad P(r, \theta, 0) = \iiint \frac{r - r' \cos \gamma}{(r^2 - 2rr' \cos \gamma + r'^2)^{3/2}} r'^2 \sin \theta'\, d\theta'\, d\phi'\, dr',$$

28. *Novi Comm. Acad. Sci. Petrop.*, 6, 1756/57, 271–311, pub. 1761 = *Opera*, (2), 12, 133–68.

29. *Misc. Taur.*, 2₂ 1760/61, 196–298, pub. 1762 = *Œuvres*, 1, 365–468.

30. *Mém. des sav. étrangers*, 10, 1785, 411–34.

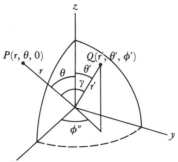

Figure 22.3

where (Fig. 22.3) r is the radius vector to the attracted point, r' is the radius vector to any point of the attracting body, and γ is the angle formed at the center of the body by the two radii vectors. The ϕ coordinate of the external point can be taken to be 0 because the solid is a figure of revolution around the z-axis. Then he expanded the integrand in powers of r'/r. This is done by writing the denominator as

$$r^3\left[1 - \left(2\frac{r'}{r}\cos\gamma - \frac{r'^2}{r^2}\right)\right]^{3/2}.$$

The quantity in the brackets can be put into the numerator and then expanded by the binomial theorem with the quantity in parentheses as the second term of the binomial. Legendre obtained for the integrand, apart from the volume element, the series

$$\frac{1}{r^2}\left\{1 + 3P_2(\cos\gamma)\frac{r'^2}{r^2} + 5P_4(\cos\gamma)\frac{r'^4}{r^4} + 7P_6(\cos\gamma)\frac{r'^6}{r^6} + \cdots\right\}.$$

The coefficients P_2, P_4, ... are rational integral functions of $\cos\gamma$. These functions are what we now call the Legendre polynomials or Laplace coefficients or zonal harmonics. Legendre gave the form of the functions so that the general P_n, namely,

(47) $$P_n(x) = \frac{(2n-1)(2n-3)\cdots 1}{n!}.$$

$$\left[x^n - \frac{n(n-1)}{2(2n-1)}x^{n-2} + \frac{n(n-1)(n-2)(n-3)}{2\cdot 4\cdot(2n-1)(2n-3)}x^{n-4} + \cdots\right],$$

could be inferred.

He could now integrate with respect to r', and he obtained

$$\frac{2}{r^2}\iint\left\{\frac{R^3}{3} + \frac{3}{5}P_2(\cos\gamma)\frac{R^5}{r^2} + \frac{5}{7}P_4(\cos\gamma)\frac{R^7}{r^4} + \cdots\right\}\sin\theta'\,d\theta'\,d\phi',$$

where $R = f(\theta')$ is the value of r' at a given θ' (it is independent of ϕ'). He then had to integrate with respect to ϕ'. For this he used[31]

$$\cos \gamma = \cos \theta \cos \theta' + \sin \theta \sin \theta' \cos \phi'.$$

Having established the subsidiary result

$$\frac{1}{\pi} \int_0^\pi P_{2n}(\cos \gamma) \, d\phi' = P_{2n}(\cos \theta) P_{2n}(\cos \theta'),$$

he obtained finally

$$P(r, \theta, 0) = \frac{3M}{r^2} \sum_{n=0}^\infty \frac{2n-3}{2n-1} P_{2n}(\cos \theta) \frac{\alpha_n}{r^{2n}},$$

where

$$\alpha_n = \frac{4\pi}{3M} \int_0^{\pi/2} R^{2n+3} P_{2n}(\cos \theta') \sin \theta' d\theta'.$$

The value of this integral depends upon the shape of the meridian curves $R = f(\theta')$.

From the above result, and on the basis of a communication from Laplace, Legendre then obtained the expression for the potential function for this problem, and from the potential derived the component of the force of attraction perpendicular to the radius vector.

In a second paper written in 1784,[32] Legendre derived some properties of the functions P_{2n}. Thus

(48) $$\int_0^1 f(x^2) P_{2n}(x) \, dx = 0$$

for each rational integral function of x^2 whose degree in x^2 is less than n. If n is any positive integer,

(49) $$\int_0^1 x^n P_{2m} \, dx = \frac{n(n-2)\cdots(n-2m+2)}{(n+1)(n+3)\cdots(n+2m+1)}.$$

If m and n are positive integers,

(50) $$\int_0^1 P_{2n}(x) P_{2m}(x) \, dx = \begin{cases} 0 & \text{for } m \gtrless n, \\ \dfrac{1}{4m+1} & \text{for } m = n. \end{cases}$$

---

31. This expression is derived as follows: In view of the equations of transformation from spherical to rectangular coordinates $x = r \sin \theta \cos \phi$, $y = r \sin \theta \sin \phi$, $z = r \cos \theta$, the rectangular coordinates of P are $(r \sin \theta \cos \phi, r \sin \theta \sin \phi, r \cos \theta)$ and the rectangular coordinates of Q are $(r' \sin \theta' \cos \phi', r' \sin \theta' \sin \phi', r' \cos \theta')$. Then using the distance formula we can express PQ. But by the law of cosines $PQ = r^2 + r'^2 + 2rr' \cos \gamma$. Equating the two expressions for PQ gives the above expression for $\cos \gamma$.

32. *Mém. de l'Acad. des Sci., Paris*, 1784, 370–89, pub. 1787.

He also proved that the zeros of each of the P_{2n} are real, different from each other, symmetric with respect to 0, and in absolute value less than 1. Also for $0 < x < 1$, $P_{2n}(x) < 1$.

Then, with the help of the orthogonality condition (50), he proves (by integration of the series term by term) that a given function of x^2 can be expressed in only one way in a series of functions $P_{2n}(x)$.

Finally, using these and other properties of his polynomials, Legendre returns to the main problem of gravitational attraction and using the expression (43) for the potential and the condition for equilibrium of a rotating fluid mass, he obtains the equation for the meridian curve of such a mass in the form of a series of his polynomials. He believed that this equation included all possible equilibrium figures for a spheroid of revolution.

Now Laplace enters the picture. He had written several papers on the force of attraction exerted by volumes of revolution (1772, pub. 1776; 1773, pub. 1776; and 1775, pub. 1778), in which he worked with the components of the force but not the potential function. The article by Legendre of 1782, published in 1785, inspired a famous and remarkable fourth paper by Laplace, "Théorie des attractions des sphéroïdes et de la figure des planètes."[33] Without mentioning Legendre, Laplace took up the problem of the attraction exerted by an arbitrary spheroid as opposed to Legendre's figures of revolution. By a spheroid Laplace meant any surface given by one equation in r, θ, and ϕ.

He starts with the theorem that the potential V of the force an arbitrary body exerts on an external point, expressed in spherical coordinates r, θ, ϕ with $\mu = \cos \theta$, satisfies the potential equation

$$(51) \qquad \frac{\partial}{\partial \mu}\left((1 - \mu^2) \frac{\partial V}{\partial \mu}\right) + \frac{1}{1 - \mu^2} \frac{\partial^2 V}{\partial \phi^2} + r \frac{\partial^2 (rV)}{\partial r^2} = 0.$$

Laplace does not say here how he obtained the equation. In a later paper[34] he gives the rectangular coordinate form (44). One may be fairly sure that he possessed the rectangular form first and derived the spherical coordinate form from it. In fact, both forms had already been given by Euler and Lagrange, but Laplace does not mention them. He may not have known their work, though this is doubtful.

In the 1782 paper Laplace sets

$$(52) \qquad V(r, \theta, \phi) = \frac{U_0}{r} + \frac{U_1}{r^2} + \frac{U_2}{r^3} + \cdots,$$

33. *Mém. de l'Acad. des Sci.*, Paris, 1782, 113–96, pub. 1785 = *Œuvres*, 10, 339–419.
34. *Mém. de l'Acad. des Sci.*, Paris, 1787, 249–67, pub. 1789 = *Œuvres*, 11, 275–92.

where $U_n = U_n(\theta, \phi)$, and substitutes this in (51). Then the individual U_n satisfy[35]

(53) $$\frac{\partial}{\partial \mu}\left[(1 - \mu^2)\frac{\partial U}{\partial \mu}\right] + \frac{1}{1 - \mu^2}\frac{\partial^2 U}{\partial \phi^2} + n(n + 1)U = 0.$$

With the help of Legendre's P_{2n} he is able to show that

(54) $$U_n(\theta, \phi) = \iiint r'^{n+2} P_{2n}(\cos \theta \cos \theta' + \sin \theta \sin \theta' \cos (\phi - \phi')) \cdot$$

$$\sin \theta' \, d\theta' \, d\phi' \, dr'.$$

Now Laplace uses this result and (52) to calculate the potential of a spheroid differing little from a sphere. He writes the equation of the surface of the spheroid as

(55) $$r = a(1 + \alpha y),$$

where α is small and y on the spheroid is a function of θ' and ϕ'. Laplace assumes that $y(\theta, \phi)$ can be expanded in a series of functions

(56) $$y = Y_0 + Y_1 + Y_2 + \cdots,$$

where the Y_n are functions of θ and ϕ and satisfy the differential equation (53). The result he obtains here is first that

(57) $$Y_n = \frac{2n + 1}{4\pi \alpha a^{n+1}} U_n.$$

These Y_n, then, may be used in (52). Also the expansion (56) may be recast as

(58) $$y(\mu, \phi) = \frac{1}{4\pi} \sum_{n=0}^{\infty} (2n + 1) \int_{-1}^{1}\int_{0}^{2\pi} Y_n(\mu', \phi')P_n(\mu, \phi, \mu', \phi') \, d\mu' \, d\phi',$$

where $\mu' = \cos \theta'$. With the value of y he now has an expression for r in (55) and with this and the U_n in (54) he obtains V in (52).

Laplace does not consider here the general problem of the development of *any* function of θ and ϕ into a series of the Y_n. In what he does do here and in later papers he presumes that such an expression is possible and is unique.

35. If we ignore the middle term (ϕ is absent) the resulting ordinary differential equation is what we now call Legendre's differential equation,

$$(1 - x^2)\frac{d^2 z}{dx^2} - 2x\frac{dz}{dx} + n(n + 1)z = 0.$$

The $P_n(x)$ satisfy this equation. On the other hand the U_n (and the Y_n in [57]) regarded as functions of the two variables $\mu = \cos \theta$ and ϕ satisfy (53). The U_n and Y_n are called by the Germans spherical functions and by Lord Kelvin spherical harmonics or spherical surface harmonics.

He deals with rational integral functions of μ, $\sqrt{1 - \mu^2}\cos\phi$, and $\sqrt{1 - \mu^2}$ sin ϕ, and so his need for the general result is limited. He does prove the basic orthogonality property

(59) $$\int_{-1}^{1}\int_{0}^{2\pi} U_n(\mu, \phi) U_m(\mu, \phi) = 0, \qquad m \neq n.$$

However in his *Mécanique céleste*, Volume 2, he shows that an arbitrary function of θ and ϕ can be expanded in a series of the U_n (or the Y_n) and shows that (59) implies uniqueness of the expansion.

Laplace wrote several more papers on the attraction of spheroids and on the shape of the earth (e.g. 1783, pub. 1786; 1787, pub. 1789); and in these papers uses expansions in spherical functions. In the last paper, in which Laplace gave the rectangular coordinate form of the potential equation, he made one mistake of consequence. He assumed that this equation holds when the point mass being attracted by a body lies inside the body. This error was corrected by Poisson (Chap. 28, sec. 4).

During the 1780s Legendre continued his investigations. His fourth paper, written in 1790,[36] introduced the $P_n(x)$ for *odd* n. The expression for $P_n(x)$ given in (47) is the correct one for all n. Legendre proves that for any positive integral m and n

(60) $$\int_{-1}^{1} P_m(x) P_n(x)\, dx = \begin{cases} 0 \text{ for } m \neq n, \\ \dfrac{2}{2n + 1} \quad \text{for } m = n. \end{cases}$$

Then he too introduces the spherical functions. That is, he lets Y_n be the coefficient of z^n in the expansion of $(1 - 2zt + z^2)^{-1/2}$ where $t = \cos\theta\cos\theta' + \sin\theta\sin\theta'\cos(\phi - \phi')$. Then, letting $\mu = \cos\theta$, $\mu' = \cos\theta'$ and $\psi = \phi - \phi'$, he shows that

$$Y_n(t) = P_n(\mu)P_n(\mu') + \frac{2}{n(n + 1)}\frac{dP_n(\mu)}{d\mu}\frac{dP_n(\mu')}{d\mu'}\sin\theta\sin\theta'\cos\psi$$

$$+ \frac{2}{(n - 1)(n)(n + 1)(n + 2)}\frac{d^2P_n(\mu)}{d\mu^2}\frac{d^2P_n(\mu')}{d\mu'^2}\sin^2\theta\sin^2\theta'\cos 2\psi + \cdots,$$

the higher terms containing higher derivatives of P_n. This equation is equivalent to

$$P_n(\cos\theta\cos\theta' + \sin\theta\sin\theta'\cos(\phi - \phi')) =$$

$$\sum_{m=1}^{n} P_n^m(\cos\theta)P_n^m(\cos\theta')\cos m(\phi - \phi'),$$

36. *Mém. de l'Acad. des Sci.*, Paris, 1789, 372–454, pub. 1793.

wherein the m is a superscript and not an exponent. The $P_n^m(x)$ satisfy

$$(61) \qquad \frac{d}{dx}\left[(1 - x^2)\frac{dP_n^m}{dx}\right] + \left[n(n + 1) - \frac{\nu^2}{1 - x^2}\right]P_n^m = 0$$

and the $P_n^m(x)$ agree up to a constant factor with

$$(1 - x^2)^{m/2}\frac{d^m P_n(x)}{dx^m}.$$

The $P_n^m(x)$ thus introduced are now called the associated Legendre polynomials. Then Legendre proves that

$$(62) \qquad \int_{-1}^{1}\int_{0}^{2\pi} U_n(\mu', \phi')P_n(\mu, \phi, \mu', \phi')\, d\mu'\, d\phi' = \frac{4\pi}{2n + 1}\, U_n(\mu, \phi)$$

and that

$$(63) \qquad \int_{-1}^{1}\int_{0}^{2\pi} \{P_n(\mu)\}^2\, d\mu\, d\phi = \frac{4\pi}{2n + 1}.$$

The fact that the $P_n(x)$ satisfy Legendre's differential equation is used in this paper.

Many other special results involving the Legendre polynomials and the spherical harmonics were obtained by Legendre, Laplace, and others. A basic result is the formula of Olinde Rodrigues (1794–1851), given in 1816,[37]

$$(64) \qquad P_n(x) = \frac{1}{2^n n!}\frac{d^n(x^2 - 1)^n}{dx^n}.$$

Laplace's work on the solution of the potential equation for the attracting force of spheroids was the beginning of a vast amount of work on this subject. Equally important was his and Legendre's work on the Legendre polynomials $P_n(x)$, the associated Legendre polynomials $P_n^m(x)$, and the spherical (surface) harmonics $Y_n(\mu, \phi)$, because rather arbitrary functions can be expressed in terms of infinite series of the P_n, P_n^m and the Y_n. These series of functions are analogous to the trigonometric functions, which Daniel Bernoulli claimed could also be used to represent arbitrary functions. The choice of class of functions depends on the differential equation being solved and on the initial and boundary conditions. Of course far more had to be done and was to be done with these functions to render them more useful in the solution of partial differential equations.

5. First Order Partial Differential Equations

Up to the time of Lagrange there was very little systematic work on first order partial differential equations. The second order equations received

37. *Corresp. sur l'Ecole Poly.*, 3, 1816, 361–85.

primary attention because the physical problems led directly to them. A few special first order equations had been solved, but these were either readily integrated or integrated by tricks. There was one exception, the equation usually called today a total differential equation and having the form

(65) $P\ dx + Q\ dy + R\ dz = 0,$

where P, Q, and R are functions of x, y, and z. Such an equation, if integrable, defines z as a function of x and y. Clairaut encountered such equations in 1739 in his work on the shape of the earth.[38] If the expression on the left side of (65) is an exact differential, that is, if there is a function $u(x, y, z) = C$ such that

(66) $du = P\ dx + Q\ dy + R\ dz,$

then Clairaut points out that

(67) $$\frac{\partial P}{\partial y} = \frac{\partial Q}{\partial x}, \qquad \frac{\partial P}{\partial z} = \frac{\partial R}{\partial x}, \qquad \frac{\partial Q}{\partial z} = \frac{\partial R}{\partial y}.$$

Clairaut showed how to solve (65) by a method still used in modern texts. The interest in equation (65) stemmed from the fact that if P, Q, R are components of velocity in fluid motion, then (65) has to be an exact differential.

If (65) is not an exact differential, then Clairaut also showed that it may be possible to find an integrating factor, that is, a function $\mu(x, y, z)$ such that when multiplied into (65), it makes the new left side an exact differential. Clairaut[39] and later d'Alembert (*Traité de l'équilibre et du mouvement des fluides*, 1744) gave a necessary condition that it be integrable (with the aid of an integrating factor). This condition (which is also sufficient) is

(68) $$P\left(\frac{\partial Q}{\partial z} - \frac{\partial R}{\partial y}\right) + Q\left(\frac{\partial R}{\partial x} - \frac{\partial P}{\partial z}\right) + R\left(\frac{\partial P}{\partial y} - \frac{\partial Q}{\partial x}\right) = 0.$$

The *general* first order partial differential equation in two independent variables is of the form

(69) $f(x, y, z, p, q) = 0,$

where $p = \partial x/\partial z$ and $q = \partial z/\partial y$. If the equation is linear in p and q then it is a linear partial differential equation and if not, then nonlinear. The significant theory was contributed by Lagrange.

Lagrange's terminology, which is still current, must be noted first to understand his work. He classified solutions of nonlinear first order equations as follows. Any solution $V(x, y, z, a, b) = 0$ containing two arbitrary constants is the *complete solution or complete integral*. By letting $b = \phi(a)$ where ϕ

38. *Mém. de l'Acad. des Sci., Paris*, 1739, 425–36.
39. *Mém. de l'Acad. des Sci., Paris*, 1740, 293–323.

is arbitrary, we obtain a one-parameter family of solutions. When $\phi(a)$ is arbitrary, the envelope of this family is called a *general integral*. When a definite $\phi(a)$ is used, the envelope is a *particular case* of the general integral. The envelope of all the solutions in the complete integral is called the *singular integral*. We shall see later what these solutions are geometrically. The complete integral is not unique in that there can be many different ones, which are not obtainable from each other by a simple change in the arbitrary constants. But from any one we can get all the solutions given by another through the particular cases and the singular integral.

Between two important papers of 1772 and 1779, which we shall take up shortly, Lagrange wrote a paper in 1774 discussing the relationships among complete, general, and singular solutions of first order partial differential equations. The general integral is obtained by eliminating a from $V(x, y, z, a, \phi(a)) = 0$ and $\partial V/\partial a = 0$ where $\phi(a)$ is arbitrary.[40] (For a particular $\phi(a)$ we get a particular solution.) The singular solution is obtained by eliminating a and b from $V(x, y, z, a, b) = 0$, $\partial V/\partial a = 0$ and $\partial V/\partial b = 0$.

Lagrange gave first the general theory of *nonlinear* first order equations. In the paper of 1772[41] he considered a general first order equation with two independent variables x and y and the dependent variable z. Here he improved on and generalized what Euler did earlier. He regarded equation (69) as given in the form where q is a function of x, y, z, and p, namely,

$$(70) \qquad q - Q(x, y, z, p) = 0$$

and sought to determine p as a function P of x, y, and z so that the two equations

$$(71) \qquad q - Q(x, y, z, p) = 0 \quad \text{and} \quad p - P(x, y, z) = 0$$

have a single infinity of common integral surfaces, or, as Lagrange put it analytically, so that the expression

$$(72) \qquad dz - p\,dx - q\,dy$$

by multiplication by a suitable factor $M(x, y, z)$ becomes an exact differential dN of $N(x, y, z) = 0$. For this to be so he must have

$$\frac{\partial N}{\partial z} = M, \qquad \frac{\partial N}{\partial x} = -Mp, \qquad \frac{\partial N}{\partial y} = -Mq.$$

For these equations the integrability conditions (67) imply

$$\frac{\partial M}{\partial x} = -\frac{\partial(Mp)}{\partial x}, \qquad \frac{\partial M}{\partial y} = -\frac{\partial(Mq)}{\partial z}, \qquad \frac{\partial(Mp)}{\partial y} = \frac{\partial(Mq)}{\partial x}.$$

40. For arbitrary ϕ it is not generally possible to actually carry out the elimination of a. The general integral is a concept and amounts to a collection of particular solutions.
41. *Nouv. Mém. de l'Acad. de Berlin*, 1772 = *Œuvres*, 3, 549–75.

If one puts into the last of these three equations the values of $\partial M/\partial x$ and $\partial M/\partial y$ from the first two, one obtains

(73) $$\frac{\partial p}{\partial y} - \frac{\partial q}{\partial x} - p\frac{\partial q}{\partial z} + q\frac{\partial p}{\partial z} = 0.$$

This is the condition (68) for the integrability of (72), a condition known before, as Lagrange remarks. In (73) q can be taken as the given function Q of x, y, z, and p so that explicitly the equation becomes

(74) $$-Q_p\frac{\partial p}{\partial x} + \frac{\partial p}{\partial y} + (Q - pQ_p)\frac{\partial p}{\partial z} - Q_x - pQ_z = 0.$$

Lagrange's plan now is to find a solution $p = P$ of this first order equation, which is *linear* in the derivatives of p and whose solution contains an arbitrary constant α. Having obtained this, he integrates the two equations

(75) $$q - Q(x, y, z, p) = 0, \qquad p - P(x, y, z, \alpha) = 0,$$

which represent $\partial z/\partial x$ and $\partial z/\partial y$ as functions of x, y, z, and finds a family of ∞^2 integral surfaces of the original equation (70); that is, he finds the complete solution. Thus far, then, Lagrange has replaced the problem of solving the nonlinear equation (70) by the problem of solving the *linear* equation (74).

In 1779 Lagrange gave his method of solving linear first order partial differential equations.[42] He considers the equation still called Lagrange's linear equation

(76) $$Pp + Qq = R,$$

where P, Q, and R are functions of x, y, and z; the equation is called non-homogeneous because of the presence of the R term. This equation is intimately related to the homogeneous equation in three independent variables

(77) $$P\frac{\partial f}{\partial x} + Q\frac{\partial f}{\partial y} + R\frac{\partial f}{\partial z} = 0.$$

What Lagrange shows readily is that if $u(x, y, z) = c$ is a solution of (76), then $f = u(x, y, z)$ is a solution of (77) and conversely. Hence the problem of solving (76) is equivalent to that of solving (77). The equation (77) in turn is related to the system of ordinary differential equations

(78) $$\frac{dx}{P} = \frac{dy}{Q} = \frac{dz}{R} \quad \text{or} \quad \frac{dy}{dx} = \frac{Q}{P} \quad \text{and} \quad \frac{dz}{dx} = \frac{R}{P}.$$

In fact, if $f = u(x, y, z)$ and $f = v(x, y, z)$ are two independent solutions of (77), then $u = c_1$ and $v = c_2$ are a solution of (78) and conversely. Hence, if we can find the solutions $u = c_1$ and $v = c_2$ of (78), $f = u$ and $f = v$ will be solutions of (77) and $u = c$ and $v = c$ will be solutions of (76). Moreover,

42. *Nouv. Mém. de l'Acad. de Berlin*, 1779 = *Œuvres*, 4, 585–634.

one can show readily that $f = \phi(u, v)$ where ϕ is an arbitrary function of u and v also satisfies (77). Then $\phi(u, v) = c$ or, since ϕ is arbitrary, $\phi(u, v) = 0$ is a general solution of (76). Lagrange gave the above scheme in 1779 and gave a proof in 1785.[43] It is perhaps worth noting incidentally that Euler was aware that the solution of (77) can be reduced to the solution of (78).

If one takes this work on linear equations in connection with the 1772 work on nonlinear equations, one sees that Lagrange had succeeded in reducing an arbitrary first order equation in x, y, and z to a system of simultaneous ordinary differential equations. He does not state the result explicitly but it follows from the above work. Curiously, in 1785 he had to solve a particular first order partial differential equation and said it was impossible with present methods; he had forgotten his earlier (1772) work.

Then Paul Charpit (d. 1784) in 1784 presumably combined the methods for nonlinear and linear equations to reduce any $f(x, y, z, p, q) = 0$ to a system of ordinary differential equations. Lacroix said in 1798 that Charpit had submitted a paper in 1784 (which was not published) in which he reduced first order partial differential equations to systems of ordinary differential equations. Jacobi found Lacroix's statement striking and expressed the wish that Charpit's work be published. But this was never done and we do not know whether Lacroix's statement is correct. Actually Lagrange had done the full job and Charpit could have added nothing. The method given in modern texts, called the Lagrange, Lagrange-Charpit, or Charpit method, is the fusion of the ideas Lagrange presented in his 1772 and 1779 papers. It states that to solve the general first order partial differential equation $f(x, y, z, p, q) = 0$ one must solve the system of ordinary differential equations (the characteristic equations of $f = 0$),

(79)
$$\frac{dx}{dt} = f_p, \qquad \frac{dy}{dt} = f_q, \qquad \frac{dz}{dt} = pf_p + qf_q,$$

$$\frac{dp}{dt} = -f_x - f_z p, \qquad \frac{dq}{dt} = -f_y - f_z q.$$

The solution is effected by finding any one integral of (79), say $u(x, y, z, p, q) = A$. One solves this and $f = 0$ simultaneously for p and q and substitutes for p and q in $dz = p\,dx + q\,dy$ (see [72]). Then one integrates by the method used for (65).

Lagrange's method is often called Cauchy's method of characteristics because the generalization to n variables of the method of arriving at (79) used by Lagrange and Charpit for a differential equation in two independent variables presents difficulties which were surmounted by Cauchy in 1819.[44]

43. *Nouv. Mém. de l'Acad. de Berlin*, 1785 = *Œuvres*, 5, 543–62.
44. *Bull. de la Société Philomathique*, 1819, 10–21; see also *Exercices d'analyse et de phys. math.*, 2, 238–72 = *Œuvres*, (2), 12, 272–309.

6. *Monge and the Theory of Characteristics*

Lagrange worked purely analytically. Gaspard Monge (1746–1818) introduced the language of geometry. His work in partial differential equations was not as great as that of Euler, Lagrange, and Legendre, but he started the movement to interpret the analytic work geometrically and introduced thereby many fruitful ideas. He saw that just as problems involving curves led to ordinary differential equations, so problems involving surfaces lead to partial differential equations. More generally, for Monge geometry and analysis were one subject, whereas for the other mathematicians of the century the two branches were distinct, with just points of contact. Monge began his work in 1770 but did not publish until much later.

It was primarily in the subject of nonlinear first order equations that Monge not only introduced the geometric interpretation but emphasized a new concept, that of characteristic curves.[45] His ideas on characteristics and on integrals as envelopes were not understood and were called a metaphysical principle by his contemporaries, but the theory of characteristics was to become a very significant theme in later work. Monge developed his ideas more fully in his lectures and in subsequent publications, notably the *Feuilles d'analyse appliquée à la géométrie* (1795). The ideas are best illustrated by his own example.

Consider the two-parameter family of spheres, all of constant radius R and centers anywhere in the XY-plane. The equation of this family is

$$(80) \qquad (x - a)^2 + (y - b)^2 + z^2 = R^2.$$

This equation is the complete integral of the nonlinear first order partial differential equation

$$(81) \qquad z^2(p^2 + q^2 + 1) = R^2,$$

because (80) contains two arbitrary constants a and b and clearly satisfies (81). Any subfamily of spheres introduced by letting $b = \phi(a)$ is a family of spheres with centers on a curve, the curve $y = \phi(x)$ in the XY-plane. The envelope of this one-parameter family of spheres (a tubular surface) is also a solution of (81). This particular solution is obtained by eliminating a from

$$(82) \qquad (x - a)^2 + (y - \phi(a))^2 + z^2 = R^2$$

and the partial derivative of (82) with respect to a, namely,

$$(83) \qquad (x - a) + (y - \phi(a))\phi'(a) = 0.$$

For each particular choice of a, equations (82) and (83) represent two particular surfaces, and therefore the two considered simultaneously repre-

45. *Hist. de l'Acad. des Sci.*, Paris, 1784, 85–117, 118–92, pub. 1787.

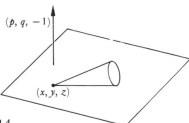

Figure 22.4

sent a curve called a characteristic curve. This curve is also the curve of intersection of two "consecutive" members of the subfamily. The set of characteristic curves fills out the envelope; that is, the envelope touches each member of the subfamily along a characteristic. The general integral is the aggregate of surfaces (envelopes of one-parameter families), each generated by a set of characteristic curves. The envelope of all the solutions of (80), that is, the singular solution, obtained by eliminating a and b from (80) and the partial derivatives of (80) with respect to a and b respectively, is $z = \pm R$.

The characteristic curve appears in another way. Consider two sub-families of spheres whose envelopes are tangential along any one sphere. We might call such envelopes consecutive envelopes. The curve of intersection of these two consecutive envelopes is the same characteristic curve on the sphere as the one obtained by considering consecutive members of either subfamily. Any one sphere may belong to an infinity of different subfamilies whose envelopes are all different, and so there will be different characteristic curves on the same sphere. All are great circles in vertical planes.

Monge gave the analytical form of the differential equations of the characteristic curves, which amounts to the fact that equations (79) determine the characteristic curves of (69). (Monge used total differential equations to express the equations of the characteristic curves.)

Monge also introduced (1784) the notion of a characteristic cone. At any point (x, y, z) of space (Fig. 22.4) one may consider a plane whose normal has the direction numbers p, q, -1. For a fixed (x, y, z), the set of p and q which satisfy

(84) $$F(x, y, z, p, q) = 0$$

determines a one-parameter family of planes all passing through (x, y, z). This set of planes envelopes a cone with vertex at (x, y, z). This is the characteristic cone or Monge cone at (x, y, z). If we now consider a surface S whose equation is $z = g(x, y)$, then the surface has a tangent plane at each (x, y, z). A necessary and sufficient condition that such a surface be an integral surface of $F = 0$ is that at each point (x, y, z) the tangent plane of F

be a tangent plane of the Monge cone at (x, y, z). A curve C on an integral surface S is called a characteristic curve if at each point of C the tangent is a generator of the Monge cone at that point. These characteristic curves are the same as those Monge deduced from the complete integral illustrated by (80) and are the solutions of the simultaneous equations (79). He also points out in a paper of 1802 which he incorporated in his *Application de l'analyse à la géométrie*[46] (1807) that each integral surface is a locus of characteristic curves and only one characteristic curve passes through each point of the integral surface.

The significance of the characteristic curves lies in the following. If one chooses a space curve $x(t), y(t), z(t)$ (for some interval of t-values) that is not a characteristic curve, then there is just one integral surface of $F = 0$ that passes through this curve; that is, there is just one $z = g(x, y)$ such that $z(t) = g(x(t), y(t))$ (for the range of t-values). On the other hand, as Monge noted in lectures of 1806, through any characteristic curve one can pass an infinity of integral surfaces. Moreover, the infinity of integral surfaces that pass through the curve are tangent to each other on that curve.

7. *Monge and Nonlinear Second Order Equations*

In addition to the second order linear equations we have already reviewed, the eighteenth-century mathematicians had occasion to consider more general linear second order equations in two independent variables and even non-linear ones. Thus they studied the linear equation

$$A\frac{\partial^2 z}{\partial x^2} + B\frac{\partial^2 z}{\partial x \partial y} + C\frac{\partial^2 z}{\partial y^2} + D\frac{\partial z}{\partial x} + E\frac{\partial z}{\partial y} + Fz + G = 0,$$

where $A, B, \cdots, G$ are functions of x and y. This equation is commonly written as

(85) $Ar + Bs + Ct + Dp + Eq + Fz + G = 0,$

where the letters $r, s, t, p,$ and q have the obvious meanings. Laplace showed in 1773[47] that equation (85) can, by a change of variables, be reduced to the form

(86) $s + ap + bq + cz + g = 0,$

where $a, b, c,$ and g are functions of x and y only, provided that $B^2 - 4AC \neq 0$. He then solved the equation in terms of an infinite series.

In his *Feuilles d'analyse*, Monge considered the nonlinear equation

(87) $Rr + Ss + Tt = V$

46. This is the title of the third edition of his *Feuilles d'analyse*.
47. *Hist. de l'Acad. des Sci., Paris,* 1773, pub. 1777 = *Œuvres,* 9, 5–68.

in which R, S, T, and V are functions of x, y, z, p, and q, so that the equation is linear only in the second derivatives r, s, and t. This type of equation arose in Lagrange's work on minimal surfaces, that is, surfaces of least area bounded by given space curves, wherein the specific differential equation is $(1 + q^2)r - 2pqs + (1 + p^2)t = 0$. (See also Chap. 24, sec. 4). Though Monge had already done some work on equation (87), in the present work (1795) he was able to solve it elegantly by the method we shall sketch.

By using the immediate facts

$$(88) \qquad\qquad dz = p\,dx + q\,dy$$

$$(89) \qquad\qquad dp = r\,dx + s\,dy$$

$$(90) \qquad\qquad dq = s\,dx + t\,dy$$

and eliminating r and t from (87), (89), and (90) he obtained the equation

$$(91) \qquad s(R\,dy^2 - S\,dx\,dy + T\,dx^2) - (R\,dy\,dp + T\,dx\,dq - V\,dx\,dy) = 0.$$

His argument then was that whenever it is possible to solve simultaneously

$$(92) \qquad\qquad R\,dy^2 - S\,dx\,dy + T\,dx^2 = 0$$

and

$$(93) \qquad\qquad R\,dy\,dp + T\,dx\,dq - V\,dx\,dy = 0$$

then (91) will be satisfied and so will (87).

Equation (92) is equivalent to two first order equations

$$(94) \quad dy - W_1(x, y, z, p, q)\,dx = 0 \quad \text{and} \quad dy - W_2(x, y, z, p, q) = 0.$$

Equations (88) and (93), together with either one of (94), constitute a system of three total differential equations in the five variables x, y, z, p, and q. When these three equations can be solved it is possible to find two solutions

$$u_1(x, y, z, p, q) = C_1 \quad \text{and} \quad u_2(x, y, z, p, q) = C_2$$

and then

$$(95) \qquad\qquad u_1 = \phi(u_2),$$

where ϕ is arbitrary, is a first order partial differential equation. The equation (95) is called an intermediate integral. Its general solution is the solution of (87). If the other equation in (94) can be used together with (88) and (93), we get another function

$$(96) \qquad\qquad u_3 = \phi(u_4).$$

In this case (95) and (96) can be solved simultaneously for p and q and these values are substituted in (88); then this total differential equation can be solved. This at least is the general scheme, though there are details we shall not take time to cover. Monge's integration of the equation for minimal surfaces was one of his claims to glory.

For the equation (87) also Monge introduced the theory of characteristics. The total differential equation of the characteristics is (92), that is,

$$R \, dy^2 - S \, dx \, dy + T \, dx^2 = 0.$$

This equation, which appears in his work as early as 1784,[48] defines at each point of an integral surface two directions that are the characteristic directions at that point. Through each point on an integral surface there pass two characteristic curves, along each of which two consecutive integral surfaces touch each other.

8. *Systems of First Order Partial Differential Equations*

Systems of partial differential equations arose first in the eighteenth-century work on fluid dynamics or hydrodynamics. The work on incompressible fluids, for example, water, was motivated by such practical problems as designing the hulls of ships to reduce resistance to motion in water and calculating the tides, the flow of rivers, the flow of water from jets, and the pressure of water on the sides of a ship. The work on compressible fluids, air in particular, sought to analyze the action of air on sails of ships, the design of windmill vanes, and the propagation of sound. The work we studied earlier on the propagation of sound was historically an application of the work on hydrodynamics specialized to waves of small amplitude.

After having treated incompressible fluids in 1752 in a paper entitled "Principles of the Motion of Fluids,"[49] Euler generalized this work in a paper of 1755, entitled "General Principles of the Motion of Fluids."[50] Here he gave the still-famous equations of fluid flow for perfect (nonviscous) compressible and incompressible fluids. The fluid is regarded as a continuum and the particles are mathematical points. He considers the force acting on a small volume of the fluid subject to the pressure p, density ρ, and external forces with components P, Q, and R per unit mass.

In one of the two approaches to fluid dynamics that Euler created, known in the literature as the spatial description, the components u, v, and w of the fluid velocity are given at every point in the fluid by

$$(97) \qquad u = u(x, y, z, t), \qquad v = v(x, y, z, t), \qquad w = w(x, y, z, t).$$

Now

$$du = \frac{\partial u}{\partial x} \, dx + \frac{\partial u}{\partial y} \, dy + \frac{\partial u}{\partial z} \, dz + \frac{\partial u}{\partial t} \, dt.$$

48. *Hist. de l'Acad. des Sci.*, Paris, 1784, 118–92, pub. 1787.
49. *Novi Comm. Acad. Sci Petrop.*, 6, 1756/57, 271–311, pub. 1761 = *Opera*, (2), 12, 133–68.
50. *Hist. de l'Acad. de Berlin*, 11, 1755, 274–315, pub. 1757 = *Opera*, (2), 12, 54–91.

In time dt, the particle at (x, y, z) will travel a distance $u\,dt$ in the x-direction, $v\,dt$ in the y-direction, and $w\,dt$ in the z-direction. Then the actual changes dx, dy, and dz in the expression for du are given by these quantities, so that

$$du = \frac{\partial u}{\partial x} u\,dt + \frac{\partial u}{\partial y} v\,dt + \frac{\partial u}{\partial z} w\,dt + \frac{\partial u}{\partial t} dt$$

or

(98)
$$\frac{du}{dt} = u \frac{\partial u}{\partial x} + v \frac{\partial u}{\partial y} + w \frac{\partial u}{\partial z} + \frac{\partial u}{\partial t},$$

and there are the corresponding expressions for dv/dt and dw/dt. These quantities give what is called now the convective rate of change of the velocity at (x, y, z) or the convective acceleration. By calculating the forces acting on the particle at (x, y, z) and applying Newton's second law, Euler obtains the system of differential equations

(99)
$$P - \frac{1}{\rho} \frac{\partial p}{\partial x} = \frac{du}{dt}$$

$$Q - \frac{1}{\rho} \frac{\partial p}{\partial y} = \frac{dv}{dt}$$

$$R - \frac{1}{\rho} \frac{\partial p}{\partial z} = \frac{dw}{dt}.$$

Euler also generalized d'Alembert's differential equation of continuity (45) and obtained for compressible flow the equation

(100)
$$\frac{\partial \rho}{\partial t} + \frac{\partial(\rho u)}{\partial x} + \frac{\partial(\rho u)}{\partial y} + \frac{\partial(\rho w)}{\partial z} = 0.$$

There are four equations and five unknowns, but the pressure p as a function of the density, the equation of state, must be specified.

In the 1755 paper Euler says, "And if it is not permitted to us to penetrate to a complete knowledge concerning the motion of fluids, it is not to mechanics, or to the insufficiency of the known principles of motion, that we must attribute the cause. It is analysis itself which abandons us here, since all the theory of the motion of fluids has just been reduced to the solution of analytic formulas." Unfortunately analysis was still too weak to do much with these equations. He then proceeds to discuss some special solutions. He also wrote other papers on the subject and dealt with the resistance encountered by ships and ship propulsion. Euler's equations are not the final ones for hydrodynamics. He neglected viscosity, which was introduced seventy years later by Navier and Stokes (Chap. 28, sec. 7).

Lagrange, too, worked on fluid motion. In the first edition of his *Mécanique analytique*, which contains some of the work, he gave Euler's basic equations and generalized them. Here he gave credit to d'Alembert but none to Euler. He too says the equations of fluid motion are too difficult to be handled by analysis. Only the cases of infinitely small movements are susceptible of rigorous calculation.

In the area of systems of partial differential equations, the equations of hydrodynamics were, in the eighteenth century, the main inspiration for mathematical research on this subject. Actually the eighteenth century accomplished little in the solution of systems.

9. *The Rise of the Mathematical Subject*

Up to 1765 partial differential equations appeared only in the solution of physical problems. The first paper devoted to purely mathematical work on partial differential equations is by Euler: "Recherches sur l'intégration de l'équation $\left(\dfrac{d\,dz}{dt^2}\right) = aa\left(\dfrac{d\,dz}{dx^2}\right) + \dfrac{b}{x}\left(\dfrac{dz}{dx}\right)$."[51] Shortly thereafter Euler published a treatise on the subject in the third volume of his *Institutiones Calculi Integralis*.[52]

Before d'Alembert's work of 1747 on the vibrating string, partial differential equations were known as equations of condition and only special solutions were sought. After this work and d'Alembert's book on the general causes of winds (1746), the mathematicians realized the difference between special and general solutions. However, once aware of this distinction, they seemed to believe that general solutions would be more important. In the first volume of his *Mécanique céleste* (1799), Laplace still complained that the potential equation in spherical coordinates could not be integrated in general form. Appreciation of the fact that general solutions such as Euler and d'Alembert obtained for the vibrating string were not as useful as particular ones satisfying initial and boundary conditions was not attained in that century.

The mathematicians did realize that partial differential equations involved no new operational techniques but differed from ordinary differential equations in that arbitrary functions might appear in the solution. These they expected to determine by reducing partial differential equations to ordinary ones. Laplace (1773) and Lagrange (1784) say clearly that they regard a partial differential equation as integrated when it is reduced to a problem of ordinary differential equations. An alternative, such as Daniel Bernoulli used for the wave equation and Laplace for the potential equation, was to seek expansions in series of special functions.

The major achievement of the eighteenth-century work on partial differential equations was to reveal their importance for problems of elasticity,

51. *Misc. Taur.*, 3_2 1762/65, 60–91, pub. 1766 = *Opera*, (1), 23, 47–73.
52. 1770 = *Opera*, (1), 13.

hydrodynamics, and gravitational attraction. Except in the case of Lagrange's work on first order equations, no broad methods were developed, nor were the potentialities in the method of expansion in special functions appreciated. The efforts were directed to solving the special equations that arose in physical problems. The theory of the solution of partial differential equations remained to be fashioned and the subject as a whole was still in its infancy.

Bibliography

Burkhardt, H.: "Entwicklungen nach oscillirenden Funktionen und Integration der Differentialgleichungen der mathematischen Physik," *Jahres. der Deut. Math.-Verein.*, 10, 1908, 1–1804.

Burkhardt, H., and W. Franz Meyer: "Potentialtheorie," *Encyk. der Math. Wiss.*, B. G. Teubner, 1899–1916, 2, A7b, 464–503.

Cantor, Moritz: *Vorlesungen über Geschichte der Mathematik*, B. G. Teubner, 1898 and 1924; Johnson Reprint Corp., 1965, Vol. 3, 858–78, Vol. 4, 873–1047.

Euler, Leonhard: *Opera Omnia*, Orell Füssli, (1), Vols. 13 (1914) and 23 (1938); (2), Vols. 10, 11, 12, and 13 (1947–55).

Lagrange, Joseph-Louis: *Œuvres*, Gauthier-Villars, 1868–70, relevant papers in Vols. 1, 3, 4, 5.

Laplace, Pierre-Simon: *Œuvres complètes*, Gauthier-Villars, 1893–94, relevant papers in Vols. 9 and 10.

Montucla, J. F.: *Histoire des mathématiques* (1802), Albert Blanchard (reprint), 1960, Vol. 3, pp. 342–52.

Langer, Rudolph E.: "Fourier Series: The Genesis and Evolution of a Theory," *Amer. Math. Monthly*, 54, No. 7, Part 2, 1947.

Taton, René: *L'Œuvre scientifique de Monge*, Presses Universitaires de France, 1951.

Todhunter, Isaac: *A History of the Mathematical Theories of Attraction and the Figure of the Earth* (1873), Dover (reprint), 1962.

Truesdell, Clifford E.: *Introduction to Leonhardi Euleri Opera Omnia Vol. X et XI Seriei Secundae*, in Euler, *Opera Omnia*, (2), 11, Part 2, Orell Füssli, 1960.

———: *Editor's Introduction* in Euler, *Opera Omnia*, (2), Vol. 12, Orell Füssli, 1954.

———: *Editor's Introduction* in Euler, *Opera Omnia*, (2), Vol. 13, Orell Füssli, 1956.

23
Analytic and Differential Geometry in the Eighteenth Century

> Geometry may sometimes appear to take the lead over analysis but in fact precedes it only as a servant goes before the master to clear the path and light him on his way.
>
> JAMES JOSEPH SYLVESTER

1. *Introduction*

The exploration of physical problems led inevitably to the search for greater knowledge of curves and surfaces, because the paths of moving objects are curves and the objects themselves are three-dimensional bodies bounded by surfaces. The mathematicians, already enthusiastic about the method of coordinate geometry and the power of the calculus, approached geometrical problems with these two major tools. The impressive results of the century were obtained in the already established area of coordinate geometry and the new field created by applying the calculus to geometrical problems, namely, differential geometry.

2. *Basic Analytic Geometry*

Two-dimensional analytic geometry was extensively explored in the eighteenth century. The improvements in elementary plane analytics are readily summarized. Whereas Newton and James Bernoulli had introduced and used what are essentially polar coordinates for special curves (Chap. 15, sec. 5), Jacob Hermann in 1729 not only proclaimed their general usefulness, but applied them freely to study curves. He also gave the transformation from rectangular to polar coordinates. Strictly Hermann used as variables p, $\cos \theta$, $\sin \theta$, which he designated by z, n, and m. Euler extended the use of polar coordinates and used trigonometric notation explicitly; with him the system is practically modern.

Though some seventeenth-century men—for example, Jan de Witt

(1625–72) in his *Elementa Curvarum Linearum* (1659)—did reduce some second degree equations in x and y to standard forms, James Stirling, in his *Lineae Tertii Ordinis Neutonianae* (1717), reduced the general second degree equation in x and y to the several standard forms.

In his *Introductio* (1748) Euler introduced the parametric representation of curves, wherein x and y are expressed in terms of a third variable. In this famous text Euler treated plane coordinate geometry systematically.

The suggestion of three-dimensional coordinate geometry can be found, as we know, in the work of Fermat, Descartes, and La Hire. The actual development was the work of the eighteenth century. Though some of the early work, Pitot's and Clairaut's for example, is tied up with the development of differential geometry, we shall consider only coordinate geometry proper at this point.

The first task was the improvement of La Hire's suggestion of a three-dimensional coordinate system. John Bernoulli, in a letter to Leibniz of 1715, introduced the three coordinate planes we use today. Through contributions too detailed to warrant space here, Antoine Parent (1666–1716), John Bernoulli, Clairaut, and Jacob Hermann clarified the notion that a surface can be represented by an equation in three coordinates. Clairaut, in his book *Recherche sur les courbes à double courbure* (Research on the Curves of Double Curvature, 1731), not only gave the equations of some surfaces but made clear that two such equations are needed to describe a curve in space. He also saw that certain combinations of the equations of two surfaces passing through a curve, the sum for example, give the equation of another surface passing through the curve. Using this fact, he explains how one can obtain the equations of the projections of these curves or, equivalently, the equations of the cylinders perpendicular to the planes of projection.

The quadric surfaces, e.g. sphere, cylinder, paraboloid, hyperboloid of two sheets, and ellipsoid, were of course known geometrically before 1700; in fact, some of them appear in Archimedes' work. Clairaut in his book of 1731 gave the equations of some of these surfaces. He also showed that an equation that is homogeneous in x, y, and z (all terms are of the same degree) represents a cone with vertex at the origin. To this result Jacob Hermann, in a paper of 1732,[1] added that the equation $x^2 + y^2 = f(z)$ is a surface of revolution about the z-axis. Both Clairaut and Hermann were primarily concerned with the shape of the earth, which by their time was believed to be some form of ellipsoid.

Though Euler had done some earlier work on the equations of surfaces, it is in Chapter 5 of the Appendix to the second volume of his *Introductio* (1748)[2] that he systematically takes up three-dimensional coordinate

1. *Comm. Acad. Sci. Petrop.*, 6, 1732/33, 36–67, pub. 1738.
2. *Opera*, (1), 9.

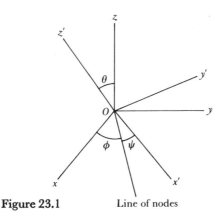

Figure 23.1   Line of nodes

geometry. He presents much of what had already been done and then studies the general second degree equation in three variables

$$(1) \qquad ax^2 + by^2 + cz^2 + dxy + exz + fyz + gx + hy + kz = l.$$

He now seeks to use change of axes to reduce this equation to the forms that result from having the principal axes of the quadric surfaces represented by (1) as the coordinate axes. He introduces the transformation from the xyz-system to the $x'y'z'$-system, whose equations are expressed (Fig. 23.1) in terms of the angles ϕ, ψ, and θ. The angle ϕ is measured in the xy-plane from the x-axis to the line of nodes, which is the line in which the $x'y'$-plane cuts the xy-plane. The angle ψ is measured in the $x'y'$-plane and locates x' with respect to the line of nodes. The angle shown is θ. Then the equations of transformation, including translation, are

$$
\begin{aligned}
x = {}& x_0 + x'(\cos\psi \cos\phi - \cos\theta \sin\psi \sin\phi) \\
& - y'(\cos\psi \sin\phi + \cos\theta \sin\psi \sin\phi) + z' \sin\theta \sin\phi \\
(2) \quad y = {}& y_0 + x'(\sin\psi \cos\phi + \cos\theta \cos\psi \sin\phi) \\
& - y'(\sin\psi \sin\phi - \cos\theta \cos\psi \sin\phi) - z' \sin\theta \sin\phi \\
z = {}& z_0 + x' \sin\theta \sin\phi + y' \sin\theta \cos\phi + z' \cos\theta.
\end{aligned}
$$

Euler uses this transformation to reduce (1) to canonical forms and obtains six distinct cases: cone, cylinder, ellipsoid, hyperboloid of one and two sheets, hyperbolic paraboloid (which he discovered), and parabolic cylinder. Like Descartes, Euler maintained that classification by the degree of the equation was the correct principle; Euler's reason was that the degree is invariant under linear transformation.

     After continuing work on this problem of change of axes, he wrote another paper,[3] in which he considers the transformation that will carry

3. *Novi Comm. Acad. Sci. Petrop.*, 15, 1770, 75–106, pub. 1771 = *Opera*, (1), 6, 287–315.

$x^2 + y^2 + z^2$ into $x'^2 + y'^2 + z'^2$. Here he—and Lagrange a little later, in a paper on the attraction of spheroids[4]—gave the symmetric form for the rotation of axes, the homogeneous linear orthogonal transformation

$$x = \lambda x' + \mu y' + \nu z'$$
$$y = \lambda' x' + \mu' y' + \nu' z'$$
$$z = \lambda'' x' + \mu'' y' + \nu'' z',$$

where

$$
\begin{array}{ll}
\lambda^2 + \lambda'^2 + \lambda''^2 = 1 & \quad \lambda\mu + \lambda'\mu' + \lambda''\mu'' = 0 \\
\mu^2 + \mu'^2 + \mu''^2 = 1 & \quad \lambda\nu + \lambda'\nu' + \lambda''\nu'' = 0 \\
\nu^2 + \nu'^2 + \nu''^2 = 1 & \quad \mu\nu + \mu'\nu' + \mu''\nu'' = 0.
\end{array}
$$

The λ, μ, and ν, unprimed and primed, are of course direction cosines in today's terminology.

Gaspard Monge's writings contain a great deal of three-dimensional analytic geometry. His outstanding contribution to analytic geometry as such is to be found in the paper of 1802 written with his pupil Jean-Nicolas-Pierre Hachette (1769–1834), "Application de l'algèbre à la géométrie."[5] The authors show that every plane section of a second degree surface is a second degree curve, and that parallel planes cut out similar and similarly placed curves. These results parallel Archimedes' geometric theorems. The authors also show that the hyperboloid of one sheet and the hyperbolic paraboloid are ruled surfaces, that is, each can be generated in two different ways by the motion of a line or each surface is formed by two systems of lines. The result on the one-sheeted hyperboloid was known by 1669 to Christopher Wren, who said that this figure could be generated by revolving a line about another not in the same plane. With the work of Euler, Lagrange, and Monge, analytic geometry became an independent and full-fledged branch of mathematics.

3. *Higher Plane Curves*

The analytic geometry described thus far was devoted to curves and surfaces of the first and second degree. It was of course natural to investigate the curves of equations of higher degree. In fact, Descartes had already discussed such equations and their curves somewhat. The study of curves of degree higher than two became known as the theory of higher plane curves, though it is part of coordinate geometry. The curves studied in the eighteenth century were algebraic; that is, their equations are given by $f(x, y) = 0$ where f is a polynomial in x and y. The degree or order is the highest degree of the terms.

4. *Nouv. Mém. de l'Acad. de Berlin*, 1773, 85–120 = *Œuvres*, 3, 619–58.
5. *Jour. de l'Ecole Poly.*, 11 cahier, 1802, 143–69.

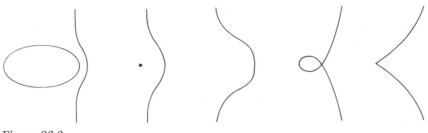

Figure 23.2

The first extensive study of higher plane curves was made by Newton. Impressed by Descartes's plan to classify curves according to the degree of their equations and then to study systematically each degree by methods suited to that degree, Newton undertook to study third degree curves. This work appeared in his *Enumeratio Linearum Tertii Ordinis*, which was published in 1704 as an appendix to the English edition of his *Opticks* but had been composed by 1676. Though the use of negative x- and y-values appears in works of La Hire and Wallis, Newton not only uses two axes and negative x- and y-values but plots in all four quadrants.

Newton showed how all curves comprised by the general third degree equation

$$(3) \quad ax^3 + bx^2y + cxy^2 + dy^3 + ex^2 + fxy + gy^2 + hx + jy + k = 0$$

can, by a change of axes, be reduced to one of the following four forms:

(a) $xy^2 + ey = ax^3 + bx^2 + cx + d$
(b) $\quad xy = ax^3 + bx^2 + cx + d$
(c) $\quad y^2 = ax^3 + bx^2 + cx + d$
(d) $\quad y = ax^3 + bx^2 + cx + d.$

The third class, which Newton called diverging parabolas, contains five species of curves whose types are shown in Figure 23.2. The species are distinguished by the nature of the roots of the cubic right-hand member, as follows: all real and distinct; two roots complex; all real but two equal and the double root greater or less than the simple root; and all three equal. Newton affirmed that every cubic curve can be obtained by projection of one of these five types from a point and then by a section of the projection.

Newton gave no proofs of many of the assertions in his *Enumeratio*. In his *Lineae*, James Stirling proved or reproved in other ways most of the assertions but not the projection theorem, which Clairaut[6] and François Nicole (1683–1758)[7] proved. Also, whereas Newton recognized seventy-two

6. *Mém. de l'Acad. des Sci., Paris*, 1731, 490–93, pub. 1733.
7. *Mém. de l'Acad. des Sci., Paris*, 1731, 494–510, pub. 1733.

species of third degree curves, Stirling added four more and abbé Jean-Paul de Gua de Malves, in a little book of 1740 entitled *Usage de l'analyse de Descartes pour découvrir sans le secours du calcul différential* . . . , added two more.

Newton's work on third degree curves stimulated much other work on higher plane curves. The topic of classifying third and fourth degree curves in accordance with one or another principle continued to interest mathematicians of the eighteenth and nineteenth centuries. The number of classes found varied with the methods of classification.

As is evident from the figures of Newton's five species of cubic curves, the curves of higher-degree equations exhibit many peculiarities not found in first and second degree curves. The elementary peculiarities, called singular points, are inflection points and multiple points. Before proceeding, let us see what some of them look like.

Inflection points are familiar from the calculus. A point at which there are two or more tangents which may coincide, is called a multiple point. At such a point two or more branches of the curve intersect. If two branches intersect at the multiple point, it is called a double point. If three branches intersect then the point is called a triple point, and so on.

If we take the equation of an algebraic curve

$$f(x, y) = 0,$$

f being a polynomial in x and y, we can by a translation always remove the constant term. If this is done and if there are first degree terms in f, say $a_1x + b_1y$, then $a_1x + b_1y = 0$ gives the equation of the tangent to the curve at the origin. The origin is not in this case a multiple point. If there are no first degree terms, and if $a_2x^2 + b_2xy + c_2y^2$ are the second degree terms, then several cases arise. The equation $a_2x^2 + b_2xy + c_2y^2 = 0$ may represent two distinct lines. These lines are tangents at the origin (this can be proven), and since there are two distinct tangents the origin is a double point; it is called a node. Thus the equation of the lemniscate (Fig. 23.3) is

$$(4) \qquad\qquad a^2(y^2 - x^2) + (y^2 + x^2)^2 = 0$$

and the second degree terms yield $y^2 - x^2 = 0$. Then $y = x$ and $y = -x$ are the equations of the tangents. Likewise, the folium of Descartes (Fig. 23.4) has the equation

$$(5) \qquad\qquad x^3 + y^3 = 3axy$$

and the tangents at the origin, which is a node, are given by $x = 0$ and $y = 0$.

When the two tangent lines are coincident, the single line is considered as a double tangent and the two branches of the curve touch each other at the point of tangency, which is called a cusp. (Sometimes cusps are

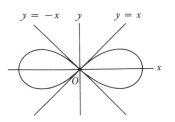

Figure 23.3. Lemniscate

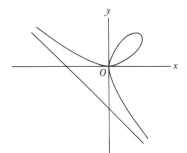

Figure 23.4. Folium of Descartes

included among the double points.) Thus the semicubical parabola (Fig. 23.5)

$$(6) \qquad\qquad ay^2 = x^3$$

has a cusp at the origin, and the equation of the two coincident tangents is $y^2 = 0$. On the curve $(y - x^2)^2 = x^5$ (Fig. 23.6) the origin is a cusp. Here both branches lie on the same side of the double tangent, which is $y = 0$. De Gua in his *Usage* had tried to show that this type of cusp could not occur, but Euler[8] gave many examples. A cusp is also called a stationary point or point of retrogression because a point moving along the curve must come to rest before continuing its motion at a cusp.

When the two tangent lines are imaginary, the double point is called a conjugate point. The coordinates of the point satisfy the equation of the curve but the point is isolated from the rest of the curve. Thus the curve (Fig. 23.7) of $y^2 = x^2(2x - 1)$ has a conjugate point at the origin. The equation of the double tangent there is $y^2 = -x^2$ and the tangents are imaginary.

Figure 23.5. Semicubical parabola　　　　　　Figure 23.6

8. *Mém. de l'Acad. de Berlin*, 5, 1749, 203-21, pub. 1751 = *Opera*, (1), 27, 236-52.

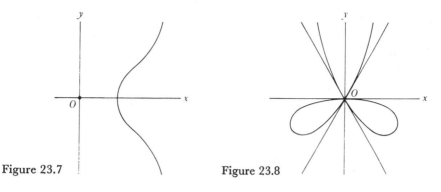

Figure 23.7   Figure 23.8

The curve of $ay^3 - 3ax^2y = x^4$ (Fig. 23.8) has a triple point at the origin. The equation of the three tangents is

$$ay^3 - 3ax^2y = 0$$

or $y = 0$ and $y = \pm x\sqrt{3}$.

The curve of $ay^4 - ax^2y^2 = x^5$ (Fig. 23.9) has a quadruple point at the origin. The origin is a combination of a node and a cusp. The tangents are $y = 0, y = 0, y = \pm x$.

Curves of the third degree (order) may have a double point (which may be a cusp) but no other multiple point. There are of course cubics with no double point.

To return to the history proper, many of these special or singular points on curves were studied by Leibniz and his successors. The analytical conditions for such points, such as that $\ddot{y} = 0$ at an inflection point and that $\dot{y}$ is indeterminate at a double point, were known even to the founders of the calculus.

Clairaut in the 1731 book referred to above assumed that a third degree curve cannot have more than three real inflection points and must have at least one. De Gua in *Usage* proved that if a third degree curve has three real inflection points, a line through two of them passes through the third

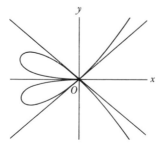

Figure 23.9

one. This theorem is often credited to Maclaurin. De Gua also investigated double points and gave the condition that if $f(x, y) = 0$ is the equation of the curve, then f_x and f_y must be 0 at a double point. A k-fold point is characterized by the vanishing of all partial derivatives through the $(k - 1)$st order. He showed that the singularities are compounded of cusps, ordinary points, and points of inflection. In addition, de Gua treated middle points of curves, the form of branches extending to infinity, and properties of such branches.

Maclaurin in his *Geometria Organica* (1720), written when he was nineteen, proved that the maximum number of double points of an irreducible curve of the nth degree is $(n - 1)(n - 2)/2$. For this purpose he counted a k-fold point as $k(k - 1)/2$ double points. He also gave upper bounds for the number of higher multiple points of each kind. He then introduced the notion of the deficiency (later called genus) of an algebraic curve as the maximum possible number of double points minus the actual number. Among curves those of deficiency 0 or possessing the maximum possible number of double points received a great deal of attention. Such curves are also called rational or unicursal. Geometrically a unicursal curve may be traversed by the continuous motion of a moving point (which may, however, pass through the point at infinity). Thus the conics, including the hyperbola, are unicursal curves.

In his *Method of Fluxions* Newton gave a method, commonly referred to as Newton's diagram or Newton's parallelogram, for determining series representations of the various branches of a curve at a multiple point (Chap. 20, sec. 2). De Gua in *Usage* replaced the Newton parallelogram by a *triangle algébrique*. Then if the origin is a singular point, for small x the equation of an algebraic curve breaks down into factors of the form $y^m - Ax^n$, where m is positive integral and n integral. The branches of the curve are given by those factors for which n is also positive. Euler noted (1749) that de Gua had neglected imaginary branches.

Gabriel Cramer (1704–52), in his *Introduction à l'analyse des lignes courbes algébriques* (1750), also tackled the expansion of y in terms of x when y and x are given in an implicit function, that is, $f(x, y) = 0$, in order to determine a series expression for each branch of the curve, particularly the branches that extend to infinity. He treated y as an ascending and as a descending series of powers of x. Like de Gua he used the triangle in place of Newton's parallelogram, and like others he neglected imaginary branches of the curves.

The conclusion that resulted from the work of obtaining series expansions for each branch of a curve issuing from a multiple point was drawn much later by Victor Puiseux (1820–83)[9] and is known as Puiseux's theorem:

9. *Jour. de Math.*, 13, 1850, 365–480.

The total neighborhood of a point (x_0, y_0) of an algebraic plane curve can be expressed by a finite number of developments

$$(7) \qquad y - y_0 = a_1(x - x_0)^{q_1/q_0} + a_2(x - x_0)^{q_2/q_0} + \cdots.$$

These developments converge in some interval about x_0 and all the q_i have no common factors. The points given by each development are called a branch of the algebraic curve.

The intersections of a curve and line and of two curves is another topic that received a great deal of attention. Stirling, in his *Lineae* of 1717, showed that an algebraic curve of the nth degree (in x and y) is determined by $n(n + 3)/2$ of its points because it has that number of essential coefficients. He also asserted that any two parallel lines cut a given curve in the same number of points, real or imaginary, and he showed that the number of branches of a curve that extend to infinity is even. Maclaurin's work, *Geometria Organica*, founded the theory of intersections of higher plane curves. He generalized on results for special cases and on this basis concluded that an equation of the mth degree and one of the nth degree intersect in mn points.

In 1748 Euler and Cramer sought to prove this result, but neither gave a correct proof. Euler[10] relied upon an argument by analogy; realizing that his argument was not complete, he said one should apply the method to particular examples. Cramer's "proof" in his book of 1750 relied entirely on examples and was certainly not acceptable. Both men took into account points of intersection with imaginary coordinates and infinitely distant common points and noted that the number mn will be attained only if both types of points are included and if any factor, such as $ax + by$, common to both curves is excluded. However, both failed to assign the proper multiplicity to several types of intersections. In 1764 Etienne Bezout (1730–83) gave a better proof of the theorem, but this was also incomplete in the count of the multiplicity assigned to points at infinity and multiple points. The proper count of the multiplicity was settled by Georges-Henri Halphen (1844–89) in 1873.[11]

In his book of 1750 Cramer took up a paradox noted by Maclaurin in his *Geometria* concerning the number of points common to two curves. A curve of degree n is determined by $n(n + 3)/2$ points. Two nth degree curves meet in n^2 points. Now if n is 3, the curve should, by the first statement, be determined by 9 points. But since two third degree curves meet in 9 points, these 9 points do not determine a unique third degree curve. A similar paradox arises when $n = 4$. Cramer's explanation of the paradox, now referred to as his, was that the n^2 equations that determine the n^2 points of intersection are not independent. All cubics that pass through 8 fixed

10. *Mém. de l'Acad. de Berlin*, 4, 1748, 234–48 = *Opera*, (1), 26, 46–59.
11. *Bull. Soc. Math. de France*, 1, 1873, 130–48; 2, 1873, 34–52; 3, 1875, 76–92 = *Œuvres*, 1, 98–157, 171–93, 337–57.

points on a given cubic must pass through the same ninth fixed point. That is, the ninth point is dependent on the first 8. Euler gave the same explanation in 1748.[12]

In 1756 Matthieu B. Goudin (1734–1817) and Achille-Pierre Dionis du Séjour (1734–94) wrote the *Traité des courbes algébriques*. Its new features are that a curve of order (degree) n cannot have more than $n(n - 1)$ tangents with a given direction, nor more than n asymptotes. As had Maclaurin, they pointed out that an asymptote cannot cut the curve in more than $n - 2$ points.

The two best eighteenth-century compendia of results on higher plane curves are the second volume of Euler's *Introductio* (1748) and Cramer's *Lignes courbes algébriques*. The latter book has a unity of viewpoint, is excellently set forth, and contains good examples. The work was often cited, even to the point of crediting Cramer with some results that were not original with him.

4. The Beginnings of Differential Geometry

Differential geometry was initiated while analytic geometry was being extended, and the two developments were often intertwined. Interest in the theory of algebraic curves waned during the latter part of the eighteenth century, and differential geometry became more important as far as geometry was concerned. This subject is the study of those properties of curves and surfaces that vary from point to point and therefore can be grasped only with the techniques of the calculus. The term "differential geometry" was first used by Luigi Bianchi (1856–1928) in 1894.

To a large extent, differential geometry was a natural outgrowth of problems of the calculus itself. Consideration of normals to curves, points of inflection, and curvature is actually the differential geometry of plane curves. However, many new problems of the late seventeenth and early eighteenth centuries, more knowledge about the curvature of plane and space curves, envelopes of families of curves, geodesics on surfaces, the study of rays of light and of wave surfaces of light, dynamical problems of motion along curves and constraints posed by surfaces, and, above all, map-making led to questions about curves and surfaces; it became evident that the calculus must be applied.

The eighteenth- and even early nineteenth-century workers in differential geometry used geometrical arguments along with analytic ones, although the latter dominated the picture. The analysis was still crude. An infinitesimal or differential of an independent variable was regarded as an extremely small constant. No real distinction was drawn between the increment of a

12. *Mém. de l'Acad. de Berlin*, 4, 1748, 219–33, pub. 1750 = *Opera*, (1), 26, 33–45.

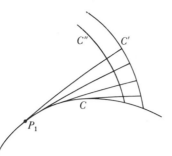

Figure 23.10

dependent variable and the differential. Differentials of higher orders were considered, but all were regarded as small and freely neglected. The mathematicians spoke of adjacent or next points on a curve as though there were no points between two adjacent points if the distance between two was sufficiently small; thus a tangent to a curve connected a point with the next one.

5. *Plane Curves*

The first applications of the calculus to curves dealt with plane curves. Some of the concepts subsequently treated by the calculus were introduced by Christian Huygens, who used purely geometrical methods. His work in this direction was motivated by his interest in light and in the design of pendulum clocks. In 1673, in the third chapter of his *Horologium Oscillatorium*, he introduced the involute of a plane curve C. Imagine a cord wrapped around C from P_1 to the right (Fig. 23.10). The end at P_1 on C is held fixed and the other unwound while the cord is kept taut. The locus C' of the free end is an involute of C. Huygens proved that at the free end the cord is perpendicular to the locus C'. Each point of the cord also describes an involute; thus C'' is also an involute, and Huygens proved that the involutes cannot touch one another. Since the cord is tangent to C at the point where it just leaves C, it follows that every orthogonal trajectory of the family of tangents to a curve is an involute of the curve.

Huygens then treated the evolute of a plane curve. Given a fixed normal at a point P on a curve, as an adjacent normal moves toward it the point of intersection of the two normals attains a limiting position on the fixed normal, which is called the center of curvature of the curve at P. The distance from the point on the curve along the fixed normal to the limiting position was shown by Huygens to be (in modern notation)

$$\frac{\left[1 + \left(\dfrac{dy}{dx}\right)^2\right]^{\frac{3}{2}}}{\dfrac{d^2y}{dx^2}}$$

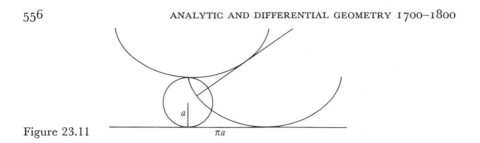

Figure 23.11

This length is the radius of the curvature of the curve at P. The locus of the centers of curvature, one on each normal, is called the evolute of the original curve. Thus the curve C (Fig. 23.10) is the evolute of any one of its involutes. In this work Huygens proved that the evolute of a cycloid is a cycloid or, more precisely, the evolute of the left half of the lower cycloid in Figure 23.11 is the right half of the upper cycloid. This theorem was proved analytically by Euler in 1764.[13] The significance of the cycloid for Huygens's work on pendulum clocks is that a pendulum bob swinging along a cycloidal arc takes exactly the same time to complete swings of large and of small amplitude. For this reason the cycloid is called the tautochrone.

Newton, too, in his *Geometria Analytica* (published in 1736 though most of it was written by 1671) introduces the center of curvature as the limiting point of intersection of a normal at P with an adjacent normal. He then states that the circle with center at the center of curvature and radius equal to the radius of curvature is the circle of closest contact with the curve at P; that is, no other circle tangent to the curve at P can come between the curve and the circle of closest contact. This circle of closest contact is called the osculating circle, the term "osculating" having been used by Leibniz in a paper of 1686.[14] The curvature of this circle is the reciprocal of its radius and is the curvature of the curve at P. Newton also gave the formula for the curvature and calculated the curvature of several curves, including the cycloid. He noted that at a point of inflection a curve has zero curvature. These results duplicate those of Huygens, but probably Newton wished to show that he could use analytical methods to establish them.

In 1691 John Bernoulli took up the subject of plane curves and produced some new results on envelopes. The caustic of a family of light rays, that is, the envelope of the family, had been introduced by Tschirnhausen in 1682. In the *Acta Eruditorum* of 1692 Bernoulli obtained the equations of some caustics, for example, the caustic of rays reflected from a spherical mirror when a beam of parallel rays strikes it.[15] Then he tackled the problem posed

13. *Novi Comm. Acad. Sci. Petrop.*, 10, 1764, 179–98, pub. 1766 = *Opera*, (1), 27, 384–400.
14. *Acta Erud.*, 1686, 289–92 = *Math. Schriften*, 7, 326–29.
15. *Opera*, 1, 52–59.

to him by Fatio de Duillier, to find the envelope of the family of parabolas that are the paths of cannon balls fired from a cannon with the same initial velocity but at various angles of elevation. Bernoulli showed that the envelope is a parabola with focus at the gun. This result had already been established geometrically by Torricelli. In the *Acta Eruditorum* of 1692 and 1694[16] Leibniz gave the general method of finding the envelope of a family of curves. If the family is given by (in our notation) $f(x, y, \alpha) = 0$, where α is the parameter of the family, the method calls for eliminating α between $f = 0$ and $\partial f/\partial \alpha = 0$. L'Hospital's text, *L'Analyse des infiniment petits* (1696), helped to perfect and spread the theory of plane curves.

6. *Space Curves*

Clairaut launched the theory of space curves, the first major development in three-dimensional differential geometry. Alexis-Claude Clairaut (1713–65) was precocious. At the age of twelve he had already written a good work on curves. In 1731 he published *Recherche sur les courbes à double courbure*, which was written in 1729 when he was but sixteen. In this book he treated the analytics of surfaces and space curves (sec. 2). Another paper by Clairaut led to his election to the Paris Academy of Sciences at the unprecedented age of seventeen. In 1743 he produced his classic work on the shape of the earth. Here he treated in more complete form than Newton or Maclaurin the shape a rotating body such as the earth assumes under the mutual gravitational attraction of its parts. He also worked on the problem of three bodies, primarily to study the moon's motion (Chap. 21, sec. 7), and wrote several papers on it, one of which won a prize from the St. Petersburg Academy in 1750. In 1763 he published his *Théorie de la lune*. Clairaut had great personal charm and was a figure in Paris society.

In his 1731 work he treated analytically fundamental problems of curves in space. He called space curves "curves of double curvature" because, following Descartes, he considered their projections on two perpendicular planes. The space curve then partakes of the curvatures of the two curves on the planes. Geometrically he thought of a space curve as the intersection of two surfaces; analytically the equation of each surface was expressed as an equation in three variables (sec. 2). Clairaut then studied tangents to curves of double curvature. He saw that a space curve can have an infinity of normals located in a plane perpendicular to the tangent. The expressions for the arc length of a space curve and the quadrature of certain areas on surfaces are also due to him.

Though Clairaut had taken a few steps in the theory of space curves, very little had been done in this subject or in the theory of surfaces by 1750.

16. Page 311; see also *Math. Schriften*, 2, 166; 3, 967, 969.

This is reflected in Euler's *Introductio* of 1748, where he presented the differential geometry of planar and spatial figures. The first was rather complete, but the second was scanty.

The next major step in the differential geometry of space curves was taken by Euler. A great deal of his work in differential geometry was motivated by his use of curves and surfaces in mechanics. His *Mechanica* (1736),[17] written when he was twenty-nine, is a major contribution to the analytical foundation of mechanics. He gave another treatment of the subject in his *Theoria Motus Corporum Solidorum seu Rigidorum* (1765).[18] In this book he derived the currently used polar coordinate formulas for the radial and normal components of acceleration of a particle moving along a plane curve, namely,

$$a_r = \frac{d^2r}{dt^2} - r\left(\frac{d\theta}{dt}\right)^2, \qquad a_\theta = r\frac{d^2\theta}{dt^2} + 2\frac{dr}{dt}\frac{d\theta}{dt}.$$

He started to write on the theory of space curves in 1774. The particular problem that very likely motivated Euler to take up this theory was the study of the skew elastica, that is, the form assumed by an initially straight band when, under pressure at the ends, it is bent and twisted into the shape of a skew curve. To treat this problem he introduced some new concepts in 1774.[19] He then gave a full treatment of the theory of skew curves in a paper presented in 1775.[20]

Euler represented space curves by the parametric equations $x = x(s)$, $y = y(s)$, $z = z(s)$, where s is arc length, and like other writers of the eighteenth century he used spherical trigonometry to carry out the analysis. From the parametric equations he has

$$dx = p\, ds, \qquad dy = q\, ds, \qquad dz = r\, ds,$$

where p, q, and r are direction cosines, varying from point to point and, of course, with $p^2 + q^2 + r^2 = 1$. The quantity ds, the differential of the independent variable, he regarded as a constant.

To study the properties of the curve he introduced the spherical indicatrix. Around any point (x, y, z) of a curve Euler describes a sphere of radius 1. The spherical indicatrix may be defined as the locus on the unit sphere of the points whose position vectors emanating from the center O are equal to the unit tangent at (x, y, z) and the unit tangents at neighboring points. Thus the two radii in Figure 23.12 represent a unit tangent at (x, y, z) and at a neighboring point of the curve. Let ds' be the arc or the angle between

17. *Opera*, (2), 1 and 2.
18. *Opera*, (2), 3 and 4.
19. *Novi Comm. Acad. Sci. Petrop.*, 19, 1774, 340–70, pub. 1775 = *Opera*, (2), 11, 158–79.
20. *Acta Acad. Sci. Petrop.*, 1, 1782, 19–57, pub. 1786 = *Opera*, (1), 28, 348–81.

Figure 23.12

the two neighboring tangents of the two points that are ds apart along the curve. Euler's definition of the radius of curvature of the curve is

$$\frac{ds'}{ds}.$$

He then derives an analytical expression for the radius of curvature:

(8) $\rho = \dfrac{ds^2}{\sqrt{(d^2x)^2 + (d^2y)^2 + (d^2z)^2}} = \dfrac{1}{\sqrt{x''^2 + y''^2 + z''^2}}.$

The plane through ds' and the center O is Euler's definition of the osculating plane at (x, y, z). John Bernoulli, who introduced the term, regarded the plane as determined by three "coincident" points. Its equation, as given by Euler, is

$$x(r\, dq - q\, dr) + y(p\, dr - r\, dp) + z(q\, dp - p\, dq) = t,$$

where t is determined by the point (x, y, z) on the curve through which the osculating plane passes. This equation is equivalent to the one we write today in vector notation as

$$(\mathbf{R} - \mathbf{r}) \cdot \mathbf{r}' \boldsymbol{\times} \mathbf{r}'' = 0,$$

where $\mathbf{r}(s)$ is the position vector with respect to some point in space of the point on the curve at which the osculating plane is determined, and $\mathbf{R}$ is the position vector of any point in the osculating plane. In vector form $\mathbf{r}$ is given by

$$x(s)\mathbf{i} + y(s)\mathbf{j} + z(s)\mathbf{k}$$

and $\mathbf{R}$ has the form $X\mathbf{i} + Y\mathbf{j} + Z\mathbf{k}$, where (X, Y, Z) are the coordinates of $\mathbf{R}$.

Clairaut had introduced the idea that a space curve has two curvatures. One of these was standardized by Euler in the manner just described. The other, now called "torsion" and representing geometrically the rate at which a curve departs from a plane at a point (x, y, z), was formulated explicitly and analytically by Michel-Ange Lancret (1774–1807), an engineer and mathematician who was a student of Monge and worked in his spirit. He singled out[21] three principal directions at any point of a

21. *Mém. divers Savans*, 1, 1806, 416–54.

curve. The first is that of the tangent. "Successive" tangents lie in a plane, the osculating plane. The normal to the curve that lies in the osculating plane is the principal normal, and the perpendicular to the osculating plane, the binormal, is the third principal direction. Torsion is the rate of change of the direction of the binormal with respect to arc length; Lancret used the terminology, flexion of successive osculating planes or successive binormals.

Lancret represented a curve by

$$x = \phi(z), \qquad y = \psi(z)$$

and called $d\mu$ the angle between successive normal planes and $d\nu$ that between successive osculating ones. Then, in modern notation,

$$\frac{d\mu}{ds} = \frac{1}{\rho}, \qquad \frac{d\nu}{ds} = \frac{1}{\tau},$$

where ρ is the radius of curvature and τ is the radius of torsion.

Cauchy improved the formulation of the concepts and clarified much of the theory of space curves in his famous *Leçons sur les applications du calcul infinitésimal à la géométrie* (1826).[22] He discarded constant infinitesimals, the ds's, and straightened out the confusion between increments and differentials. He pointed out that when one writes

$$ds^2 = dx^2 + dy^2 + dz^2$$

one should mean

$$\left(\frac{ds}{dt}\right)^2 = \left(\frac{dx}{dt}\right)^2 + \left(\frac{dy}{dt}\right)^2 + \left(\frac{dz}{dt}\right)^2.$$

Cauchy preferred to write a surface as $w(x, y, z) = 0$ instead of the unsymmetric form $z = f(x, y)$, and he wrote the equation of a straight line through the point (ξ, η, ζ) as

$$\frac{\xi - x}{\cos \alpha} = \frac{\eta - y}{\cos \beta} = \frac{\zeta - z}{\cos \gamma},$$

where $\cos \alpha$, $\cos \beta$, and $\cos \gamma$ are the direction cosines of the line, though more often he used direction numbers instead of direction cosines.

Cauchy's development of the geometry of curves is practically modern. He got rid of the spherical trigonometry in the proofs, but he, too, took the arc length as the independent variable. He obtains for the direction cosines of the tangent at any point

$$\frac{dx}{ds}, \frac{dy}{ds}, \frac{dz}{ds}, \quad \text{or} \quad x'(s), y'(s), z'(s).$$

22. *Œuvres*, (2), 5.

The direction numbers of the principal normal are shown to be

$$\frac{d^2x}{ds^2}, \frac{d^2y}{ds^2}, \frac{d^2z}{ds^2}, \quad \text{or} \quad x''(s), y''(s), z''(s)$$

and the curvature k of the curve is

$$k = \frac{1}{\rho} = \sqrt{(x'')^2 + (y'')^2 + (z'')^2}.$$

Then he proves that, if the direction cosines of the tangent are $\cos \alpha$, $\cos \beta$, and $\cos \gamma$,

(9)
$$x'' = \frac{d(\cos \alpha)}{ds} = \frac{\cos \lambda}{\rho}, \qquad y'' = \frac{d(\cos \beta)}{ds} = \frac{\cos \mu}{\rho},$$

$$z'' = \frac{d(\cos \gamma)}{ds} = \frac{\cos \nu}{\rho},$$

where ρ is the radius of curvature already introduced and $\cos \lambda$, $\cos \mu$, and $\cos \nu$ are the direction cosines of a normal, which he takes to be the principal one. He shows next that

$$\frac{1}{\rho} = \frac{d\omega}{ds}$$

where ω is the angle between adjacent tangents.

He introduces the osculating plane as the plane of the tangent and principal normal. The normal to this plane is the binormal, and its direction cosines $\cos L$, $\cos M$, and $\cos N$ are given by the formulas

$$\frac{\cos L}{dy\, d^2z - dz\, d^2y} = \frac{\cos M}{dz\, d^2x - dx\, d^2z} = \frac{\cos N}{dx\, d^2y - dy\, d^2x}.$$

He can then prove that

(10)
$$\frac{d\cos L}{ds} = \frac{\cos \lambda}{\tau}, \qquad \frac{d\cos M}{ds} = \frac{\cos \mu}{\tau}, \qquad \frac{d\cos N}{ds} = \frac{\cos \nu}{\tau},$$

where $1/\tau$ is the torsion, and that the torsion equals $d\Omega/ds$, where Ω is the angle between osculating planes.

Formulas (9) and (10) are two of the three famous Serret-Frénet formulas, the third being

(11)
$$\frac{d\cos \lambda}{ds} = -\frac{\cos \alpha}{\rho} - \frac{\cos L}{\tau}, \qquad \frac{d\cos \mu}{ds} = -\frac{\cos \beta}{\rho} - \frac{\cos M}{\tau},$$

$$\frac{d\cos \nu}{ds} = -\frac{\cos \gamma}{\rho} - \frac{\cos N}{\tau},$$

where $1/\tau$ is the torsion and $1/\rho$ is the curvature. These formulas (9), (10), and (11), which give the derivatives of the direction cosines of the tangent, binormal, and normal respectively, were published by Joseph Alfred Serret (1819–85) in 1851[23] and Fréderic-Jean Frénet (1816–1900) in 1852.[24] The significance of curvature and torsion is that they are the two essential properties of space curves. Given the curvature and torsion as functions of arc length along the curve, the curve is completely determined except for position in space. This theorem is readily proven on the basis of the Serret-Frénet formulas.

7. The Theory of Surfaces

Like the theory of space curves, the theory of surfaces made a slow start. It began with the subject of geodesics on surfaces, with geodesics on the earth as the main concern. In the *Journal des Sçavans* of 1697, John Bernoulli posed the problem of finding the shortest arc between two points on a convex surface.[25] He wrote to Leibniz in 1698 to point out that the osculating plane (the plane of the osculating circle) at any point of a geodesic is perpendicular to the surface at that point. In 1698 James Bernoulli solved the geodesic problem on cylinders, cones, and surfaces of revolution. The method was a limited one, though in 1728 John Bernoulli[26] did have some success with the method and found geodesics on other kinds of surfaces.

In 1728 Euler[27] gave differential equations for geodesics on surfaces. Euler used the method he introduced in the calculus of variations (see Chap. 24, sec. 2). In 1732 Jacob Hermann[28] also found geodesics on particular surfaces.

Clairaut in 1733 and again in 1739[29] in his work on the shape of the earth treated more fully geodesics on surfaces of revolution. He proved in the 1733 paper that for any surface of revolution, the sine of the angle made by a geodesic curve and any meridian (any position of the generating curve) it crosses varies inversely as the length of the perpendicular from the point of intersection to the axis. In another paper[30] he also proved the nice theorem that if at any point M of a surface of revolution a plane be passed normal to the surface and to the plane of the meridian through M, then the curve cut out on the surface has a radius of curvature at M equal to the

23. *Jour. de Math.*, 16, 1851, 193–207.
24. *Jour. de Math.*, 17, 1852, 437–47.
25. *Opera*, 1, 204–5.
26. *Opera*, 4, 108–28.
27. *Comm. Acad. Sci. Petrop.*, 3, 1728, 110–24, pub. 1732 = *Opera*, (1), 25, 1–12.
28. *Comm. Acad. Sci. Petrop.*, 6, 1732/3, 36–67.
29. *Hist. le l'Acad. des Sci.*, Paris, 1733, 186–94, pub. 1735 and 1739, 83–96, pub. 1741.
30. *Mém. de l'Acad. des Sci.*, Paris, 1735, 117–22, pub. 1738.

length of the normal between M and the axis of revolution. Clairaut's methods were analytical, but like most of his predecessors he did not employ the ideas we now associate with the calculus of variations.

In 1760, in his *Recherches sur la courbure des surfaces*[31] Euler established the theory of surfaces. This work is Euler's most important contribution to differential geometry and a landmark in the subject. He represents a surface by $z = f(x, y)$ and introduces the now standard symbolism

$$p = \frac{\partial z}{\partial x}, \qquad q = \frac{\partial z}{\partial y}, \qquad r = \frac{\partial^2 z}{\partial x^2}, \qquad s = \frac{\partial^2 z}{\partial x \, \partial y}, \qquad t = \frac{\partial^2 z}{\partial y^2}.$$

He then says, "I begin by determining the radius of curvature of any plane section of a surface; then I apply this solution to sections which are perpendicular to the surface at any given point; and finally I compare the radii of curvature of these sections with respect to their mutual inclination, which puts us in a position to establish a proper idea of the curvature of surfaces."

He obtains first a rather complex expression for the radius of curvature of any curve made by cutting the surface with a plane. He then particularizes the result by applying it to normal sections (sections containing a normal to the surface). For normal sections the general expression for the radius of curvature simplifies a little. Next he defines the principal normal section as that normal section perpendicular to the *xy*-plane. (This use of "principal" is not followed today.) The radius of curvature of a normal section whose plane makes an angle ϕ with the plane of the principal normal section has the form

$$(12) \qquad \frac{1}{L + M \cos 2\phi + N \sin 2\phi},$$

where L, M, and N are functions of x and y. To obtain the greatest and least curvature of all normal sections through one point on the surface (or, when the form of the denominator in [12] is indefinite, to get the two greatest curvatures), he sets the derivative with respect to ϕ of the denominator equal to zero and (in both cases) obtains $\tan 2\phi = N/M$. There are two roots differing by 90° so that there are two mutually perpendicular normal planes. We call the corresponding curvatures the principal curvatures κ_1 and κ_2.

It follows from Euler's results that the curvature κ of any other normal section making an angle α with one of the sections with principal curvature is

$$(13) \qquad \kappa = \kappa_1 \cos^2 \alpha + \kappa_2 \sin^2 \alpha.$$

This result is called Euler's theorem.

The same results were obtained in 1776 in a more elegant manner by a student of Monge, Jean-Baptiste-Marie-Charles Meusnier de La Place (1754–93), who also worked in hydrostatics and in chemistry with Lavoisier.

31. *Mém. de l'Acad. de Berlin*, 16, 1760, 119–43, pub. 1767 = *Opera*, (1), 28, 1–22.

Meusnier[32] then took up the curvature of non-normal sections, for which Euler had obtained a very complex expression. Meusnier's result, still called Meusnier's theorem, is that the curvature of a plane section of a surface at a point P is the curvature of the normal section through the same tangent at P divided by the sine of the angle that the original plane makes with the tangent plane at P. There follows the beautiful result that if one considers the family of planes through the same tangent line MM' to a surface, the centers of curvature of the sections made by these planes lie on a circle in a plane perpendicular to MM' and whose diameter is the radius of curvature of the normal section. Meusnier then proves the theorem that the only surfaces for which the two principal curvatures are everywhere equal are planes or spheres. His paper was remarkably simple and fertile; it helped to make intuitive a number of results reached in the eighteenth century.

A major concern of the theory of surfaces, motivated by the needs of map-making, is the study of developable surfaces, that is, surfaces that can be flattened out on a plane without distortion. Since the sphere cannot be cut and then so flattened, the problem was to find surfaces with shapes close to that of a sphere that could be unrolled without distortion. Euler was the first to consider the subject. This work is contained in his "De Solidis Quorum Superficiém in Planum Explicare Licet." [33] Surfaces were regarded as boundaries of solids in the eighteenth century; this is why Euler speaks of solids whose surfaces may be unfolded on a plane. In this paper he introduces the parametric representation of surfaces, that is,

$$x = x(t, u), \qquad y = y(t, u), \qquad z = z(t, u),$$

and asks what conditions these functions must satisfy so that the surface may be developable on a plane. His approach is to represent t and u as rectangular coordinates in a plane and then to form a small right-angled triangle (t, u), $(t + dt, u)$, $(t, u + du)$. Because the surface is developable this triangle must be congruent to a small triangle on the surface. If we denote the partial derivatives of x, y, and z with respect to t by l, m, and n, and those with respect to u by λ, μ, ν, then the corresponding triangle on the surface is (x, y, z), $(x + l\,dt, y + m\,dt, z + n\,dt)$, and $(x + \lambda\,du, y + \mu\,du, z + \nu\,du)$. From the congruence of the two triangles Euler derives

$$(14) \quad l^2 + m^2 + n^2 = 1, \qquad \lambda^2 + \mu^2 + \nu^2 = 1, \qquad l\lambda + m\mu + n\nu = 0.$$

These are the analytic necessary and sufficient conditions for developability. The condition is equivalent to requiring that the line element on the surface equal the line element in the plane. Analytically the problem of determining whether a surface is developable is the problem of finding a parametric

32. *Mém. divers Savans*, 10, 1785, 477–85.
33. *Novi Comm. Acad. Sci. Petrop.*, 16, 1771, 3–34, pub. 1772 = *Opera*, (1), 28, 161–86.

representation $x(t, u)$, $y(t, u)$, and $z(t, u)$ such that the partial derivatives satisfy the conditions (14).

Euler then investigated the relationship between space curves and developable surfaces and showed that the family of tangents to any space curve fills out or constitutes a developable surface. He tried unsuccessfully to show that every developable surface is a ruled surface, that is, one generated by a moving straight line, and conversely. The converse is, in fact, not true.

The subject of developable surfaces was taken up independently by Gaspard Monge. With Monge geometry and analysis supported each other. He embraced simultaneously both aspects of the same problem and showed the usefulness of thinking both geometrically and analytically. Because analysis had dominated the eighteenth century, despite some analytic geometry and differential geometry by Euler and Clairaut, the effect of Monge's double view was to put geometry on at least an equal basis with analysis and then to inspire the revival of pure geometry. He is the first real innovator in synthetic geometry after Desargues.

His extensive work in descriptive geometry (which primarily serves architecture), analytic geometry, differential geometry, and ordinary and partial differential equations won the admiration and envy of Lagrange. The latter, after listening to a lecture by Monge, said to him, "You have just presented, my dear colleague, many elegant things. I wish I could have done them." Monge contributed much to physics, chemistry, metallurgy (problems of forges), and machinery. In chemistry he worked with Lavoisier. Monge saw the need for science in the development of industry and advocated industrialization as a path to the betterment of life. He was inspired by an active social concern, perhaps because he knew the hardships of humble origins. For this reason, too, he supported the French Revolution and served as Minister of the Navy and as a member of the Committee on Public Health in the succeeding governments. He designed armaments and instructed governmental staff in technical matters. His admiration for Bonaparte seduced him into following the latter in his counter-revolutionary measures.

Monge helped to organize the Ecole Polytechnique and as a professor there founded a school of geometers. He was a great teacher and a force in inspiring nineteenth-century mathematical activity. His vigorous and fertile lectures communicated enthusiasm to his students, among whom were at least a dozen of the most famous men of the early nineteenth century.

Monge created results in three-dimensional differential geometry which go far beyond Euler's. His paper of 1771, *Mémoire sur les développées, les rayons de courbure, et les différents genres d'inflexions des courbes à double courbure*, published much later,[34] was followed by his *Feuilles d'analyse appliquée à la géométrie* (1795, second ed. 1801). The *Feuilles* was as much differential geometry as

34. *Mém. divers Savans*, 10, 1785, 511–50.

analytic geometry and partial differential equations. Based on lecture notes, it offered a systematization and extension of old results, new results of some importance, and the translation of various properties of curves and surfaces into the language of partial differential equations. In pursuing the correspondence between the ideas of analysis and those of geometry, Monge recognized that a family of surfaces having a common geometric property or defined by the same method of generation should satisfy a partial differential equation (cf. Chap. 22, sec. 6).

Monge's first major work, the paper on the developable surfaces of curves of double curvature, takes up space curves and surfaces associated with them. At this time Monge did not know of Euler's work on developable surfaces. He treats space curves either as the intersections of two surfaces or by means of their projections on two perpendicular planes, given by $y = \phi(x)$ and $z = \psi(x)$. At any ordinary point there is an infinity of normals (perpendiculars to the tangent line) lying on one plane, the normal plane. In this plane there is a line he calls the (polar) axis, which is the limit of the intersection with a neighboring normal plane. As one moves along the curve the normal planes envelop a developable surface, called the polar developable. The surface is also swept out by the axes of the normal planes. The perpendicular from P to the axis in the normal plane at P is the principal normal and the foot Q of the perpendicular is the center of curvature.

To get the equation of the polar developable, he finds the equation of the normal plane and eliminates x between this equation and the partial derivative of the equation with respect to x. He also gives a rule for finding the envelope of any one-parameter family of planes that is the one we use today. The rule, which applies equally well to a one-parameter family of surfaces, is this: In our notation, he takes $F(x, y, z, \alpha) = 0$ as the family. To find where this meets "an infinitely near surface" he finds $\partial F/\partial \alpha = 0$. The curve of intersection of the two surfaces is called the characteristic curve. The equation of the envelope is found by eliminating α between $F = 0$ and $\partial F/\partial \alpha = 0$. He applies this method to the study of other developable surfaces, each of which he regards as the envelope of a one-parameter family of planes.

Monge also considered the edge of regression (*arête de rebroussement*) of a developable surface. The edge (curve) is formed by a set of lines generating the surface. The intersection of any two neighboring lines is a point and the locus of such points is the edge of regression Γ. Then the tangents to Γ are the generators, or Γ is the envelope of the family of generating lines. The edge of regression separates the developable surface into two nappes, just as a cusp separates a plane curve into two parts. Monge obtained the equations of the edge of regression. In the case of the polar developable surface of a space curve, the edge of regression is the locus of the centers of curvature of the original curve.

In 1775 Monge presented to the Académie des Sciences another paper on surfaces, particularly developable surfaces that occur in the theory of shadows and penumbras.[35] Using the definition that a developable surface can be flattened out on a plane without distortion, he argues intuitively that a developable surface is a ruled surface (but not conversely) on which two consecutive lines are concurrent or parallel and that any developable is equivalent to that formed by the tangents to a space curve. It is in this paper of 1775 that he gives a general representation of developable surfaces. Their equations are always of the form

$$z = x[F(q) - qF'(q)] + f(q) - qf'(q),$$

where $q = \partial z/\partial y$ and the partial differential equation which such surfaces, except for cylinders perpendicular to the xy-plane, always satisfy is

$$z_{xx}z_{yy} - z_{xy}^2 = 0.$$

Monge then studies ruled surfaces and gives a general representation for them. Also, he gives the third order partial differential equation which they satisfy and integrates this differential equation. He then proves that developable surfaces are a special case of ruled surfaces.

In the "Mémoire sur la théorie des déblais et des remblais" (Excavation and Fill) of 1776,[36] he considers the problem arising in building fortifications which involves the transport of earth and other materials from one place to another and seeks to do it in the most efficient way, that is, the material transported times the distance transported should be a minimum. Only part of this work is important for the differential geometry of surfaces; in fact, the results of the practical problem are not too realistic and, as Monge says, he published the paper for the sake of the geometrical results in it. Here he initiates the treatment of the subject of a family of lines depending upon two parameters or a congruence of lines. Then, following up on the work of Euler and Meusnier, he considers the family of normals to a surface S, which also forms a congruence of lines. Consider in particular the normals along a line of curvature, a line of curvature being a curve on the surface that has a principal curvature at each point on the curve. The surface normals at these points of a line of curvature form a developable surface, called a normal developable. Likewise, the surface normals along the line of curvature perpendicular to the first one form a developable surface. Since there are two families of lines of curvature on the surface, there are two families of developable surfaces, and the members of one intersect the members of the other at right angles. The intersection of any two in fact takes place along a surface normal. On each normal there are

35. *Mém. divers Savans*, 9, 1780, 382–440.
36. *Mém. de l'Acad. des Sci.*, Paris, 1781, 666–704, pub. 1784.

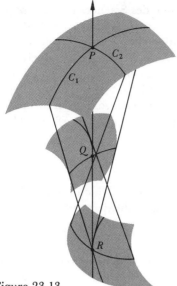

Figure 23.13

two points distant from the surface by the amount of the two principal curvatures. Each set of points determined by one of the principal curvatures, one on each of the normals to a line of curvature, lies on the edge of regression of the developable surface formed by that set of normals, so that the normals are tangent to that edge. The edges of regression of one family of developable surfaces all constitute a surface called a center surface. There are then two such center surfaces (Fig. 23.13). The envelope of each family of developable surfaces is called a focal surface.

The further details of Monge's work on families of surfaces, families which satisfy nonlinear and linear first, second, and even third order partial differential equations, are more significant for the subject of partial differential equations.

Monge tended to treat fully a number of concrete curves and surfaces through which he explained his ideas. The generalizations and the exploitation of his ideas were carried out by nineteenth-century men. Always practically oriented, Monge concluded his *Feuilles* with a projection of how his theory could be applied to architecture and in particular to the construction of large meeting halls.

Some additional contributions to the theory of surfaces were made by Charles Dupin (1784–1873), a pupil of Monge. Dupin was graduated from the Ecole Polytechnique as a naval engineer and, like Monge, constantly kept applications of geometry in mind. His text *Développements de géométrie*

(1813) was subtitled "With applications to the stability of ships, excavation and fill, fortifications, optics, and so forth," and in other writings, notably his *Applications de géométrie et de mécanique* (1822), he made many applications to geometry and to mechanics. The first few results we shall describe are in the 1813 book.

One of Dupin's contributions, which sums up and clarifies prior results of Euler and Meusnier, is called the Dupin indicatrix. Given the tangent plane of a surface at a point M, he laid off in each direction and starting from M a segment whose length is equal to the square root of the radius of curvature of the normal section of the surface in that direction. The locus of the endpoints of these segments is a conic, the indicatrix, which gives a first approximation to the form of the surface around M (because it is almost similar to a section of the surface made by a plane near M and parallel to the tangent plane at M). The lines of curvature of a surface at a point, that is, the curves on the surface through M having extreme (maximum or minimum) curvature, are the curves having as tangents at M the axes of the indicatrix.

In the three-dimensional rectangular coordinate system, the coordinate surfaces are the three families of surfaces $x =$ const., $y =$ const., and $z =$ const. Suppose one has three families of surfaces, each given by an equation in x, y, and z. If each member of one family cuts the members of the other two families orthogonally, then the three families are called orthogonal. Dupin's foremost result in this subject, which is in his *Développements*, is the theorem that the three families of orthogonal surfaces cut each other along the lines of curvature (the curves of maximum or minimum normal curvature) on each surface.

Dupin also extended Monge's results on congruences of lines. If the congruence, the two-parameter family, is cut orthogonally by a family of surfaces, as is the case in optics where the lines are light rays and the surfaces are wave fronts, then the congruence is called normal. Apparently using Monge's results, though he does not refer to them, Etienne-Louis Malus (1775–1812), a French physicist, proved [37] that a normal congruence of lines emanating from one point (a homocentric set) remains such after reflection or refraction (according to the laws of optics) at a surface. In 1816 [38] Dupin proved that this is true for any normal congruence after any number of reflections. Lambert A. J. Quetelet (1796–1874) then gave a proof that a normal congruence remains normal after any number of refractions.[39] The subject of congruences of lines and complexes of lines, families introduced by Malus and depending on three parameters, was pursued by many men in the nineteenth century.

37. *Jour. de l'Ecole Poly.*, Cahier 14, 1808, 1–44, 84–129.
38. *Annales de Chimie et de Physique*, 5, 1817, 85–88. Also in his *Applications* of 1822, 195–97.
39. *Correspondance mathématique et physique*, 1, 1825, 147–49.

8. *The Mapping Problem*

A great deal of the differential geometry of the eighteenth century was motivated by problems of geodesy and map-making. However, the map-making problem involves special considerations and difficulties that generated mathematical developments, notably conformal transformations or conformal mapping, running across many mathematical subjects. Map-making is of course much older than differential geometry, and even mathematical methods of mapping go far back. Of these, stereographic projection and other methods (Chap. 7, sec. 5) stem from Ptolemy, and Mercator's projection (Chap. 12, sec. 2) dates from the sixteenth century. It may have been intuitively clear prior to the eighteenth century that no true map of a sphere could be made on a plane, that is, one cannot map a sphere on a plane and still preserve lengths. If this were possible all the geometric properties would be preserved. Only developable surfaces can be so mapped and these, the eighteenth-century work revealed, are cylinders (not necessarily circular), cones, and any surface generated by the tangent lines of a space curve. Since a map of a sphere on a plane cannot preserve all the geometric properties, attention was directed toward maps that preserve angles.

In such maps, if two curves on one surface meet at an angle α and the corresponding curves on the other meet at the same angle, and if the direction of the angles is preserved, the map is said to be conformal. Stereographic projection and the Mercator projection are conformal. Conformality does not mean that two corresponding *finite* figures are similar, because the equality of the angles is a property holding at a point.

J. H. Lambert initiated a new epoch in theoretical cartography. He was the first to consider the conformal mapping of the sphere on the plane in full generality; and in a book of 1772, *Anmerkungen und Zusätze zur Entwerfung der Land-und Himmelscharten* (Notes and Additions on Designing Land Maps and Maps of the Heavens), he obtained the formulas for this mapping. Euler, too, made many contributions to the subject and actually made a map of Russia. In a paper presented to the St. Petersburg Academy in 1768,[40] Euler, by using complex functions, devised a method of representing conformal transformations from one plane to another. But he did not capitalize on it. Then, in two papers presented in 1775,[41] he showed that the sphere cannot be mapped congruently into a plane. Here again he used complex functions and treated rather general conformal representations. He also gave a full analysis of the Mercator and stereographic projections. Lagrange in 1779[42] obtained all conformal transformations of a portion of

40. *Novi Comm. Acad. Sci. Petrop.*, 14, 1769, 104–28, pub. 1770 = *Opera*, (1), 28, 99–119.
41. *Acta Acad. Sci. Petrop.*, 1, 1777, 107–32 and 133–42, pub. 1778 = *Opera*, (1), 28, 248–75 and 276–87.
42. *Nouv. Mém. de l'Acad. de Berlin*, 1779, 161–210, pub. 1781 = *Œuvres*, 4, 637–92.

the earth's surface onto a plane area that transform latitude and longitude circles into circular arcs.

Further progress in the mapping problem and in conformal mapping in particular awaited the extension of differential geometry and complex function theory.

Bibliography

Ball, W. W. Rouse: "On Newton's Classification of Cubic Curves," *Proceedings of the London Mathematical Society*, 22, 1890, 104–43.

Bernoulli, John: *Opera Omnia*, 4 vols., 1742, reprint by Georg Olms, 1968.

Berzolari, Luigi: "Allgemeine Theorie der höheren ebenen algebraischen Kurven," *Encyk. der Math. Wiss.*, B. G. Teubner, 1903–15, III C4, 313–455.

Boyer, Carl B.: *History of Analytic Geometry*, Scripta Mathematica, 1956, Chaps. 6–8.

————: A History of Mathematics, John Wiley and Sons, 1968, Chaps. 20 and 21.

Brill, A., and Max Noether: "Die Entwicklung der Theorie der algebraischen Funktionen in älterer und neuerer Zeit," *Jahres. der Deut. Math.-Verein.*, 3, 1892/3, 107–56.

Cantor, Moritz: *Vorlesungen über Geschichte der Mathematik*, B. G. Teubner, 1898 and 1924; Johnson Reprint Corp., 1965, Vol. 3, 18–35, 748–829; Vol. 4, 375–88.

Chasles, M.: *Aperçu historique sur l'origine et le développement des méthodes en géométrie* (1837), 3rd ed., Gauthier-Villars, 1889, pp. 142–252.

Coolidge, Julian L.: "The Beginnings of Analytic Geometry in Three Dimensions," *Amer. Math. Monthly*, 55, 1948, 76–86.

————: A History of Geometrical Methods, Dover (reprint), 1963, pp. 134–40, 318–46.

————: *The Mathematics of Great Amateurs.* Dover (reprint), 1963, Chap. 12.

————: *A History of Conic Sections and Quadric Surfaces*, Dover (reprint), 1968.

Euler, Leonhard: *Opera Omnia*, Orell Füssli, Series 1, Vol. 6 (1921), Vols. 26–29 (1953–56); Series 2, Vols. 3 and 4 (1948–50), Vol. 9 (1968).

Huygens, Christian: *Horologium Oscillatorium* (1673), reprint by Dawsons, 1966; also in Huygens, *Œuvres Complètes*, 18, 27–438.

Hofmann, Jos. E.: "Über Jakob Bernoullis Beiträge zur Infinitesimalmathematik," *L'Enseignement Mathématique*, (2), 2, 61–171, 1956; published separately by Institut de Mathématiques, Geneva, 1957.

Kötter, Ernst: "Die Entwickelung der synthetischen Geometrie von Monge bis auf Staudt," *Jahres. der Deut. Math.-Verein.*, 5, Part II, 1896, 1–486; also as a book, B. G. Teubner, 1901.

Lagrange, Joseph-Louis: *Œuvres*, Gauthier-Villars, 1867–69, Vol. 1, 3–20; Vol. 3, 619–92.

Lambert, J. H.: *Anmerkungen und Zusätze zur Entwerfung der Land-und Himmelscharten* (1772), Ostwald's Klassiker No. 54, Wilhelm Engelmann, Leipzig, 1896.

Loria, Gino: *Spezielle algebraische und tranzendente ebenen Kurven, Theorie und Geschichte*, 2 vols., 2nd ed., B. G. Teubner, 1910–11.

Montucla, J. F.: *Histoire des mathématiques* (1802), Albert Blanchard (reprint), 1960, Vol. 3, pp. 63–102.

Struik, D. J.: *A Source Book in Mathematics, 1200–1800*, Harvard University Press 1969, pp. 168–78, 180–83, 263–69, 413–19.

Taton, René: *L'Œuvre scientifique de Monge*, Presses Universitaires de France, 1951, Chap. 4.

Whiteside, Derek T.: "Patterns of Mathematical Thought in the Later Seventeenth Century," *Archive for History of Exact Sciences*, 1, 1961, 179–388. See pp. 202–5 and 270–311.

————: *The Mathematical Works of Isaac Newton*, Johnson Reprint Corp., 1967, Vol. 2, 137–61. This contains Newton's *Enumeratio* in English.

24

The Calculus of Variations in the Eighteenth Century

> For since the fabric of the universe is most perfect and the work of a most wise Creator, nothing at all takes place in the universe in which some rule of maximum or minimum does not appear. LEONHARD EULER

1. *The Initial Problems*

As in the areas of series and differential equations, the early work on the calculus of variations could hardly be distinguished from the calculus proper. But within a few years after Newton's death in 1727 it was clear that a totally new branch of mathematics, with its own characteristic problems and methodology, had come into being. This new subject, almost comparable in importance with differential equations for mathematics and science, supplied one of the grandest principles in all of mathematical physics.

To gain some preliminary notion of the nature of the calculus of variations, let us consider the problems that launched the mathematicians into the subject. Historically the first significant problem was posed and solved by Newton. In Book II of his *Principia*, he studied the motion of objects in water; then, in the Scholium to Proposition 34 of the third edition, he considered the shape that a surface of revolution moving at a constant velocity in the direction of its axis must have if it is to offer the least resistance to the motion. Newton assumed that the resistance of the fluid at any point on the surface of the body is proportional to the component of the velocity normal to the surface. In the *Principia* itself he gave only a geometrical characterization of the desired shape, but in a letter presumably written to David Gregory in 1694 he gave his solution.

In modern form, Newton's problem is to find the minimum value of the integral

$$J = \int_{x_1}^{x_2} \frac{y(x)[y'(x)]^3}{1 + [y'(x)]^2} \, dx$$

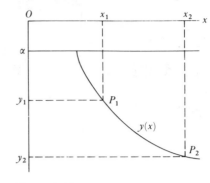

Figure 24.1   Figure 24.2

by choosing the proper function $y(x)$ for the shape of the curve that is to be rotated around the x-axis (Fig. 24.1). The peculiar feature of this problem (and of calculus of variations problems generally) is that it poses an integral whose value depends upon an unknown function $y(x)$ which appears in the integrand and which is to be determined so as to make the integral a minimum or a maximum.

Newton's solution, though it did use the idea of introducing a change in the shape of a part of the meridian arc $y(x)$, which is almost what the essential method of the calculus of variations involves, is not typical of the technique of the subject and so we shall not look into it. It may be of interest that the parametric equations of the proper $y(x)$ are

$$x = \frac{c}{p} (1 + p^2)^2, \qquad y = a + c\left(- \log p + p^2 + \frac{3}{4} p^4\right),$$

where p is the parameter. Of this work Newton says, "This proposition I conceive may be of use in the building of ships." Problems of this nature have become important in the design not only of ships but submarines and airplanes.

In the *Acta Eruditorum* of June 1696[1] John Bernoulli proposed as a challenge to other mathematicians the now famous brachistochrone problem. The problem is to determine the path down which a particle will slide from one given point to another not directly below in the shortest time. The initial velocity v_1 at P_1 (Fig. 24.2) is given; friction and air resistance are to be neglected. In modern form, this problem is to minimize the integral J which represents the time of descent where

$$J = \frac{1}{\sqrt{2g}} \int_{x_1}^{x_2} \sqrt{\frac{1 + [y'(x)]^2}{y(x) - \alpha}} \, dx.$$

1. Page 269 = *Opera*, 1, 161.

Here g is the gravitational acceleration and $\alpha = y_1 - v_1^2/2g$. Again the $y(x)$ in the integrand must be chosen so as to make J a minimum. The problem had been formulated and incorrectly solved by Galileo (1630 and 1638), who gave the arc of a circle as the answer. The correct answer is the arc of the unique cycloid joining P_1 and the second point P_2, which is concave upward; the line l on which the generating circle rolls must be at just the proper height, $y = \alpha$, above the given initial point of fall. Then there is one and only one cycloid through the two points.

Newton, Leibniz, L'Hospital, John Bernoulli, and his elder brother James found the correct solution. All these were published in the May issue of the *Acta Eruditorum* of 1697. The solutions of the two Bernoullis warrant further comment. John's method [2] was to see that the path of quickest descent is the same as the path of a ray of light in a medium with a suitably selected variable index of refraction, $n(x, y) = c/\sqrt{y - \alpha}$. The law of refraction at a sharp discontinuity (Snell's law) was known; so John broke up the medium into a finite number of layers with a sharp change in index from layer to layer and then let the number of layers go to infinity. James's method [3] was much more laborious and more geometrical. But it was also more general and was a bigger step in the direction of method for the calculus of variations.

The cycloid was well known through the work of Huygens and others on the pendulum problem (Chap. 23, sec. 5). When the Bernoulli brothers found that it was also the solution of the brachistochrone problem, they were amazed. John Bernoulli said, [4] "With justice we admire Huygens because he first discovered that a heavy particle traverses a cycloid in the same time, no matter what the starting point may be. But you will be struck with astonishment when I say that this very same cycloid, the tautochrone of Huygens, is the brachistochrone we are seeking."

Another important class of problems calls for geodesics, that is, paths of minimum length between two points on a surface. If the surface is a plane then the integral involved is

$$J = \int_{x_1}^{x_2} \sqrt{1 + [y'(x)]^2} \, dx,$$

and the answer is of course a line segment. In the eighteenth century the geodesic problem of most interest concerned the shortest paths on the surface of the earth, whose precise shape was not known, though the mathematicians believed it was some form of ellipsoid and most likely a figure of revolution. The early work on geodesics already noted (Chap. 23, sec. 7) did not use the method of the calculus of variations but it was clear that special devices would not be powerful enough to treat the general geodesic problem.

2. *Acta Erud.*, 1697, 206–11 = *Opera*, 1, 187–93.
3. *Acta Erud.*, 1697, 211–17 = *Opera*, 2, 768–78.
4. *Opera*, 1, 187–93.

Analytically the problems thus far formulated are of the form

$$J = \int_{x_1}^{x_2} f(x, y, y') \, dx$$

and call for finding the $y(x)$ that extends from (x_1, y_1) to (x_2, y_2) and that minimizes or maximizes J. Another class of problems, called isoperimetrical problems, also entered the history of the calculus of variations at the end of the seventeenth century. The progenitor of this class of problems, of all closed plane curves with a given perimeter to find the one that bounds maximum area, may date back to pre-Greek times. There is a story that Princess Dido of the ancient Phoenician city of Tyre ran away from her home to settle on the Mediterranean coast of North Africa. There she bargained for some land and agreed to pay a fixed sum for as much land as could be encompassed by a bull's hide. The shrewd Dido cut the hide into very thin strips, tied the strips end to end and proceeded to enclose an area having the total length of these strips as its perimeter. Moreover, she chose land along the sea so that no hide would be needed along the shore. According to the legend Dido decided that the length of hide should form a semicircle—the correct shape to enclose maximum area.

Apart from the work of Zenodorus (Chap. 5, sec. 7), there was practically no work on isoperimetrical problems until the end of the seventeenth century. In a move to challenge and embarrass his brother, James Bernoulli posed a rather complicated isoperimetrical problem involving several cases in the *Acta Eruditorum* of May 1697.[5] James even offered John a prize of fifty ducats for a satisfactory solution. John gave several solutions, one of which was obtained in 1701,[6] but all were incorrect. James gave a correct solution.[7] The brothers quarreled about the correctness of each other's solutions. Actually James's method, as in the case of the brachistochrone problem, was a major step toward the general technique soon to be fashioned. In 1718 John[8] considerably improved his brother's solution.

Analytically the basic isoperimetric problem is formulated thus. The possible curves are represented parametrically by

$$x = x(t), \qquad y = y(t), \qquad t_1 \leq t \leq t_2$$

and because they are closed curves, $x(t_1) = x(t_2)$ and $y(t_1) = y(t_2)$. Moreover no curve must intersect itself. The problem then calls for determining the $x(t)$ and $y(t)$ such that the length

$$L = \int_{t_1}^{t_2} \sqrt{(x')^2 + (y')^2} \, dt$$

5. Page 214.
6. *Mém. de l'Acad. des. Sci.*, Paris, 1706, 235 = *Opera*, 1, 424.
7. *Acta Erud.*, 1701, 213 ff. = *Opera*, 2, 897–920.
8. *Mém. de l'Acad. des Sci.*, Paris, 1718, 100 ff. = *Opera*, 2, 235–69.

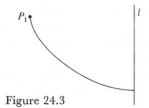

Figure 24.3

is a given constant and such that the area integral

$$J = \int_{t_1}^{t_2} (xy' - x'y)\, dt$$

is a maximum. There are two new features in this isoperimetrical problem. One, the use of the parametric representation, is incidental. The other is the presence of the auxiliary condition that L must be a constant.

Another problem that James posed in the May 1697 issue of the *Acta* was to determine the shape of the curve along which a particle slides from a given point P_1 with given initial velocity v_1 to any point of a line l (Fig. 24.3) so as to make the time of sliding from P_1 to l a minimum. This problem differs from the preceding ones in that the possible curves do not extend from one fixed point to another but from a fixed point to some line. The answer, given by James in the *Acta* of 1698 (though possessed by John in 1697 but not published by him), is an arc of a cycloid that cuts the line l at right angles. This problem was later generalized to the cases where l can be any given curve and where in place of P_1 another curve is given, so that the problem is to find the path requiring least time to slide from some point on one given curve to some point on another. This class of problems is described by the phrase "problems with variable endpoints."

2. *The Early Work of Euler*

In 1728 John Bernoulli proposed to Euler the problem of obtaining geodesics on surfaces by using the property that the osculating planes of geodesics cut the surface at right angles (Chap. 23, sec. 7). This problem started Euler off on the calculus of variations. He solved it in 1728.[9] In 1734 Euler generalized the brachistochrone problem to minimize quantities other than time and to take into account resisting media.[10]

Then Euler undertook to find a more general approach to problems in the subject. His method, which was a simplification of James Bernoulli's, was to replace the integral of a problem by a finite sum and to replace

9. *Comm. Acad. Sci. Petrop.*, 3, 1728, 110–24, pub. 1732 = *Opera*, (1), 25, 1–12.
10. *Comm. Acad. Sci. Petrop.*, 7, 1734/35, 135–49, pub. 1740 = *Opera*, (1), 25, 41–53.

derivatives in the integrand by difference quotients, thus making the integral a function of a finite number of ordinates of the arc $y(x)$. He then varied one or more arbitrarily selected ordinates and calculated the variation in the integral. By equating the variation of the integral to zero and by using a crude limiting process to transform the resulting difference equation, he obtained the differential equation which must be satisfied by the minimizing arc.

By the above described method applied to integrals of the form

$$(1) \qquad J = \int_{x_1}^{x_2} f(x, y, y') \, dx,$$

Euler succeeded in showing that the function $y(x)$ that minimizes or maximizes the value of J must satisfy the ordinary differential equation

$$(2) \qquad f_y - \frac{d}{dx} (f_{y'}) = 0.$$

This notation must be understood in the following sense. The integrand $f(x, y, y')$ is to be regarded as a function of the independent variables x, y, and y' insofar as f_y and $f_{y'}$ are concerned. However $df_{y'}/dx$ must be taken to be the derivative of $f_{y'}$ wherein $f_{y'}$ depends on x through x, y, and y'. That is, Euler's differential equation is equivalent to

$$(3) \qquad f_y - f_{y'x} - f_{y'y}y' - f_{y'y'}y'' = 0.$$

Since f is known, this equation is a second order, generally nonlinear, ordinary differential equation in $y(x)$. This famous equation, which Euler published[11] in 1736, is still the basic differential equation of the calculus of variations. It is, as we shall see more clearly later, a necessary condition that the minimizing or maximizing function $y(x)$ must satisfy.

Euler then tackled more difficult problems that involved special side conditions, as in the isoperimetric problems, but his procedure was still to solve the differential equation (3) in order to get first the possible minimizing or maximizing arcs and then to determine from the number of constants in the general solution of (2) or (3) what side conditions he could apply. One of the problems he tackled was called to his attention by Daniel Bernoulli in a letter of 1742. Bernoulli proposed to find the shape of an elastic rod subject to pressure at both ends by assuming that the square of the curvature along the curve taken by the bent rod, that is, $\int_0^L ds/R^2$, where s is arc length and R is the radius of curvature, is a minimum. This condition amounts to assuming that the potential energy stored up in the shape taken by the rod is a minimum.

---

11. *Comm. Acad. Sci. Petrop.*, 8, 1736, 159–90, pub. 1741 = *Opera*, (1), 25, 54–80.

The differential equation (3) is not the proper one when the integrands of the integrals to be minimized or maximized are more complicated than in (1). In the years from 1736 to 1744 Euler improved his methods and obtained the differential equations analogous to (3) for a large number of problems. These results he published in 1744 in a book, *Methodus Inveniendi Lineas Curvas Maximi Minimive Proprietate Gaudentes* (The Art of Finding Curved Lines Which Enjoy Some Maximum or Minimum Property).[12] Euler's work in his *Methodus* was cumbersome because he used geometric considerations, successive differences, and series and he changed derivatives to difference quotients and integrals to finite sums. He failed, in other words, to make most effective use of the calculus. But he ended with simple and elegant formulas applicable to a large variety of problems; and he treated a large number of examples to show the convenience and generality of his method. One example deals with minimal surfaces of revolution. Here the problem is to determine the plane curve $y = f(x)$ lying between (x_0, y_0) and (x_1, y_1) such that when revolved around the x-axis it generates the surface of least area. The integral to be minimized is

$$(4) \qquad\qquad A = \int_{x_0}^{x_1} 2\pi y \sqrt{1 + y'^2} \, dx.$$

Euler proved that the function $f(x)$ must be an arc of a catenary; the surface so generated is called a catenoid. In an appendix to his 1744 book Euler also gave a definitive solution of the elastic-rod problem referred to above. He not only deduced that the shape of the rod took the form of an elliptic integral, but also gave solutions for different kinds of end-conditions. This book brought him immediate fame and recognition as the greatest living mathematician.

With this work the calculus of variations came into existence as a new branch of mathematics. However, geometrical arguments were used extensively, and the combined analytical and geometrical arguments were not only complicated but hardly provided a systematic general method. Euler was fully aware of these limitations.

3. *The Principle of Least Action*

While progress in the solution of problems of the calculus of variations was being made, a new motivation for work in the subject came directly from physics. The contemporary development was the Principle of Least Action. To explain the basis for this principle we must go back a bit. Euclid had proved in his *Catoptrica* (Chap. 7, sec. 7) that light traveling from P to a mirror (Fig. 24.4) and then to Q takes the path for which $\angle 1 = \angle 2$. Then the Alexandrian Heron proved that the path PRQ, which light actually

12. *Opera*, (1), 24.

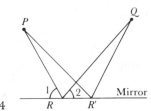

Figure 24.4

takes, is shorter than any other path, such as $PR'Q$, which it could conceivably take. Since the light takes the shortest path, if the medium on the upper side of the line RR' is homogeneous, then the light travels with constant velocity and so takes the path requiring least time. Heron applied this principle of shortest path and least time to problems of reflection from concave and convex spherical mirrors.

Basing their case on this phenomenon of reflection and on philosophic, theological, and aesthetic principles, philosophers and scientists after Greek times propounded the doctrine that nature acts in the shortest possible way or, as Olympiodorus (6th cent. A.D.) said in his *Catoptrica*, "Nature does nothing superfluous or any unnecessary work." Leonardo da Vinci said nature is economical and her economy is quantitative, and Robert Grosseteste believed that nature always acts in the mathematically shortest and best possible way. In medieval times it was commonly accepted that nature behaved in this manner.

The seventeenth-century scientists were at least receptive to this idea but, as scientists, tried to tie it to phenomena that supported it. Fermat knew that under reflection light takes the path requiring least time and, convinced that nature does indeed act simply and economically, affirmed in letters of 1657 and 1662 [13] his Principle of Least Time, which states that light always takes the path requiring least time. He had doubted the correctness of the law of refraction of light (Chap. 15, sec. 4) but when he found in 1661 [14] that he could deduce it from his Principle, he not only resolved his doubts about the law but felt all the more certain that his Principle was correct.

Fermat's Principle is stated mathematically in several equivalent forms. According to the law of refraction

$$\frac{\sin i}{\sin r} = \frac{v_1}{v_2},$$

where v_1 is the velocity of light in the first medium and v_2 in the second. The ratio of v_1 to v_2 is denoted by n and is called the index of refraction of the

13. *Œuvres*, 2, 354–59, 457–63.
14. *Œuvres*, 2, 457–63.

second medium relative to the first, or, if the first is a vacuum, n is called the absolute index of refraction of the nonvacuous medium. If c denotes the velocity of light in a vacuum, then the absolute index $n = c/v$ where v is the velocity of light in the medium. If the medium is variable in character from point to point, then n and v are functions of x, y, and z. Hence the time required for light to travel from a point P_1 to a point P_2 along a curve $x(\sigma)$, $y(\sigma)$, $z(\sigma)$ is given by

$$(5) \qquad J = \int_{\sigma_1}^{\sigma_2} \frac{ds}{v} = \int_{\sigma_1}^{\sigma_2} \frac{n}{c}\, ds = \frac{1}{c} \int_{\sigma_1}^{\sigma_2} n(x, y, z) \sqrt{\dot{x}^2 + \dot{y}^2 + \dot{z}^2}\, d\sigma,$$

where σ_1 is the value of σ at P_1 and σ_2 the value at P_2. Thus the Principle states that the path light actually takes in traveling from P_1 to P_2 is given by the curve which makes J a minimum.[15]

By the early eighteenth century the mathematicians had several impressive examples of the fact that nature does attempt to maximize or minimize some important quantities. Huygens, who had at first objected to Fermat's Principle, showed that it does hold for the propagation of light in media with variable indices of refraction. Even Newton's first law of motion, which states that the straight line or shortest distance is the natural motion of a body, showed nature's desire to economize. These examples suggested that there might be some more general principle. The search for such a principle was undertaken by Maupertuis.

Pierre-Louis Moreau de Maupertuis (1698–1759), while working with the theory of light in 1744, propounded his famous Principle of Least Action in a paper entitled "Accord des différentes lois de la nature qui avaient jusqu'ici paru incompatibles."[16] He started from Fermat's Principle, but in view of disagreements at that time as to whether the velocity of light was proportional to the index of refraction as Descartes and Newton believed, or inversely proportional as Fermat believed, Maupertuis abandoned least time. In fact he did not believe that it was always correct.

Action, Maupertuis said, is the integral of the product of mass, velocity, and distance traversed, and any changes in nature are such as to make the action least. Maupertuis was somewhat vague because he failed to specify the time interval over which the product of m, v, and s was to be taken and because he assigned a different meaning to action in each of the applications he made to optics and some problems of mechanics.

Though he had some physical examples to support his Principle, Maupertuis advocated it also for theological reasons. The laws of behavior of matter had to possess the perfection worthy of God's creation; and the

15. There are instances, as for example in the reflection of light from a concave mirror, where light takes the path requiring maximum time. This fact was known to Fermat and was explicitly stated by William R. Hamilton.
16. *Mém. de l'Acad. des Sci., Paris*, 1744.

least action principle seemed to satisfy this criterion because it showed that nature was economical. Maupertuis proclaimed his principle to be a universal law of nature and the first scientific proof of the existence of God. Euler, who had corresponded with Maupertuis on this subject between 1740 and 1744, agreed with Maupertuis that God must have constructed the universe in accordance with some such basic principle and that the existence of such a principle evidenced the hand of God.

In the second appendix to his 1744 book Euler formulated the Principle of Least Action as an exact dynamical theorem. He limited himself to the motion of a single particle moving along plane curves. Moreover, he supposed that the speed is dependent upon position or, in modern terms, that the force is derivable from a potential. Whereas Maupertuis wrote

$$mvs = \text{min.},$$

Euler wrote

$$\partial \int v \, ds = 0,$$

by which he meant that the rate of change of the integral for a change in the path must be zero. He also wrote that, since $ds = v \, dt$,

$$\partial \int v^2 \, dt = 0.$$

Just what Euler meant by the rate of change of the integral was still vague here even though he applied the principle correctly in specific problems by using his technique of the calculus of variations. At least he showed that Maupertuis's action was least for motions along plane curves.

Euler went further than Maupertuis in believing that all natural phenomena behave so as to maximize or minimize some function, so that the basic physical principles should be expressed to the effect that some function is maximized or minimized. In particular this should be true in dynamics, which studies the motions of bodies propelled by forces. Euler was not too far from the truth.

4. The Methodology of Lagrange

Euler's work attracted the attention of Lagrange, who began to concern himself with problems of the calculus of variations in 1750 when he was nineteen. He discarded the geometric-analytic arguments of the Bernoullis and Euler and introduced purely analytical methods. In 1755 he obtained a general procedure, systematic and uniform for a wide variety of problems, and worked on it for several years. His famous publication on the subject was the "Essai d'une nouvelle méthode pour déterminer les maxima et les minima des formules intégrales indéfinies."[17] In a letter to Euler of August

17. *Misc. Taur.*, 2, 1760/61, 173–95, pub. 1762 = *Œuvres*, 1, 333–62.

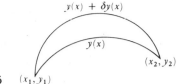

Figure 24.5 (x_1, y_1)

1755 he described the method, which he termed the method of variations, but which Euler in a paper presented to the Berlin Academy in 1756[18] named the calculus of variations.

Let us note Lagrange's method for the basic problem of the calculus of variations, namely, to minimize or maximize the integral

(6)
$$J = \int_{x_1}^{x_2} f(x, y, y') \, dx,$$

where $y(x)$ is to be determined. One of Lagrange's innovations was not to vary individual ordinates of the minimizing or maximizing curve $y(x)$ but to introduce new curves running between the endpoints (x_1, y_1) and (x_2, y_2). These new curves (Fig. 24.5) were represented by Lagrange in the form $y(x) + \delta y(x)$, the δ being a special symbol introduced by Lagrange to indicate a variation of the *entire curve* $y(x)$. The introduction of a new curve in the integrand of (6) of course changes the value of J. The increment in J, which we shall denote by ΔJ, is then

$$\Delta J = \int_{x_1}^{x_2} \{ f(x, y + \delta y, y' + \delta y') - f(x, y, y') \} \, dx.$$

Now Lagrange regards f as a function of three independent variables, but since x is not changed, the integrand can be expanded by means of Taylor's theorem applied to a function of two variables. The expansion gives first degree terms in δy and $\delta y'$, second degree terms in these increments, and so forth. Lagrange then writes

(7)
$$\Delta J = \delta J + \frac{1}{2} \delta^2 J + \frac{1}{3!} \delta^3 J + \cdots$$

where δJ indicates the integral of first degree terms in δy and $\delta y'$, $\delta^2 J$ indicates the integral of the second degree terms, and so forth. Thus

$$\delta J = \int_{x_1}^{x_2} (f_y \, \delta y + f_{y'} \, \delta y') \, dx$$

$$\delta^2 J = \int_{x_1}^{x_2} \{ f_{yy} (\delta y)^2 + 2 f_{yy'} (\delta y)(\delta y') + f_{y'y'} (\delta y')^2 \} \, dx.$$

18. "Elementa Calculi Variationum," *Novi Comm. Acad. Sci. Petrop.*, 10, 1764, 51–93, pub. 1766 = *Opera*, (1), 25, 141–76.

δJ is called the first variation of J; $\delta^2 J$, the second variation; and so forth.

Lagrange now argues that the value of δJ, since it contains the first order terms in the small variations δy and $\delta y'$, dominates the right side of (7), so that when δJ is positive or negative ΔJ will be positive or negative. But at a maximum or minimum of J, ΔJ must have the same sign, as in the case of ordinary maxima and minima of a function $f(x)$ of one variable, so that for $y(x)$ to be a maximizing function, δJ must be 0. Moreover, Lagrange says,

$$(8) \qquad\qquad \delta y' = \frac{d(\delta y)}{dx};$$

that is, the order of the operations d and δ can be interchanged. This is correct, though the reason was not clear to Lagrange's contemporaries and Euler clarified it later. [It is easily seen to be correct if we write $y + \delta y$ as $y + n(x)$, where $n(x)$ is the variation of $y(x)$. Then $\delta y = y + n(x) - y = n(x)$ and $\delta y' = y' + n'(x) - y' = n'(x)$. But $n'(x) = dn(x)/dx = d(\delta y)/dx$.] Using (8) Lagrange writes the first variation as

$$\delta J = \int_{x_1}^{x_2} \left[f_y \, \delta y + f_{y'} \frac{d}{dx}(\delta y) \right] dx.$$

Integrating the second term by parts and using the fact that δy must vanish at x_1 and at x_2 yields

$$(9) \qquad\qquad \delta J = \int_{x_1}^{x_2} \left(f_y \, \delta y - \left(\frac{d}{dx} f_{y'}\right) \delta y \right) dx.$$

Now δJ must be 0 for every variation δy. Hence Lagrange concludes that the coefficient of δy must be 0,[19] or that

$$(10) \qquad\qquad f_y - \frac{d}{dx}(f_{y'}) = 0.$$

Thus Lagrange arrived at the same ordinary differential equation for $y(x)$ that Euler had obtained. Lagrange's method of deriving (10) (except for his use of differentials) and even his notation are used today. Of course, (10) is a necessary condition on $y(x)$ but not sufficient.

In this paper of 1760/61 Lagrange also deduced for the first time end-conditions that must be satisfied by a minimizing curve for problems with variable endpoints. He found the transversality conditions that must hold at the intersections of the minimizing curve with the fixed curves or

19. The fact that the coefficient of δy must be 0 was intuitively accepted or incorrectly proven by every writer on the subject for one hundred years after Lagrange's work. Even Cauchy's proof was inadequate. The first correct proof was given by Pierre Frédéric Sarrus (1798–1861) (*Mém. divers Savans*, (2), 10, 1848, 1–128). The result is now known as the fundamental lemma of the calculus of variations.

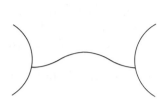

Figure 24.6

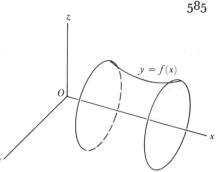

Figure 24.7

surfaces on which the endpoints of the comparison curves are allowed to vary (Fig. 24.6).

Though much more remains to be said about maximizing and minimizing integrals of the form (6), historically the next step, made by Lagrange in his 1760/61 paper and in a following one,[20] was to consider problems leading to multiple integrals. The integral to be maximized or minimized is of the form

$$(11) \qquad J = \iint f(x, y, z, p, q) \, dx \, dy$$

wherein z is a function of x and y, $p = \partial z/\partial x$, and $q = \partial z/\partial y$. The integration is over some area in the xy-plane. The problem, then, is to find the function $z(x, y)$ that maximizes or minimizes the value of J. One of the most important problems that comes under this class of double integrals is to find the surface of least area among all surfaces whose boundary is fixed in some way. Thus one might be given two closed non-intersecting curves in space and seek the surface of minimum area bounded by these two curves. As a special case of the minimal surface problem, the two curves can be circles parallel to the yz-plane (Fig. 24.7) and with centers on the x-axis. Then the possible minimal surfaces are necessarily surfaces of revolution bounded by the two curves and the problem is to find the surface of revolution of minimum area. This last problem, as we noted above, had already been solved by Euler in 1744. However, this special case of a surface of revolution can be treated by the theory applicable to the integral (11).

By a method similar to the one he had used for the simpler integral (6), Lagrange obtained the differential equation that the function $z(x, y)$ minimizing (11) must satisfy. If we use the common notation

$$\frac{\partial z}{\partial x} = p, \qquad \frac{\partial z}{\partial y} = q, \qquad \frac{\partial^2 z}{\partial x^2} = r, \qquad \frac{\partial^2 z}{\partial x \partial y} = s, \qquad \frac{\partial^2 z}{\partial y^2} = t,$$

20. *Misc. Taur.*, 4, 1766/69 = *Œuvres*, 2, 37–63.

then the equation is

(12) $Rr + Ss + Tt = U$

wherein R, S, T, and U are functions of x, y, z, p, and q. This nonlinear second order partial differential equation, called the equation of Monge, is not easy to solve; equations of this form had been the subject of research from the days of Euler onward (Chap. 22, sec. 7).

In the case of the minimal surface problem, the integral (11) becomes

(13) $$\iint (1 + p^2 + q^2)^{1/2} \, dx \, dy,$$

and for this special class of problems the partial differential equation (12) becomes

(14) $(1 + q^2)r - 2pqs + (1 + p^2)t = 0.$

This equation was given by Lagrange in his 1760/61 paper (though not quite in this form) and is a major analytical result in the theory of minimal surfaces. Geometrically, as Meusnier pointed out in a paper of 1785,[21] this partial differential equation expresses the fact that at any point on the minimizing surface the principal radii of curvature are equal and opposite or that the mean curvature, that is, the average of the principal curvatures, is zero.

In a later paper (1770)[22] Lagrange also considered single and multiple integrals in which higher derivatives than the first ones appear in the integrand. This topic has been well developed since Lagrange's time and is now standard material in the calculus of variations. However, since the principles are not basically different from the cases already considered, we shall not go into this extension of the subject. The content of Lagrange's papers on the calculus of variations is incorporated in his *Mécanique analytique*.

The calculus of variations was not well understood by the contemporaries of Lagrange and Euler. Euler explained Lagrange's method in numerous writings and used it to re-prove a number of old results. Though he realized that the calculus of variations was a new branch or technique, which he says is symbolized by the new operational symbol δ, he, like Lagrange, tried to base the logic of the calculus of variations on the ordinary calculus. Euler's idea[23] was to introduce a parameter t such that the curves of the family considered in a variations problem would vary with t, that is, for each t of some range there would be a curve $y_t(x)$. Then, says Euler, whereas $dy = (dy/dx) \, dx$, $\delta y = (dy/dt) \, dt$. Hence the variation δy is expressed by a partial differentiation with respect to t. He then formulated the technique

21. *Mém. divers Savans*, 10, 1785, 477–85.
22. *Nouv. Mém. de l'Acad. de Berlin*, 1770 = *Œuvres*, 3, 157–86.
23. *Novi Comm. Acad. Sci. Petrop.*, 16, 1771, 35–70, pub. 1772 = *Opera*, (1), 25, 208–35.

of the calculus of variations in terms of this new concept of differentiation with respect to t. His final results were, of course, the same as those already obtained.

Euler proceeded (1779)[24] to consider space curves with maximum or minimum properties and (1780) extensions of the brachistochrone problem when the applied force (which is gravity in the usual problem) operates in three dimensions or when a resisting medium is present.[25]

5. *Lagrange and Least Action*

Lagrange applied the calculus of variations to dynamics. He took over from Euler the Principle of Least Action and became the first to express the principle in concrete form, namely, that for a single particle the integral of the product of mass, velocity, and distance taken between two fixed points is a maximum or a minimum; that is, $\int mv\,ds$ must be a maximum or a minimum for the actual path taken by the particle. Alternatively, since $ds = v\,dt$, then $\int mv^2\,dt$ must be a maximum or a minimum. The quantity mv^2 [today $(1/2)mv^2$] is called the kinetic energy; in Lagrange's day it was called living force. Lagrange also asserted that the principle is true for a collection of particles and even for extended masses, though he was not clear on the last case.

Using the Principle of Least Action and the method of the calculus of variations, Lagrange obtained his famous equations of motion. Let us consider the case where the kinetic energy is a function of x, y, and z. Then for a single particle the kinetic energy T is given by

$$(15) \qquad\qquad T = \frac{1}{2}\,m(\dot{x}^2 + \dot{y}^2 + \dot{z}^2).$$

Lagrange also supposed that the forces acting to cause the motion are derivable from a potential function V, which depends on x, y, and z. An additional condition, then, is that $T + V = $ const., that is, the total energy is constant. Lagrange's action is

$$(16) \qquad\qquad \int_{t_0}^{t_1} T\,dt$$

and his Principle of Least Action states that this action must be a minimum or a maximum, that is

$$(17) \qquad\qquad \delta \int_{t_0}^{t_1} T\,dt = 0.$$

24. *Mém. de l'Acad. des Sci. de St. Peters.*, 4, 1811, 18–42, pub. 1813 = *Opera*, (1), 25, 293–313.

25. *Mém. de l'Acad. des Sci. de St. Peters.*, 8, 1817/18, 17–45, pub. 1822 = *Opera*, (1), 25, 314–42.

In a minimizing or maximizing action, even though the motion takes place between two fixed points in space and the two fixed time values t_0 and t_1, the space and time variables must be varied.

By applying the method of the calculus of variations to the action integral, Lagrange derived equations analogous to Euler's equation (2), namely,

$$(18) \qquad \frac{d}{dt}\left(\frac{\partial T}{\partial x}\right) + \frac{\partial V}{\partial x} = 0$$

and the two corresponding equations with y and z. These equations are the equivalent of Newton's second law of motion.

Lagrange made the further step of introducing what are now called generalized coordinates. That is, in place of rectangular coordinates one may use polar coordinates or, in fact, any set of coordinates q_1, q_2, q_3, which are needed to fix the position of the particle (or extended mass). Then

$$x = x(q_1, q_2, q_3)$$
$$y = y(q_1, q_2, q_3)$$
$$z = z(q_1, q_2, q_3),$$

where the q_i are now functions of t. In terms of the new coordinates, T becomes a function of the q_i and $\dot{q}_i$ while V becomes a function of the q_i. Then the equations (18) become

$$(19) \qquad \frac{d}{dt}\left(\frac{\partial T}{\partial \dot{q}_i}\right) - \frac{\partial T}{\partial q_i} + \frac{\partial V}{\partial q_i} = 0, \qquad i = 1, 2, 3.$$

This is a set of 3 simultaneous second order ordinary differential equations in the q_i. They are the Euler (characteristic) equations for the action integral. If n coordinates are needed to fix the position of the moving object, for example, 2 particles require 6 coordinates, then equations (19) are replaced by n equations.[26]

These generalized coordinates need not have either geometrical or physical significance. One speaks of them today as coordinates in a configuration space and then the $q_i(t)$ are the equations of a path in configuration space. Thus Lagrange recognized that the variational principle, namely, that the action must be a minimum or maximum, can be used with any set

26. Lagrange is explicit that the number of variables in T and in V are the number required to determine the position of the mechanical system. Thus if there are N independent particles and each requires three coordinates (x_i, y_i, z_i) to describe its path in space, then $3N$ coordinates are required. In this case there will be $3N$ coordinates q_i, $3N$ equations relating the Cartesian coordinates to the q_i and $3N$ equations (19). The number of independent coordinates or the number of degrees of freedom, as physicists put it, depends on the system being treated and on the constraints in the motion.

of coordinates and the Lagrangian equations of motion (19) are invariant in form with respect to any coordinate transformation.

Lagrange's Principle, though it amounts to Newton's second law of motion, has several advantages over Newton's formulation. First of all, any convenient coordinate system is already, so to speak, built into the formulation. Secondly, it is easier to handle problems with constraints on the motion. Thirdly, instead of a series of separate differential equations, which may be numerous if many particles are involved, there is—to start with, at least—one principle from which the differential equations follow. Finally, though his principle supposes a knowledge of the kinetic and potential energies of a problem, it does not require a knowledge of the forces acting. With his principle Lagrange deduced major laws of mechanics and solved new problems, though it was not broad enough to include all the problems mechanics deals with. His work on the action principle is fully treated in his *Mécanique analytique*. He also started the movement to deduce the laws of other branches of physics from variational principles that would be the analogues of least action. He himself gave a variational principle for a broad class of hydrodynamical problems. This subject will be resumed in our study of the nineteenth-century work.

From the mathematical standpoint Lagrange's work on least action gave major importance to the calculus of variations. In particular Lagrange had derived the Euler equations for an integral whose integrand contains one independent variable but several dependent variables and their derivatives. This is an extension of the original calculus of variations problem, which contains only one dependent variable and its derivative. In this extended case the Euler equations are a system of second order ordinary differential equations in the q_i.

6. *The Second Variation*

The Euler differential equation, as Euler and Lagrange realized, is only a necessary condition for the solution to furnish a maximum or a minimum. They used the differential equation to find the solution and then decided on intuitive or physical grounds whether it furnished a maximum or a minimum. The role of the Euler equation is entirely analogous to the condition $f'(x) = 0$ in the ordinary calculus. A value of x that maximizes or minimizes $y = f(x)$ must satisfy $f'(x) = 0$, but the converse is not necessarily true.

The question of what additional conditions a solution of the Euler equation must satisfy to actually furnish a maximum or a minimum value of an integral depending on $y(x)$ was tackled unsuccessfully by Laplace in 1782. It was then taken up by Legendre in 1786.[27] Guided by the fact that

27. *Hist. de l'Acad. des Sci., Paris*, 1786, 7–37, pub. 1788.

in the ordinary calculus the sign of $f''(x)$ at a value of x for which $f'(x) = 0$ determines whether $f(x)$ has a maximum or a minimum, Legendre considered the second variation $\delta^2 J$, recast its form, and concluded that J is a maximum for the curve $y(x)$ which satisfies Euler's equation and passes through (x_0, y_0) and (x_1, y_1) provided that $f_{y'y'} \leq 0$ at each x along $y(x)$. Likewise, J is a minimum for a $y(x)$ satisfying the first two conditions provided that $f_{y'y'} \geq 0$ at each x along $y(x)$. Legendre then extended this result to more general integrals than (6). However, Legendre realized in 1787 that the condition on $f_{y'y'}$ was just a necessary condition on $y(x)$ in order that it be a maximizing or minimizing curve. The problem of finding sufficient conditions that a curve $y(x)$ maximize or minimize an integral such as (6) was not solved in the eighteenth century.

Bibliography

Bernoulli, James: *Opera*, 2 vols., 1744, reprint by Birkhaüser, 1968.

Bernoulli, John: *Opera Omnia*, 4 vols., 1742, Georg Olms (reprint), 1968.

Bliss, Gilbert A.: *The Calculus of Variations*, Open Court, 1925.

————: "The Evolution of Problems in the Calculus of Variations," *Amer. Math. Monthly*, 43, 1936, 598–609.

Cantor, Moritz: *Vorlesungen über Geschichte der Mathematik*, B. G. Teubner, 1898 and 1924, Vol. 3, Chap. 117, and Vol. 4, 1066–74.

Caratheodory, C.: Introduction to Series (1), Vol. 24 of Euler's *Opera Omnia*, viii–lxii, Orell Füssli, 1952. Also in C. Caratheodory: *Gesammelte mathematische Schriften*, C. H. Beck, 1957, Vol. 5, pp. 107–74.

Darboux, Gaston: *Leçons sur la théorie générale des surfaces*, 2nd ed., Gauthier-Villars, 1914, Vol. 1, Book III, Chaps. 1–2.

Euler, Leonhard: *Opera Omnia*, (1), Vols. 24–25, Orell Füssli, 1952.

Hofmann, Joseph E.: "Über Jakob Bernoullis Beiträge zur Infinitesimalmathematik," *L'Enseignement Mathématique*, (2), 2, 1956, 61–171; published separately by Institut de mathématiques, Geneva, 1957.

Huke, Aline: *An Historical and Critical Study of the Fundamental Lemma in the Calculus of Variations*, University of Chicago Contributions to the Calculus of Variations, University of Chicago Press, Vol, 1, 1930, pp. 45–160.

Lagrange, Joseph-Louis: *Œuvres de Lagrange*, Gauthier-Villars, 1867–69, relevant papers in Vols. 1–3.

————: *Mécanique analytique*, 2 vols., 4th ed., Gauthier-Villars, 1889.

Lecat, Maurice: *Bibliographie du calcul des variations depuis les origines jusqu'à 1850*, Gand, 1916.

Montucla, J. F.: *Histoire des mathématiques*, 1802, Albert Blanchard (reprint), 1960, Vol. 3, 643–58.

Porter, Thomas Isaac: "A History of the Classical Isoperimetric Problem," *University of Chicago Contributions to the Calculus of Variations*, University of Chicago Press, 1933, Vol. 2, pp. 475–517.

Smith, David E.: *A Source Book in Mathematics*, Dover (reprint), 1959, pp. 644–55.

Struik, D. J.: *A Source Book in Mathematics, 1200–1800*, Harvard University Press, 1969, pp. 391–413.

Todhunter, Isaac: *A History of the Calculus of Variations during the Nineteenth Century*, 1861, Chelsea (reprint), 1962.

Woodhouse, Robert: *A History of the Calculus of Variations in the Eighteenth Century*, 1810, Chelsea (reprint), 1964.

25
Algebra in the Eighteenth Century

I present higher analysis as it was in its childhood but you
are bringing it to man's estate.

JOHN BERNOULLI, IN A LETTER TO EULER

1. Status of the Number System

Though algebra and analysis were hardly distinguished from each other in
the eighteenth century because the significance of the limit concept was still
obscure, it is desirable from our modern point of view to separate the two
fields of activity. In the seventeenth century, algebra was a major center of
interest; in the eighteenth century it became subordinate to analysis, and
except in the theory of numbers the motivation for work in it came largely
from analysis.

Since the basis of algebra is the number system, let us note the status of
this subject. By 1700 all of the familiar members of the system—whole
numbers, fractions, irrationals, and negative and complex numbers—were
known. However, opposition to the newer types of numbers was expressed
throughout the century. Typical are the objections of the English mathematician Baron Francis Masères (1731–1824), a Fellow of Clare College in
Cambridge and a member of the Royal Society. Masères, who did write
acceptable papers in mathematics and a substantial treatise on the theory of
life insurance, published in 1759 his *Dissertation on the Use of the Negative Sign
in Algebra.* He shows how to avoid negative numbers (except to indicate the
subtraction of a larger quantity from a smaller one), and especially negative
roots, by carefully segregating the types of quadratic equations so that those
with negative roots are considered separately; and, of course, the negative
roots are to be rejected. He does the same with cubics. Then he says of
negative roots,

> ...they serve only, as far as I am able to judge, to puzzle the whole
> doctrine of equations, and to render obscure and mysterious things that
> are in their own nature exceeding plain and simple.... It were to be
> wished therefore that negative roots had never been admitted into

algebra or were again discarded from it: for if this were done, there is good reason to imagine, the objections which many learned and ingenious men now make to algebraic computations, as being obscure and perplexed with almost unintelligible notions, would be thereby removed; it being certain that Algebra, or universal arithmetic, is, in its own nature, a science no less simple, clear, and capable of demonstration, than geometry.

Certainly negative numbers were not really well understood until modern times. Euler, in the latter half of the eighteenth century, still believed that negative numbers were greater than ∞. He also argued that $(-1) \cdot (-1) = +1$ because the product must be $+1$ or -1 and since $1 \cdot (-1) = -1$, then $(-1) \cdot (-1) = +1$. Carnot, the noted French geometer, thought the use of negative numbers led to erroneous conclusions. As late as 1831 Augustus De Morgan (1806–71), professor of mathematics at University College, London, and a famous mathematical logician and contributor to algebra, in his *On the Study and Difficulties of Mathematics*, said,"The imaginary expression $\sqrt{-a}$ and the negative expression $-b$ have this resemblance, that either of them occurring as the solution of a problem indicates some inconsistency or absurdity. As far as real meaning is concerned, both are equally imaginary, since $0 - a$ is as inconceivable as $\sqrt{-a}$."

De Morgan illustrated this by means of a problem. A father is 56; his son is 29. When will the father be twice as old as the son? He solves $56 + x = 2(29 + x)$ and obtains $x = -2$. Thus the result, he says, is absurd. But, he continues, if we change x to $-x$ and solve $56 - x = 2(29-x)$, we get $x = 2$. He concludes that we phrased the original problem wrongly and thus were led to the unacceptable negative answer. De Morgan insisted that it was absurd to consider numbers less than zero.

Though nothing was done in the eighteenth century to clarify the concept of irrational numbers, some progress was made in this subject. In 1737 Euler showed, substantially, that e and e^2 are irrational and Lambert showed that π is irrational (Chap. 20, sec. 6). The work on the irrationality of π was motivated largely by the desire to solve the problem of squaring the circle. Legendre's conjecture that π may not be a root of an algebraic equation with rational coefficients led to a distinction between types of irrationals. Any root, real or complex, of any algebraic (polynomial) equation with rational coefficients is called an algebraic number. Thus the roots of

$$a_0 x^n + a_1 x^{n-1} + \cdots + a_{n-1} x + a_n = 0,$$

where the a_i are rational numbers, are called algebraic numbers. Consequently, every rational number and some irrationals are algebraic numbers, for any rational number c is a root of $x - c = 0$, and $\sqrt{2}$ is a root of $x^2 - 2 = 0$. Those numbers that are not algebraic are called transcendental because, as Euler put it, "they transcend the power of algebraic methods."

This distinction between algebraic and transcendental numbers was recognized by Euler at least as early as 1744. He conjectured that the logarithm to a rational base of a rational number must be either rational or transcendental. However, no number was known to be transcendental in the eighteenth century and the problem of showing that there are transcendental numbers remained open.

Complex numbers were more of a bane to the eighteenth-century mathematicians. These numbers were practically ignored from their introduction by Cardan until about 1700. Then (Chap. 19, sec. 3) complex numbers were used to integrate by the method of partial fractions, which was followed by the lengthy controversy about complex numbers and the logarithms of negative and complex numbers. Despite his correct resolution of the problem of the logarithms of complex numbers, neither Euler nor the other mathematicians were clear about these numbers.

Euler tried to understand what complex numbers really are, and in his *Vollständige Anleitung zur Algebra* (Complete Introduction to Algebra), which first appeared in Russian in 1768–69 and in German in 1770 and is the best algebra text of the eighteenth century, says,

> Because all conceivable numbers are either greater than zero or less than 0 or equal to 0, then it is clear that the square roots of negative numbers cannot be included among the possible numbers [real numbers]. Consequently we must say that these are impossible numbers. And this circumstance leads us to the concept of such numbers, which by their nature are impossible, and ordinarily are called imaginary or fancied numbers, because they exist only in the imagination.

Euler made mistakes with complex numbers. In this *Algebra* he writes $\sqrt{-1} \cdot \sqrt{-4} = \sqrt{4} = 2$ because $\sqrt{a}\sqrt{b} = \sqrt{ab}$. He also gives $i^i = 0.2078795763$ but misses other values of this quantity. He gave this number originally in a letter to Goldbach of 1746 and also in his article of 1749 on the controversy between Leibniz and John Bernoulli (Chap. 19, sec. 3). Though he calls complex numbers impossible numbers, Euler says they are of use. The use he has in mind occurs when we tackle problems about which we do not know whether there is or is not an answer. Thus if we are asked to separate 12 into two parts whose product is 40, we would find that the parts are $6 + \sqrt{-4}$ and $6 - \sqrt{-4}$. Thereby, he says, we recognize that the problem cannot be solved.

Some positive steps beyond Euler's correct conclusion on the logarithms of complex numbers were made, but their influence on the eighteenth century was limited. In his *Algebra* (1685, Chaps. 66–69), John Wallis showed how to represent geometrically the complex roots of a quadratic equation with real coefficients. Wallis said, in effect, that complex numbers are no more absurd than negative numbers; and since the latter can be represented on a direct line, it should be possible to represent complex numbers in the plane. He started with an axis on which real numbers were marked out in relation to an origin; a distance from the origin along this axis represented the real part of the root, the distance being measured in the positive or negative direction of the axis according as this part of the root was positive or negative. From the point on the real axis thus determined, a line was drawn perpendicular to the axis of reals whose length represented the number which, multiplied by $\sqrt{-1}$, gave the imaginary part of the root, the line being drawn in one direction or the opposite according as the number was positive or negative. (He failed to introduce the y-axis itself as the axis of imaginaries.) Wallis then gave a geometrical construction for the roots of $ax^2 + bx + c = 0$ when the roots are real and when they are complex. His work was correct, but it was not a useful representation of $x + iy$ for other purposes. There were other attempts to represent complex numbers geometrically in the eighteenth century, but they were not broadly useful. Nor did geometrical representation at this time make complex numbers more acceptable.

Early in the century most mathematicians believed that different roots of complex numbers would introduce different types or orders of complex numbers and that there might be ideal roots whose nature they could not specify but which might somehow be calculated. But d'Alembert, in his prize work *Réflexions sur la cause générale des vents* (1747), affirmed that every expression built up from complex numbers by means of algebraic operations (in which he included raising to an arbitrary power), is a complex number of the form $A + B\sqrt{-1}$. The one difficulty he had in proving this assertion was the case of $(a + bi)^{g+hi}$. His demonstration of this fact had to be mended by Euler, Lagrange, and others. In the *Encyclopédie*, d'Alembert maintained a discrete silence on complex numbers.

Throughout the eighteenth century complex numbers were used effectively enough for mathematicians to acquire some confidence in them (Chap. 19, sec. 3; Chap. 27, sec. 2). Where used in intermediate stages of mathematical arguments, the results proved to be correct; this fact had telling effect. Yet there were doubts as to the validity of the arguments and often even of the results.

In 1799 Gauss gave his first proof of the fundamental theorem of algebra, and since this necessarily depended on the recognition of complex numbers, Gauss solidified the position of these numbers. The nineteenth century then plunged forward boldly with complex functions. But even long after this theory was developed and employed in hydrodynamics, the Cambridge University professors preserved "an invincible repulsion to the objectionable $\sqrt{-1}$, cumbrous devices being adopted to avoid its occurrence or use wherever possible."

The general attitude toward complex numbers even as late as 1831 may be learned from De Morgan's book *On the Study and Difficulties of Mathematics*. He says that this book contains nothing that could not be found in the best works then in use at Oxford and Cambridge. Turning to complex numbers, he states,

> We have shown the symbol $\sqrt{-a}$ to be void of meaning, or rather self-contradictory and absurd. Nevertheless, by means of such symbols, a part of algebra is established which is of great utility. It depends upon the fact, which must be verified by experience, that the common rules of algebra may be applied to these expressions [complex numbers] without leading to any false results. An appeal to experience of this nature appears to be contrary to the first principles laid down at the beginning of this work. We cannot deny that it is so in reality, but it must be recollected that this is but a small and isolated part of an immense subject, to all other branches of which these principles apply in their fullest extent.

The "principles" he refers to are that mathematical truths should be derived by deductive reasoning from axioms.

He then compares negative roots and complex roots.

> There is, then, this distinct difference between the negative and the imaginary result. When the answer to a problem is negative, by changing the sign of x in the equation which produced that result, we may either discover an error in the method of forming that equation or show that the question of the problem is too limited, and may be extended so as to admit of a satisfactory answer. When the answer to a problem is imaginary this is not the case.... We are not advocates for stopping the progress of the student by entering fully into all the arguments for and against such questions, as the use of negative quantities, etc., which he could not understand, and which are inconclusive on both sides; but he might be made aware that a difficulty does exist, the nature of which might be pointed out to him, and he might then, by the consideration of a sufficient number of examples, treated separately, acquire confidence in the results to which the rules lead.

By the time De Morgan wrote the above lines, the concepts of complex numbers and complex functions were well on the way to clarification. But the diffusion of the new knowledge was slow. Certainly throughout the eighteenth and the first half of the nineteenth centuries the meaning of complex numbers was debated hotly. All the arguments of John Bernoulli, d'Alembert, and Euler were continually rehashed. Even twentieth-century textbooks on trigonometry supplemented presentations that employed complex numbers by proofs not involving $\sqrt{-1}$.

We should note here another point whose significance is almost in inverse proportion to the conciseness with which it can be stated: In the eighteenth century no one worried about the logic of the real or complex number systems. What Euclid had done in Book V of the *Elements* to establish the properties of incommensurable magnitudes was disregarded. That this exposition had been tied to geometry, whereas by now arithmetic and algebra were independent of geometry, accounts in part for this disregard. Moreover this logical development, even if suitably modified to free it of geometry, could not establish the logical foundations of negative and complex numbers; and this fact too may have caused mathematicians to desist from any attempt to found the number system rigorously. Finally, the century was concerned primarily to use mathematics in science, and since the rules of operation were intuitively secure at least for real numbers no one really worried about the foundations. Typical is the statement of d'Alembert in his article on negative numbers in the *Encyclopédie*. The article is not at all clear and d'Alembert concludes that "the algebraic rules of operation with negative numbers are generally admitted by everyone and acknowledged as exact, whatever idea we may have about these quantities." The various types of numbers, never properly introduced to the world, nevertheless gained a firmer place in the eighteenth-century mathematical community.

2. *The Theory of Equations*

One of the investigations continued from the seventeenth century with hardly a break was the solution of polynomial equations. The subject is fundamental in mathematics and so the interest in obtaining better methods of solving equations of any degree, getting better methods of approximating roots of equations, and completing the theory—in particular by proving that every nth degree polynomial equation has n roots—was natural. In addition, the use of the method of partial fractions for integration raised the question of whether any polynomial with real coefficients can be decomposed into a product of linear factors or a product of linear and quadratic factors with real coefficients to avoid the use of complex numbers.

As we saw in Chapter 19 (sec. 4), Leibniz did not believe that every polynomial with real coefficients could be decomposed into linear and quadratic factors with real coefficients. Euler took the correct position. In a letter to Nicholas Bernoulli (1687–1759) of October 1, 1742, Euler affirmed without proof that a polynomial of arbitrary degree with real coefficients could be so expressed. Nicholas did not believe the assertion to be correct and gave the example of

$$x^4 - 4x^3 + 2x^2 + 4x + 4$$

with the zeroes $1 + \sqrt{2 + \sqrt{-3}}$, $1 - \sqrt{2 + \sqrt{-3}}$, $1 + \sqrt{2 - \sqrt{-3}}$, $1 - \sqrt{2 - \sqrt{-3}}$, which he said contradicts Euler's assertion. On December 15, 1742, Euler, writing to Goldbach (Fuss, Vol. 1, 169–71), pointed out that complex roots occur in conjugate pairs, so that the product of $x - (a + b\sqrt{-1})$ and $x - (a - b\sqrt{-1})$, wherein $a + b\sqrt{-1}$ and $a - b\sqrt{-1}$ are a conjugate pair, gives a quadratic expression with real coefficients. Euler then showed that this was true for Bernoulli's example. But Goldbach, too, rejected the idea that every polynomial with real coefficients can be factored into real factors and gave the example $x^4 + 72x - 20$. Euler then showed Goldbach that the latter had made a mistake and that he (Euler) had proved his theorem for polynomials up to the sixth degree. However, Goldbach was not convinced, because Euler did not succeed in giving a general proof of his assertion.

The kernel of the problem of factoring a real polynomial into linear and quadratic factors with real coefficients was to prove that every such polynomial had at least one real or complex root. The proof of this fact, called the fundamental theorem of algebra, became a major goal.

Proofs offered by d'Alembert and Euler were incomplete. In 1772[1] Lagrange, in a long and detailed argument, "completed" Euler's proof. But Lagrange, like Euler and his contemporaries, applied freely the ordinary properties of numbers to what were supposedly the roots without establishing that the roots must at worst be complex numbers. Since the nature of the roots was unknown, the proof was actually incomplete.

The first substantial proof of the fundamental theorem, though not rigorous by modern standards, was given by Gauss in his doctoral thesis of 1799 at Helmstädt.[2] He criticized the work of d'Alembert, Euler, and Lagrange and then gave his own proof. Gauss's method was not to calculate a root but to demonstrate its existence. He points out that the complex roots $a + ib$ of $P(x + iy) = 0$ correspond to points (a, b) of the plane, and if $P(x + iy) = u(x, y) + iv(x, y)$ then (a, b) must be an intersection of the curves $u = 0$ and $v = 0$. By a qualitative study of the curves he shows that a continuous arc of one joins the points of two distinct regions separated by the other. Then the curve $u = 0$ must cut the curve $v = 0$. The argument was highly original. However, he depended on the graphs of these curves, which were somewhat complicated, to show that they must cross. In this same paper Gauss proved that the nth degree polynomial can be expressed as a product of linear and quadratic factors with real coefficients.

Gauss gave three more proofs of the theorem. In the second proof[3] he dispensed with geometrical arguments. In this proof he also showed that the product of the differences of each two roots (which we, following Sylvester, call the discriminant) can be expressed as a linear combination of the polynomial and its derivative, so that a necessary and sufficient condition that the polynomial and its derivative have a common root is that the discriminant

1. *Nouv. Mém. de l'Acad. de Berlin*, 1772, 222 ff. = *Œuvres*, 3, 479–516.
2. *Werke*, 3, 1–30; also reproduced in Euler, *Opera*, (1), 6, 151–69.
3. *Comm. Soc. Gott.*, 3, 1814/15, 107–42 = *Werke*, 3, 33–56.

vanish. However, this second proof assumed that a polynomial cannot change signs at two different values of x without vanishing in between. The proof of this fact lay beyond the rigor of the time.

The third proof[4] used in effect what we now call the Cauchy integral theorem (Chap. 27, sec. 4).[5] The fourth proof[6] is a variation of the first as far as method is concerned. However, in this proof Gauss uses complex numbers more freely because, he says, they are now common knowledge. It is worth noting that the theorem was, in many proofs, not proved in all generality. Gauss's first three proofs and later proofs by Cauchy, Jacobi, and Abel assumed that the (literal) coefficients represented real numbers, whereas the full theorem includes the case of complex coefficients. Gauss's fourth proof did allow the coefficients of the polynomial to be complex numbers.

Gauss's approach to the fundamental theorem of algebra inaugurated a new approach to the entire question of mathematical existence. The Greeks had wisely recognized that the existence of mathematical entities must be established before theorems about them can be entertained. Their criterion of existence was constructibility. In the more explicit formal work of the succeeding centuries, existence was established by actually obtaining or exhibiting the quantity in question. For example, the existence of the solutions of a quadratic equation is established by exhibiting quantities that satisfy the equation. But in the case of equations of degree higher than four, this method is not available. Of course a proof of existence such as Gauss's may be of no help at all in computing the object whose existence is being established.

While the work that ultimately showed that every polynomial equation with real coefficients has at least one root was under way, the mathematicians also pushed hard to solve equations of degree higher than four by algebraic processes. Leibniz and his friend Tschirnhausen were the first to make serious efforts. Leibniz[7] reconsidered the irreducible case of the third degree equation and convinced himself that one could not avoid the use of complex numbers to solve this type. He then tackled the solution of the fifth degree equation but without success. Tschirnhausen[8] thought he had solved this problem by transforming the given equation into a new one by means of a transform $y = P(x)$, where $P(x)$ is a suitable fourth degree polynomial.

4. *Comm. Soc. Gott.*, 3, 1816 = *Werke*, 3, 59–64.

5. For a discussion of Gauss's third proof see M. Bocher, "Gauss's Third Proof of the Fundamental Theorem of Algebra," *Amer. Math. Soc., Bull.*, 1, 1895, 205–9. A translation of the third proof can be found in H. Meschkowski, *Ways of Thought of Great Mathematicians*, Holden-Day, 1964.

6. *Abhand. der Ges. der Wiss. zu Gött.*, 4, 1848/50, 3–34 = *Werke*, 3, 73–102.

7. *Der Briefwechsel von Gottfried Wilhelm Leibniz mit Mathematikern*, Georg Olms (reprint), 1961, Vol. 1, 547–64.

8. *Acta Erud.*, 2, 1683, 204–07.

This transformation eliminated all but the x^5 and constant terms of the equation. But Leibniz showed that to determine the coefficients of $P(x)$ one had to solve equations of degree higher than five, and so the method was useless.

For a while the problem of solving the nth degree equation centered on the special case $x^n - 1 = 0$, called the binomial equation. Cotes and De Moivre showed, through the use of complex numbers, that the solution of this problem amounts to the division of the circumference of a circle into n equal parts. To obtain the roots by radicals (trigonometric solutions are not necessarily algebraic) it is sufficient to consider the case of n an odd prime, because if $n = pm$, where p is a prime, then one can consider $(x^m)^p - 1 = 0$. If this equation can be solved for x^m, then $x^m - A$ can be considered where A is any of the roots of the preceding equation. Alexandre-Théophile Vandermonde (1735–96) affirmed in a paper of 1771 [9] that every equation of the form $x^n - 1 = 0$, where n is a prime, is solvable by radicals. However, Vandermonde merely verified that this is so for values of n up to 11. The decisive work on binomial equations was done by Gauss (Chap. 31, sec. 2).

The major effort toward solving equations of degree higher than four concentrated on the general equation, and toward this end some subsidiary work on symmetric functions proved important. The expression $x_1x_2 + x_2x_3 + x_3x_1$ is a symmetric function of x_1, x_2, and x_3 because replacement throughout of any x_i by an x_j and x_j by x_i leaves the entire expression unaltered. The interest in symmetric functions arose when the seventeenth-century algebraists noted and Newton proved that the various sums of the products of the roots of a polynomial equation can be expressed in terms of the coefficients. For example, for $n = 3$, the sum of the products taken two at a time,

$$a_1a_2 + a_2a_3 + a_3a_1$$

is an elementary symmetric function; and if the equation is written as

$$x^3 - c_1x^2 + c_2x - c_3 = 0,$$

the sum equals c_2. The progress made by Vandermonde in the 1771 paper was to show that any symmetric function of the roots can be expressed in terms of the coefficients of the equation.

The outstanding work of the eighteenth century on the problem of the solution of equations by radicals, after efforts by many men including Euler,[10] was by Vandermonde in the 1771 paper and Lagrange, in his massive paper "Réflexions sur la résolution algébrique des équations."[11] Vandermonde's

9. *Hist. de l'Acad. des Sci.*, Paris, 1771, 365–416, pub. 1774.
10. *Comm. Acad. Sci. Petrop.*, 6, 1732/3, 216–31, pub. 1738 = *Opera*, (1), 6, 1–19 and *Novi Comm. Acad. Sci. Petrop.*, 9, 1762/63, 70–98, pub. 1764 = *Opera*, (1), 6, 170–96.
11. *Nouv. Mém. de l'Acad. de Berlin*, 1770, 134–215, pub. 1772 and 1771, 138–254, pub. 1773 = *Œuvres*, 3, 205–421.

ideas are similar but not so extensive nor so clear. We shall therefore present Lagrange's version. Lagrange set himself the task of analyzing the methods of solving third and fourth degree equations, to see why these worked and to see what clue these methods might furnish for solving higher-degree equations.

For the third degree equation

(1) $$x^3 + nx + p = 0$$

Lagrange noted that if one introduces the transformation (Chap. 13, sec. 4)

(2) $$x = y - (n/3y),$$

one obtains the auxiliary equation

(3) $$y^6 + py^3 - n^3/27 = 0.$$

This equation is also known as the reduced equation because it is quadratic in y^3 and with $r = y^3$ becomes

(4) $$r^2 + pr - n^3/27 = 0.$$

Now we see that we can calculate the roots r_1 and r_2 of this equation in terms of the coefficients of the original one. But to go back to y from r we must introduce cube roots or solve

$$y^3 - r = 0.$$

Then if we let ω be the particular cube root of unity $(-1 + \sqrt{-3})/2$, the values of y are

$$\sqrt[3]{r_1}, \; \omega\sqrt[3]{r_1}, \; \omega^2\sqrt[3]{r_1}, \; \sqrt[3]{r_2}, \; \omega\sqrt[3]{r_2}, \; \omega^2\sqrt[3]{r_2}$$

and the distinct solutions of (1) are

$$x_1 = \sqrt[3]{r_1} + \sqrt[3]{r_2}, \qquad x_2 = \omega\sqrt[3]{r_1} + \omega^2\sqrt[3]{r_2}, \qquad x_3 = \omega^2\sqrt[3]{r_1} + \omega\sqrt[3]{r_2}.$$

Thus the solutions of the original equation are obtained in terms of those of the reduced equation.

Lagrange showed that the different procedures utilized by his predecessors all amount to the above method. Then he pointed out that we should turn our attention not to x as a function of the y-values but to y as a function of x because it is the reduced equation that permits us to solve at all, and the secret must lie in the relation that expresses the solutions of the reduced equation in terms of the solutions of the proposed equation.

Lagrange noted that each of the y-values can be written (since $1 + \omega + \omega^2 = 0$) in the form

(5) $$y = \frac{1}{3}(x_1 + \omega x_2 + \omega^2 x_3)$$

when x_1, x_2, and x_3 are taken in particular orders. Examination of this expression enables us to perceive two properties of the reduced equation in y. First, the roots x_1, x_2, and x_3 in the expression for y are not one fixed choice for x_1, x_2, and x_3, and the expression is, so to speak, ambiguous. Thus any of the three x-values can be x_1, any of the other two x_2, etc. But there are 3! permutations of the x's. Hence there are six values for y, and y should satisfy a sixth degree equation. Thus the degree of the reduced equation is determined by the number of permutations of the roots in the proposed equation.

In the second place, the relation (5) also shows why one can reduce the sixth degree reduced equation to the second degree equation. For among the six permutations, three (including the identity) come from interchanging all the x_i and three from interchanging just two and keeping one fixed. But then in view of the value of ω, the six values of y that result are related by

(6) $$y_1 = \omega^2 y_2 = \omega y_3; \qquad y_4 = \omega^2 y_5 = \omega y_6$$

and by cubing,

$$y_1^3 = y_2^3 = y_3^3, \qquad y_4^3 = y_5^3 = y_6^3.$$

Another way of putting this result is that the function

$$(x_1 + \omega x_2 + \omega^2 x_3)^3$$

can take on only two values under all six permutations of x_1, x_2, and x_3, and this is why the equation that y satisfies proves to be a quadratic in y^3. Moreover the coefficients of the sixth degree equation that y satisfies are rational functions of the coefficients of the original cubic.

In the case of the general fourth degree equation in x, Lagrange considers

$$y = x_1 x_2 + x_3 x_4.$$

This function of the four roots takes on only three distinct values under all 24 possible permutations of the four roots. Hence there should be a third degree equation that y satisfies and the coefficients of this equation should be rational functions of those of the original equation. These statements do apply to the fourth degree equation.

Lagrange then takes up the general nth degree equation

(7) $$x^n + a_1 x^{n-1} + \cdots + a_{n-1} x + a_n = 0.$$

The coefficients of this equation are assumed to be independent; that is, there must be no relation holding among the a_i. Then the roots must also be independent, because if there should be a relation among the roots one could show that this must be true of the coefficients (since, essentially, the coefficients are symmetric functions of the roots). Thus the n roots of the general equation must be considered as independent variables, and each function of them is a function of independent variables.

To understand Lagrange's plan of solution let us consider first

$$x^2 + bx + c = 0.$$

We know two functions of the roots, namely, $x_1 + x_2$ and x_1x_2. These are symmetric functions, that is, they do not change when the roles of the roots are exchanged. When a function does not change under a permutation performed on its variables, one says that the function admits the permutation. Thus the function $x_1 + x_2$ admits the permutation of x_1 and x_2 but the function $x_1 - x_2$ does not.

Lagrange then demonstrated two important propositions. If a function $\phi(x_1, x_2, \ldots, x_n)$ of the roots of the general equation of degree n admits all the permutations of the x_i that another function $\psi(x_1, x_2, \ldots, x_n)$ admits (and possibly other permutations that ψ does not admit), the function ϕ can be expressed rationally in terms of ψ and the coefficients of the general equation (7). Thus the function x_1 of the quadratic admits all the permutations which $x_1 - x_2$ admits, (there is just one, the identity); then

$$x_1 = \frac{-b + (x_2 - x_1)}{2}.$$

Lagrange's proof of this proposition also shows how to express ϕ as a rational function of ψ.

Lagrange's second proposition states: If a function $\phi(x_1, x_2, \ldots, x_n)$ of the roots of the general equation does not admit all the permutations admitted by a function $\psi(x_1, x_2, \ldots, x_n)$, but takes on under the permutations that ψ admits r different values, then ϕ is a root of an equation of degree r whose coefficients are rational functions of ψ and the coefficients of the given general equation of degree n. This equation of degree r can be constructed. Thus $x_1 - x_2$ does not admit all the permutations $x_1 + x_2$ admits but takes on the two values $x_1 - x_2$ and $x_2 - x_1$. Then $x_1 - x_2$ is a root of a second degree equation whose coefficients are rational functions of $x_1 + x_2$ and of b and c. As a matter of fact, $x_1 - x_2$ is a root of

$$t^2 - (b^2 - 4c) = 0$$

because $b^2 - 4ac = (x_1 - x_2)^2$. With the value of this root, namely $\sqrt{b^2 - 4c}$, we can find x_1 by means of the preceding equation for x_1.

Likewise for the equation $x^3 + px + q = 0$ the expression (the ϕ)

$$(x_1 + \omega x_2 + \omega^2 x_3)^3,$$

where $\omega = (-1 + i\sqrt{3})/2$, takes on two values under the six possible permutations of the roots, whereas $x_1 + x_2 + x_3$ (the ψ) admits all six permutations. If the two values are denoted by A and B, one can show that A and B are roots of a second degree equation whose coefficients are rational in p and q (since $x_1 + x_2 + x_3 = 0$). If we solve the second degree equation and the roots are A and B, we can then find x_1, x_2, and x_3 from

$$x_1 + x_2 + x_3 = 0$$
$$x_1 + \omega x_2 + \omega^2 x_3 = \sqrt[3]{A}$$
$$x_1 + \omega^2 x_2 + \omega x_3 = \sqrt[3]{B}.$$

For the fourth degree equation, Lagrange started with the function

(8) $x_1 x_2 + x_3 x_4,$

which takes on three different values under the 24 possible permutations of the roots, whereas $x_1 + x_2 + x_3 + x_4$ admits all 24. Hence (8) is a root of a third degree equation whose coefficients are rational functions of those of the original equation. And in fact the auxiliary or reduced equation for the general quartic is of the third degree.

For the nth degree equation with general coefficients, Lagrange's idea was to start with a symmetric function ϕ_0 of the roots which admits all $n!$ permutations of the roots. Such a function, he points out, could be $x_1 + x_2 + \cdots + x_n$. Then he would choose a function ϕ_1 which admits only some of the substitutions. Suppose ϕ_1 takes on under the $n!$ permutations, say, r different values. ϕ_1 will then be a root of an equation of degree r whose coefficients are rational functions of ϕ_0 and the coefficients of the given general equation. The equation of degree r can be constructed. Further, if ϕ_0 is taken to be one of the symmetric functions that relates roots and coefficients, then the coefficients of the equation of degree r are known entirely in terms of coefficients of the given general equation. If the equation of degree r can be solved algebraically, then ϕ_1 will be known in terms of the coefficients of the original equation. Then a function ϕ_2 is chosen that admits only some of the permutations ϕ_1 admits. ϕ_2 may take on, say, s different values under the substitutions that ϕ_1 admits. Then ϕ_2 will be a root of an equation of degree s whose coefficients are rational functions of ϕ_1 and of the coefficients of the given general equation. The coefficients of this sth degree equation will be known if the rth degree equation which has ϕ_1 as one of its roots can be solved. If the sth degree equation can be solved algebraically, then ϕ_2 will be known in terms of the coefficients of the original equation.

One continues thus to ϕ_3, ϕ_4, ... until the final function which is chosen to be x_1. If then the equations of degree r, s, ... can be solved algebraically, one will know x_1 in terms of the coefficients of the given general equation. The other roots x_2, x_3, ..., x_n come out of the same process. The equations of degree r, s, ..., are today called resolvent equations.[12]

12. Lagrange used the word "resolvent" for the special forms of the ϕ_i functions and not for the equations which the ϕ_i satisfy. Thus for the cubic equation, $x_1 + \omega x_2 + \omega^2 x_3$ is one form of Lagrange's resolvent.

Lagrange's method worked for the general second, third, and fourth degree equations. He tried to solve the fifth degree equation in this way but found the work so difficult that he abandoned it. Whereas for the cubic he had but to solve an equation of the second degree, for the quintic he had to solve a sixth degree equation. Lagrange sought in vain to find a resolving function (in his sense of the term) that would satisfy an equation of degree less than five. However his work did not give any criterion for picking ϕ_i's that would satisfy algebraically solvable equations. Also his method applied only to the general equation because his two basic propositions assume that the roots are independent.

Lagrange was drawn to the conclusion that the solution of the general higher-degree equation (for $n > 4$) by algebraic operations was likely to be impossible. (For special equations of higher degree he offered little.) He decided that either the problem was beyond human capacities or the nature of the expressions for the roots must be different from all those thus far known. Gauss, too, in his *Disquisitiones* of 1801, declared that the problem could not be solved.

Lagrange's method, despite its lack of success, does give insight into the reason for the successes when $n \leq 4$ and failures when $n > 4$; this insight was capitalized on by Abel and Galois (Chap. 31). Moreover, Lagrange's idea that one must consider the number of values that a rational function takes on when its variables are permuted led to the theory of permutation or substitution groups. In fact, he had in effect the theorem that the order of a subgroup (the number of elements) must be a divisor of the order of the group. In Lagrange's work, which preceded any work on group theory, this result takes the form that the number r of values that ϕ_1 takes on is a divisor of $n!$.

Influenced by Lagrange, Paolo Ruffini (1765–1822), a mathematician, doctor, politician, and an ardent disciple of Lagrange, made several attempts during the years 1799 to 1813 to prove that the general equation of degree higher than the fourth could not be solved algebraically. In his *Teoria generale delle equazioni*,[13] Ruffini succeeded in proving by Lagrange's own method that no resolving function (in Lagrange's sense) existed that would satisfy an equation of degree less than five. In fact he demonstrated that no rational function of n elements existed that took on 3 or 4 values under permutations of the n elements when $n > 4$. Then he boldly undertook to prove, in his *Riflessioni intorno alla soluzione delle equazioni algebraiche generali*,[14] that the algebraic solution of the general equation of degree $n > 4$ was impossible. This effort was not conclusive, though at first Ruffini believed it was correct. Ruffini used but did not prove the auxiliary theorem, now known as Abel's theorem, that if an equation is solvable by radicals, the

13. 1799 = *Opere Mat.*, 1, 1–324.
14. 1813 = *Opere Mat.*, 2, 155–268.

expressions for the roots can be given such a form that the radicals in them are rational functions with rational coefficients of the roots of the given equation and the roots of unity.

3. Determinants and Elimination Theory

The study of a system of linear equations which we write today as

$$(9) \qquad x_i = \sum_{j=1}^{n} a_{ij} y_j, \qquad i = 1, 2, \ldots, m$$

wherein the x_i are known and the y_j are the unknowns, was initiated before 1678 by Leibniz. In 1693 Leibniz[15] used a systematic set of indices for the coefficients of a system of three linear equations in two unknowns x and y. He eliminated the two unknowns from the system of three linear equations and obtained a determinant, now called the resultant of the system. The vanishing of this determinant expresses the fact that there is an x and y satisfying all three equations.

The solution of simultaneous linear equations in two, three, and four unknowns by the method of determinants was created by Maclaurin, probably in 1729, and published in his posthumous *Treatise of Algebra* (1748). Though not as good in notation, his rule is the one we use today and which Cramer published in his *Introduction à l'analyse des lignes courbes algébriques* (1750). Cramer gave the rule in connection with determining the coefficients of the general conic, $A + By + Cx + Dy^2 + Exy + x^2 = 0$, passing through five given points. His determinants were, as at present, the sum of the products formed by taking one and only one element from each row and column, with the sign of each product determined by the number of derangements of the elements from a standard order, the sign being positive if this number was even and negative if odd. In 1764, Bezout[16] systematized the process of determining the signs of the terms of a determinant. Given n homogeneous linear equations in n unknowns, Bezout showed that the vanishing of the determinant of the coefficients (the vanishing of the resultant) is the condition that non-zero solutions exist.

Vandermonde[17] was the first to give a connected and logical exposition of the theory of determinants as such,—that is, apart from the solution of linear equations,—though he, too, did apply them to the solution of linear equations. He also gave a rule for expanding a determinant by using second order minors and their complementary minors. In the sense that he concentrated on determinants, he is the founder of the theory.

15. *Math. Schriften,* 2, 229, 238–40, 245.
16. *Hist. de l'Acad. des Sci. Paris,* 1764, 288–388.
17. *Mém. de l'Acad. des Sci., Paris,* 1772, 516–32, pub. 1776.

In a paper of 1772, "Recherches sur le calcul intégral et sur le système du monde,"[18] Laplace, who referred to Cramer's and Bezout's work, proved some of Vandermonde's rules and generalized his method of expanding determinants by using a set of minors of r rows and the complementary minors. This method is still known by his name.[19]

The condition that, say, a set of three linear nonhomogeneous equations in two unknowns have a common solution, the vanishing of the resultant, also expresses the result of eliminating x and y from the three equations. But the problem of elimination extends in other directions. Given two polynomials

$$f = a_0 x^m + \cdots + a_n$$
$$g = b_0 x^n + \cdots + b_n,$$

one can ask for the condition that $f = 0$ and $g = 0$ have a common root. Since the condition involves the fact that at least one value of x satisfying $f = 0$ also satisfies $g = 0$, the substitution of that value of x obtained from $f = 0$ in $g = 0$ should result in a condition on the a_i and b_i. This condition, or eliminant, or resultant, was first investigated by Newton. In his *Arithmetica Universalis* he gave rules for eliminating x from two equations, which could be of degrees two to four.

Euler, in Chapter 19 of the second volume of his *Introductio*, gives two methods of elimination. The second is the forerunner of Bezout's multiplier method and was better described by Euler in a paper of 1764.[20] Bezout's method proved to be the most widely accepted one, and we shall therefore examine it. In his *Cours de mathématique* (1764–69), he considers two equations of degree n,

(10)
$$f(x) = a_n x^n + a_{n-1} x^{n-1} + \cdots + a_0 = 0$$
$$\phi(x) = b_n x^n + b_{n-1} x^{n-1} + \cdots + b_0 = 0.$$

One multiplies f by b_n and ϕ by a_n and subtracts; next one multiplies f by $b_n x + b_{n-1}$ and ϕ by $a_n x + a_{n-1}$ and subtracts; then f is multiplied by $b_n x^2 + b_{n-1} x + b_{n-2}$ and ϕ by $a_n x^2 + a_{n-1} x + a_{n-2}$ and again one subtracts; etc. Each of the equations so obtained is of degree $n - 1$ in x. One can regard this set of equations as a system of n linear homogeneous equations in the unknowns x^{n-1}, x^{n-2}, $\ldots$, 1. The resultant of this system of linear equations, which is the determinant of the coefficients of the unknowns, is the resultant of the original two, $f = 0$ and $\phi = 0$. Bezout also gave a method of finding the resultant when the two equations are not of the same degree.[21]

18. *Mém. de l'Acad. des Sci.*, Paris, 1772, 267–376, pub. 1776 = *Œuvres*, 8, 365–406.
19. See M. Bocher, *Introduction to Higher Algebra*, Dover (reprint), 1964, p. 26.
20. *Mém. de l'Acad. de Berlin*, 20, 1764, 91–104, pub. 1766 = *Opera*, (1), 6, 197–211.
21. An exposition can be found in W. S. Burnside and A. W. Panton, *The Theory of Equations*, Dover (reprint), 1960, Vol. 2, p. 76.

Elimination theory was also applied to two equations, $f(x, y) = 0$ and $g(x, y) = 0$, of degree higher than 1. The motivation was to establish the number of common solutions of the two equations or, geometrically, to find the number of intersections of the curves corresponding to the equations. The outstanding method for eliminating one unknown from $f(x, y) = 0$ and $g(x, y) = 0$, sketched first by Bezout in a paper of 1764, was presented by him in his *Théorie générale des équations algébriques* (1779). Bezout's idea was that by multiplying $f(x, y)$ and $g(x, y)$ by suitable polynomials, $F(x)$ and $G(x)$ respectively, he could form

(11) $$R(y) = F(x)f(x, y) + G(x)g(x, y).$$

Moreover he sought an F and G such that the degree of $R(y)$ would be as small as possible.

The question of the degree of the eliminant was also answered by Bezout in his *Théorie* (and independently by Euler in the 1764 paper). Both gave the answer as mn, the product of the two degrees of f and g, and both proved this theorem by reducing the problem to one of elimination from an auxiliary set of linear equations. This product is the number of points of intersection of the two algebraic curves. Jacobi[22] and Minding[23] also gave Bezout's method of elimination for the two equations, but neither mentions Bezout. It could be that Bezout's work was not known to them.

4. *The Theory of Numbers*

The theory of numbers in the eighteenth century remained a series of disconnected results. The most important works in the subject were Euler's *Anleitung zur Algebra* (German edition, 1770) and Legendre's *Essai sur la théorie des nombres* (1798). The second edition appeared in 1808 under the title *Théorie des nombres*, and an expanded third edition in two volumes appeared in 1830. The problems and results to be described are a small sample of what was done.

In 1736 Euler established[24] the minor theorem of Fermat, namely, if p is a prime and a any positive integer, $a^p - a$ is divisible by p. Many proofs of this same theorem were given by other men in the eighteenth and nineteenth centuries. In 1760 Euler generalized this theorem[25] by introducing the ϕ function or totient of n; $\phi(n)$ is the number of integers $< n$ and prime to n so that if n is a prime, $\phi(n)$ is $n - 1$. [The notation $\phi(n)$ was introduced by Gauss.] Then Euler proved that if a is relatively prime to n,

$$a^{\phi(n)} - 1$$

is divisible by n.

22. *Jour. für Math.*, 15, 1836, 101–24 = *Gesam. Werke*, 3, 297–320.
23. *Jour. für Math.*, 22, 1841, 178–83.
24. *Comm. Acad. Sci. Petrop.*, 8, 1736, 141–46, pub. 1741 = *Opera*, (1), 2, 33–37.
25. *Novi Comm. Acad. Sci. Petrop.*, 8, 1760/1, 74–104, pub. 1763 = *Opera*, (1), 3, 531–55.

As for the famous Fermat conjecture on $x^n + y^n = z^n$, Euler proved[26] this was correct for $n = 3$ and $n = 4$; the case $n = 4$ had been proven by Frénicle de Bessy. This work of Euler had to be completed by Lagrange, Legendre, and Gauss. Legendre then proved[27] the conjecture for $n = 5$. As we shall see, the history of the efforts to prove Fermat's conjecture is extensive.

Fermat had also conjectured (Chap. 13, sec. 7) that the formula

$$2^{2^n} + 1$$

yielded primes for an unspecified set of values of n. This is true for $n = 0$, 1, 2, 3, and 4. Fermat did not claim it was always true, and Euler showed in 1732[28] that 641 was a factor when $n = 5$. In fact the formula is now known not to yield primes for many other values of n, whereas no other value beyond 4 has been found for which the formula does yield primes. However, the formula is of interest in that it reappears in the work of Gauss on the constructibility of regular polygons (Chap. 31, sec. 2).

A topic with many subdivisions concerns the decomposition of integers of various types into other classes of integers. Fermat had affirmed that each positive integer is the sum of at most four squares (repetition of a square as in $8 = 4 + 4$ is permitted if it is counted the number of times it appears). Over a period of forty years, Euler kept trying to prove this theorem and gave partial results.[29] Using some of Euler's work, Lagrange[30] proved the theorem. Neither Euler nor Lagrange obtained the number of representations.

Euler, in the 1754/5 paper just referred to and in another paper in the same journal[31] proved Fermat's assertion that every prime of the form $4n + 1$ is decomposable uniquely into a sum of two squares. However, Euler's proof did not follow the method of descent that Fermat had sketched for this theorem. In another paper[32] Euler also proved that every divisor of the sum of two relatively prime squares is the sum of two squares.

Edward Waring (1734–98) stated, in his *Meditationes Algebraicae* (1770), the theorem known now as "Waring's theorem," that every integer is either a cube or the sum of at most nine cubes; also every integer is either a fourth power or the sum of at most 19 fourth powers. He conjectured also that every positive integer can be expressed as the sum of at most r kth powers, the r depending upon k. These theorems were not proven by him.[33]

26. *Algebra*, Part II, Second Section, 509–16 = *Opera*, (1), 1, 484–89 (for $n = 3$); and *Comm. Acad. Sci. Petrop.*, 10, 1738, 125–46, pub. 1747 = *Opera*, (1), 2, 38–59 (for $n = 4$).
27. *Mém. de l'Acad. des Sci.*, Paris, 6, 1823, 1–60, pub. 1827.
28. *Comm. Acad. Sci Petrop.*, 6, 1732/3, 103–07 = *Opera*, (1), 2, 1–5.
29. *Novi Comm. Acad. Sci. Petrop.*, 5, 1754/5, 13–58, pub. 1760 = *Opera*, (1), 2, 338–72.
30. *Nouv. Mém. de l'Acad. de Berlin*, 1, 1770, 123–33, pub. 1772 = *Œuvres*, 3, 189–201.
31. 5, 1754/5, 3–13, pub. 1760 = *Opera*, (1), 2, 328–37.
32. *Novi Comm. Acad. Sci. Petrop.*, 4, 1752/53, 3–40, pub. 1758 = *Opera*, (1) 2, 295–327.
33. The general theorem was proven by David Hilbert (*Math. Ann.*, 67, 1909, 281–300).

In a letter to Euler of June 7, 1742, Christian Goldbach, a Prussian envoy to Russia, stated without proof that every even integer is the sum of two primes and every odd integer is either a prime or a sum of three primes. The first part of the assertion is now known as Goldbach's conjecture and is still an unsolved problem. The second assertion actually follows from the first, because if n is odd substract any prime p from it. Then $n - p$ is even.

Among somewhat more specialized results dealing with the decomposition of numbers is Euler's proof that $x^4 - y^4$ and $x^4 + y^4$ cannot be squares.[34] Euler and Lagrange proved many of Fermat's assertions to the effect that certain primes can be expressed in particular ways. For example Euler proved[35] that a prime of the form $3n + 1$ can be expressed uniquely in the form $x^2 + 3y^2$.

Amicable and perfect numbers continued to interest the mathematicians. Euler[36] gave 62 pairs of amicable numbers, including the three pairs already known. Two of his pairs were incorrect. He also proved in a posthumously published paper[37] the converse of Euclid's theorem: Every even perfect number is of the form $2^{p-1}(2^p - 1)$ where the second factor is prime.

John Wilson (1741–93), who had been a prize student of mathematics at Cambridge but became a lawyer and judge, stated a theorem still named after him: For every prime p, the quantity $(p - 1)! + 1$ is divisible by p; in addition, if the quantity is divisible by p, then p is a prime. Waring published the statement in his *Meditationes Algebraicae* and Lagrange proved it in 1773.[38]

The problem of solving the equation $x^2 - Ay^2 = 1$ in integers has already been discussed (Chap. 13, sec. 7). Euler, in a paper of 1732/33, erroneously called it Pell's equation, and the name has stuck. He became interested in this equation because he needed its solutions to solve $ax^2 + bx + c = y^2$ in integers; he wrote several papers on this last theme. In 1759 Euler gave a method of solving the Pellian equation by expressing $\sqrt{A}$ as a continued fraction.[39] Euler's idea was that the values x and y which satisfy the equation are such that the ratios x/y are convergents (in the sense of continued fractions) to $\sqrt{A}$. He failed to prove that his method always gives solutions and that all its solutions are given by the continued fraction

34. *Comm. Acad. Sci. Petrop.*, 10, 1738, 125–46, pub. 1747 = *Opera*, (1), 2, 38–59; also in *Algebra* (1770), Part II, Ch. 13, arts. 202–8 = *Opera*, (1), 1, 436–43.

35. *Novi Comm. Acad. Sci. Petrop.*, 8, 1760/1, 105–28, pub. 1763 = *Opera*, (1), 2, 556–75.

36. "De numeris amicabilibus," *Opuscula varii argumenti*, 2, 1750, 23–107 = *Opera*, (1), 2, 86–162.

37. "De numeris amicabilibus," *Comm. Arith.*, 2, 1849, 627–36 = *Opera postuma*, 1, 1862, 85–100 = *Opera*, (1), 5, 353–65.

38. *Nouv. Mém. de l'Acad. de Berlin*, 2, 1771, 125 ff., pub. 1773 = *Œuvres*, 3, 425–38.

39. *Novi Comm. Acad. Sci. Petrop.*, 11, 1765, 28–66, pub. 1767 = *Opera*, (1), 3, 73–111.

development of $\sqrt{A}$. The existence of solutions of Pell's equation was shown in 1766 by Lagrange[40] and then more simply in later papers.[41]

Fermat had asserted that he could determine when the more general equation $x^2 - Ay^2 = B$ was solvable in integers and that he could solve it when solvable. The equation was solved by Lagrange in the two papers just mentioned.

The problem of giving all integral solutions of the general equation

$$ax^2 + 2bxy + cy^2 + 2dx + 2ey + f = 0,$$

where the coefficients are integers, was also tackled. Euler gave incomplete classes of solutions; then Lagrange[42] gave the complete solution. In the next volume of the *Mémoires*[43] he gave a simpler proof.

The most original and perhaps the most consequential discovery of the eighteenth century in the theory of numbers is the law of quadratic reciprocity. It uses the notion of quadratic remainders or residues. In the language introduced by Euler in the 1754/55 paper and adopted by Gauss, if there exists an x such that $x^2 - p$ is divisible by q, then p is said to be a quadratic residue of q; if there is no such x, p is said to be a quadratic nonresidue of q. Legendre (1808) invented a symbol that is now used to represent either state of affairs. The symbol is (p/q) and means the following: For any number p and any prime q,

$$(p/q) = \begin{cases} 1 & \text{if } p \text{ is a quadratic residue of } q \\ -1 & \text{if } p \text{ is a quadratic nonresidue of } q. \end{cases}$$

It is also understood that $(p/q) = 0$ if p divides evenly into q.

The law of quadratic reciprocity, in symbolic form, states that if p and q are distinct odd primes, then

$$(p/q)(q/p) = (-1)^{(p-1)(q-1)/4}.$$

This means that if the exponent of (-1) is even, p is a quadratic residue of q and q is a quadratic residue of p or neither is a quadratic residue of the other. When the exponent is odd, which occurs when p and q are of the form $4k + 3$, one prime will be a quadratic residue of the other but not the second of the first.

The history of this law is detailed. Euler in a paper of 1783[44] gave four theorems and a fifth summarizing theorem that states the law of quadratic reciprocity very clearly. However, he did not prove these theorems. The

40. *Misc. Taur.*, 4, 1766/69, 19 ff. = *Œuvres*, 1, 671–731.
41. *Mém. de l'Acad. de Berlin*, 23, 1767, 165–310, pub. 1769, and 24, 1768, 181–256, pub. 1770 = *Œuvres*, 2, 377–535 and 655–726; also in Lagrange's additions to his translation of Euler's *Algebra*; see the bibliography.
42. *Mém. de l'Acad. de Berlin*, 23, 1767, 165–310, pub. 1769 = *Œuvres*, 2, 377–535.
43. 24, 1768, 181–256, pub. 1770 = *Œuvres*, 2, 655–726.
44. *Opuscula Analytica*, 1, 1783, 64–84 = *Opera*, (1), 3, 497–512.

work on this paper dates from 1772 and incorporates even earlier work. Kronecker observed in 1875[45] that the statement of the law is actually contained in a much earlier paper by Euler.[46] However, Euler's "proof" was based on calculations. In 1785 Legendre announced the law independently, though he cites another paper by Euler, in the same volume of the *Opuscula*, in his own paper on the subject. His proof[47] was incomplete. In his *Théorie des nombres*[48] he again stated the law and gave another proof. However, this one, too, was incomplete, because he assumed that there are an infinite number of primes in certain arithmetic progressions. The desire to find what is behind the law and to derive its many implications has been a key theme in investigations since 1800 and has led to important discoveries, some of which we shall consider in a later chapter.

The eighteenth-century work on the theory of numbers closes with Legendre's classic *Théorie* of 1798. Though it contains a number of interesting results, in this domain as in others (such as the subject of elliptic integrals), Legendre made no great innovations. One could reproach him for presenting a collection of propositions from which general conceptions could have been extracted but were not. This was done by his successors.

Bibliography

Cajori, Florian: "Historical note on the Graphical Representation of Imaginaries Before the Time of Wessel," *Amer. Math. Monthly*, 19, 1912, 167–71.

Cantor, Moritz: *Vorlesungen über Geschichte der Mathematik*, B. G. Teubner, 1898 and 1924; Johnson Reprint Corp., 1965, Vol. 3, Chap. 107; Vol. 4, pp. 153–98.

Dickson, Leonard E.: *History of the Theory of Numbers*, 3 vols., Chelsea (reprint), 1951.

——: "Fermat's Last Theorem," *Annals of Math.*, 18, 1917, 161–87.

Euler, Leonhard: *Opera Omnia*, (1), Vols. 1–5, Orell Füssli, 1911–44.

——: *Vollständige Anleitung zur Algebra* (1770) = *Opera Omnia*, (1), 1.

Fuss, Paul H. von, ed.: *Correspondance mathématique et physique de quelques célèbres géomètres du XVIIIème siècle*, 2 vols. (1843), Johnson Reprint Corp., 1967.

Gauss, Carl Friedrich: *Werke*, Königliche Gesellschaft der Wissenschaften zu Göttingen, 1876, Vol. 3, pp. 3–121.

Gerhardt, C. I.: *Der Briefwechsel von Gottfried Wilhelm Leibniz mit Mathematikern*, Mayer und Müller, 1899; Georg Olms (reprint), 1962.

Heath, Thomas L.: *Diophantus of Alexandria*, 1910, Dover (reprint), 1964, pp. 267–380.

Jones, P. S.: "Complex Numbers: An Example of Recurring Themes in the

45. *Werke*, 2, 3–10.
46. *Comm. Acad. Sci. Petrop.*, 14, 1744/46, 151–81, pub. 1751 = *Opera*, (1), 2, 194–222.
47. *Hist. de l'Acad. des Sci.*, Paris, 1785, 465–559, pub. 1788.
48. 1798, 214–26; 2nd ed., 1808, 198–207.

Development of Mathematics," *The Mathematics Teacher*, 47, 1954, 106–14, 257–63, 340–45.

Lagrange, Joseph-Louis: *Œuvres*, Gauthier-Villars, 1867–69, Vols. 1–3, relevant papers.

————: "Additions aux éléments d'algèbre d'Euler," *Œuvres*, Gauthier-Villars, 1877, Vol. 7, pp. 5–179.

Legendre, Adrien-Marie, *Théorie des nombres*, 4th ed., 2 vols., A. Blanchard (reprint), 1955.

Muir, Thomas: *The Theory of Determinants in the Historical Order of Development*, 1906, Dover (reprint), 1960, Vol. 1, pp. 1–52.

Ore, Oystein: *Number Theory and its History*, McGraw-Hill, 1948.

Pierpont, James: "Lagrange's Place in the Theory of Substitutions," *Amer. Math. Soc. Bulletin*, 1, 1894/5, 196–204.

————: "Zur Geschichte der Gleichung des V. Grades (bis 1858)," *Monatshefte für Mathematik und Physik*, 6, 1895, 15–68.

Smith, H. J. S.: *Report on the Theory of Numbers*, 1867, Chelsea (reprint), 1965; also in Vol. 2 of the *Collected Mathematical Papers of H. J. S. Smith*, 1894, Chelsea (reprint), 1965.

Smith, David Eugene: *A Source Book in Mathematics*, 1929, Dover (reprint), 1959, Vol. 1, relevant selections. One of the selections is an English translation of Gauss's second proof of the fundamental theorem of algebra.

Struik, D. J.: *A Source Book in Mathematics, 1200–1800*, Harvard University Press, 1969, pp. 26–54, 99–122.

Vandiver, H. S.: "Fermat's Last Theorem," *Amer. Math. Monthly*, 53, 1946, 555–78.

Whiteside, Derek T.: *The Mathematical Works of Isaac Newton*, Johnson Reprint Corp., 1967, Vol. 2, pp. 3–134. This section contains Newton's *Universal Arithmetic* in English.

Wussing, H. L.: *Die Genesis der abstrakten Gruppenbegriffes*, VEB Deutscher Verlag der Wissenschaften, 1969.

26
Mathematics as of 1800

> When we cannot use the compass of mathematics or the torch
> of experience...it is certain that we cannot take a single step
> forward.   VOLTAIRE

1. *The Rise of Analysis*

If the seventeenth century has correctly been called the century of genius,
then the eighteenth may be called the century of the ingenious. Though both
centuries were prolific, the eighteenth-century men, without introducing
any concept as original and as fundamental as the calculus, but by exercising
virtuosity in technique, exploited and advanced the power of the calculus
to produce what are now major branches: infinite series, ordinary and
partial differential equations, differential geometry, and the calculus of
variations. In extending the calculus to these several areas, they built what
is now the most extensive domain of mathematics, which we call analysis
(though the word now includes a couple of other branches upon which the
eighteenth-century men barely touched). The progress in coordinate geometry and algebra, on the other hand, was hardly more than a minor
extension of what the seventeenth century had initiated. Even the major
problem of algebra, the solution of the nth degree equation, received attention because it was needed for analysis, as for example, in integration by the
method of partial fractions.

   During the first third or so of the century, geometrical methods were
used freely; but Euler and Lagrange, in particular, recognizing the greater
effectiveness of analytic methods, deliberately and gradually replaced
geometrical arguments by analytic ones. Euler's many texts showed how
analysis could be used. Toward the end of the century Monge did revive
pure geometry, though he used it largely to give intuitive meaning and
guidance to the work in analysis. Monge is often referred to as a geometer,
but this is because, working at a time when geometry had dried up, he put
new life into it by showing its importance, at least for the purposes just
described. In fact, in a paper published in 1786, he implicitly attached
greater importance to analysis, when he observed that geometry could make
progress because analysis could be used to study it. He, as well as the others,
did not, in the main, look for new geometrical ideas. The primary interest
and the end results were in analytical work.

The classic statement of the importance of analysis was made by Lagrange in his *Mécanique analytique*. He writes in his preface,

> We already have various treatises on Mechanics but the plan of this one is entirely new. I have set myself the problem of reducing this science [mechanics], and the art of solving the problems appertaining to it, to general formulas whose simple development gives all the equations necessary for the solutions of each problem.... No diagrams will be found in this work. The methods which I expound in it demand neither constructions nor geometrical or mechanical reasonings, but solely algebraic [analytic] operations subjected to a uniform and regular procedure. Those who like analysis will be pleased to see mechanics become a new branch of it, and will be obliged to me for having extended its domain.

Laplace, too, stressed the power of analysis. In his *Exposition du système du monde* he says,

> The algebraic analysis soon makes us forget the main object [of our researches] by focusing our attention on abstract combinations and it is only at the end that we return to the original objective. But in abandoning oneself to the operations of analysis, one is led by the generality of this method and the inestimable advantage of transforming the reasoning by mechanical procedures to results often inaccessible to geometry. Such is the fecundity of the analysis that it suffices to translate into this universal language particular truths in order to see emerge from their very expression a multitude of new and unexpected truths. No other language has the capacity for the elegance that arises from a long sequence of expressions linked one to the other and all stemming from one fundamental idea. Therefore the geometers [mathematicians] of this century convinced of its [analysis] superiority have applied themselves primarily to extending its domain and pushing back its bounds.[1]

A few features of the analysis warrant note. Newton's emphasis on the derivative and antidifferentiation was retained, so that the summation concept was rarely employed. On the other hand, Leibniz's *concept*, that is, the differential form of the derivative, and Leibniz's notation became standard, despite the fact that throughout the century the Leibnizian differentials had no precise meaning. The first differentials dy and dx of a function $y = f(x)$ were legitimized in the nineteenth century (Chap. 40, sec. 3), but the higher-order differentials, which the eighteenth-century men used freely, have not been put on a rigorous basis even today. The eighteenth century also continued the Leibnizian tradition of formal manipulation of analytical expressions and, in fact, accentuated this practice.

The importance assigned to analysis had implications that the eighteenth-century men did not appreciate. It furthered the separation of number

1. Book V, Chap. 5 = *Œuvres*, 6, 465–66.

from geometry and implicitly underscored the issue of the proper foundations for the number system, algebra, and analysis itself. This problem was to become critical in the nineteenth century. Peculiarly the eighteenth-century mathematicians still referred to themselves as geometers, the term in vogue during the preceding ages in which geometry had dominated mathematics.

2. *The Motivation for the Eighteenth-Century Work*

Far more than in any other century, the mathematical work of the eighteenth was directly inspired by physical problems. In fact, one can say that the goal of the work was not mathematics, but rather the solution of physical problems; mathematics was a means to physical ends. Laplace, though perhaps an extreme case, certainly regarded mathematics as just a tool for physics and was himself entirely concerned with its value for astronomy.

The major physical field was, of course, mechanics, and particularly celestial mechanics. Mechanics became the paradise for mathematics, much as Leonardo had predicted, because it suggested so many directions of research. So extensive was the concern of mathematics with the problems of mechanics that d'Alembert in the *Encyclopédie* and Denis Diderot (1713–84) in his *Pensées sur l'interprétation de la nature* (1754) wrote of a transition from the seventeenth-century age of mathematics to an age of mechanics. They believed, in fact, that the mathematical work of men such as Descartes, Pascal, and Newton was passé, and that mechanics was to be the major interest of mathematicians. Mathematics generally was useful, in their opinion, only so far as it served physics. The eighteenth century concentrated on the mechanics of discrete systems of masses and of continuous media. Optics was temporarily pushed into the background.

What was actually happening, however, was the opposite of what Diderot and d'Alembert maintained. The truer interpretation, certainly in light of future developments, was expressed by Lagrange when he said that those who liked analysis would be pleased to see mechanics become a new branch of it. Put more broadly, the program initiated by Galileo, and consciously pursued by Newton, of expressing the basic physical principles as quantitative mathematical statements and deducing new physical results by mathematical arguments, had been advanced immeasurably. Physics was becoming more and more mathematical, at least in those areas where the physical principles had become sufficiently well understood. The increasing incorporation of major branches of physics into mathematical frameworks established mathematical physics.

Mathematics not only began to embrace science, but, partly because there was no sharp separation between science and what we would call engineering, the mathematicians undertook technological problems as a matter of course. Euler, for example, worked on the design of ships, the action of sails, ballistics, cartography, and other practical problems. Monge took his work on excavation and fill and the design of windmill vanes as seriously as any problem of differential geometry or differential equations.

J. F. (Jean-Etienne) Montucla (1725–99), in his *Histoire des mathématiques* (2nd ed., 1799–1802) divided mathematics into two parts, the one "comprising those things that are pure and abstract, the other those that one calls compound, or more ordinarily physico-mathematics." His second part comprised fields that can be approached and treated mathematically, that is, mechanics, optics, astronomy, military and civil architecture, insurance, acoustics, and music. Under optics he included dioptrics and even optometry, catoptrics, and perspective. Mechanics included dynamics and statics, hydrodynamics and hydrostatics. Astronomy covered geography, theoretical astronomy, spherical astronomy, gnomonics (e.g. sundials), chronology, and navigation. Montucla also included astrology, the constructon of observatories, and the design of ships.

3. *The Problem of Proof*

That physical problems motivated most mathematical work was of course not peculiar to the eighteenth century, but in that century the merger of mathematics and physics was crucial. The major development, as we have noted, was analysis. However, the very basis of the calculus not only was not clear, but had been under attack almost from the beginning of the seventeenth-century work on the subject. Eighteenth-century thinking was certainly loose and intuitive. Any delicate questions of analysis, such as the convergence of series and integrals, the interchange of the order of differentiation and integration, the use of differentials of higher order, and questions of existence of integrals and solutions of differential equations, were all but ignored. That the mathematicians were able to proceed at all was due to the fact that the rules of operation were clear. Having formulated the physical problems mathematically, the virtuosos got to work, and new methodologies and conclusions emerged. Certainly the mathematics itself was purely formal. Euler was so fascinated by formulas that he could hardly refrain from manipulating them. How could the mathematicians have dared merely to apply rules and yet assert the reliability of their conclusions?

The physical meaning of the mathematics guided the mathematical steps and often supplied partial arguments to fill in nonmathematical steps. The reasoning was in essence no different from a proof of a theorem of geometry, wherein some facts entirely obvious in the figure are used even though no axiom or theorem supports them. Finally, the physical correctness of the conclusions gave assurance that the mathematics must be correct.

Another element in eighteenth-century mathematical thought supported the arguments. The men trusted the symbols far more than they did logic. Because infinite series had the same symbolic form for all values of x, the distinction between values of x for which the series converged and values for which it diverged did not seem to demand attention. And even though they recognized that some series, such as $1 + 2 + 3 + \ldots$, had an infinite sum, they preferred to try to give meaning to the sum rather than question the summation.

Likewise the somewhat free use of complex numbers rested on the confidence in symbols. Since second degree expressions, $ax^2 + bx + c$, could be expressed as a product of linear factors when the zeros were real, it was equally clear that there should be linear factors when the zeros were complex. The formal operations of the calculus, differentiation and anti-differentiation, were extended to new functions despite the fact that the mathematicians were conscious of their own lack of clarity about the ideas. This reliance upon formalism blinded them somewhat. Thus the difficulty they had in broadening their notion of function was caused by their adherence to the conviction that functions must be expressible by formulas.

The eighteenth-century men were fully aware of the mathematical requirement of proof. We have seen that Euler did try to justify his use of divergent series and Lagrange, among others, did offer a foundation for the calculus. But the few efforts to achieve rigor, significant because they show that standards of rigor vary with the times, did not logicize the work of the century; and the men almost willingly took the position that what cannot be cured must be endured. They were so intoxicated with their physical successes that most often they were indifferent to the missing rigor. Very striking is the extreme confidence in the conclusions while the theory was so ill-assured. Because the eighteenth-century mathematicians were willing to plunge ahead so boldly without logical support, this period has been called the heroic age in mathematics.

Perhaps because the few efforts to rigorize the calculus were unsuccessful and additional questions of rigor raised by subsequent analytical work were hopelessly beyond resolution, some mathematicians abandoned the effort to secure it and, like the fox with the grapes, consciously derided the rigor of the Greeks. Sylvestre-François Lacroix (1765–1843), in the second edition (1810–19) of his three-volume *Traité du calcul différentiel et du calcul intégral*, says, in the preface to Volume 1 (p. 11), "Such subtleties as the Greeks worried about we no longer need." The typical attitude of the century was: Why go to the trouble of proving by abstruse reasoning things which one never doubts in the first place, or of demonstrating what is more evident by means of what is less evident?

Even Euclidean geometry was criticized, on the ground that it offered proofs where none were deemed to be needed. Clairaut said in his *Eléments de géométrie* (1741),

> It is not surprising that Euclid goes to the trouble of demonstrating that two circles which cut one another do not have the same center,

that the sum of the sides of a triangle which is enclosed within another is smaller than the sum of the sides of the enclosing triangle. This geometer had to convince obstinate sophists who glory in rejecting the most evident truths; so that geometry must, like logic, rely on formal reasoning in order to rebut the quibblers. . . . But the tables have turned. All reasoning concerned with what common sense knows in advance, serves only to conceal the truth and to weary the reader and is today disregarded.

This attitude of the eighteenth century was also expressed by Josef Maria Hoene-Wronski (1778–1853), who was a great algorithmist but was not concerned with rigor. A paper of his was criticized by a Commission of the Paris Academy of Sciences as lacking in rigor; Wronski replied that this was "pedantry which prefers means to the end."

The men were, on the whole, aware of their own indifference to rigor. D'Alembert said in 1743, "Up to the present . . . more concern has been given to enlarging the building than to illuminating the entrance, to raising it higher than to giving proper strength to the foundations." Accordingly, the eighteenth century broke new ground as has not been done since. The great creations, in view of the limited number of men involved, were more numerous than in any other century. The nineteenth- and twentieth-century men, inclined to look down on the crude, often brash, inductive work of the eighteenth, stressed its excesses and errors in order to belittle its triumphs.

4. The Metaphysical Basis

Though the mathematicians did recognize that their creations had not been reformulated in terms of the deductive model of Euclid, they were confident of the truth of the mathematics. This confidence rested in part, as noted, on the physical correctness of the conclusions, but also on philosophical and theological grounds. Truth was assured because mathematics was simply unearthing the mathematical design of the universe. The keynote of late seventeenth- and eighteenth-century philosophy, expressed especially by Thomas Hobbes, John Locke, and Leibniz, was the pre-established harmony between reason and nature. This doctrine had really been unchallenged since the time of the Greeks. Need one quibble, then, if the mathematical laws that so clearly applied to nature lacked the precision of purely mathematical proof? Though the eighteenth century uncovered only pieces, they were pieces of underlying truths. The remarkable accuracy of the mathematical deductions, especially in celestial mechanics, were glorious confirmation of the century's confidence in the mathematical design of the universe.

The eighteenth-century men were likewise convinced that certain mathematical principles must be true because the mathematical design must have incorporated them. Thus, since a perfect universe would not tolerate waste, its action was the least required to achieve its purposes. Hence the Principle of Least Action was, as Maupertuis affirmed and Euler seconded in his *Methodus Inveniendi*, unquestionable.

The conviction that the world was mathematically designed derived from the earlier linking of science and theology. We may recall that most of the leading figures of the sixteenth and seventeenth centuries were not only deeply religious but found in their theological views the inspiration and conviction vital to their scientific work. Copernicus and Kepler were certain that the heliocentric theory must be true because God must have preferred the mathematically simpler theory. Descartes's belief that our innate ideas, among them the axioms of mathematics, are true and that our reasoning is sound rested on the conviction that God would not deceive us, so that to deny the truth and clarity of mathematics would be to deny God. Newton, as already mentioned, regarded the chief value of his scientific efforts to be the study of God's work and the support of revealed religion. Many passages in his works are glorifications of God, and the General Scholium at the end of Book III of his *Principia* is largely a tribute to God reminiscent of Kepler. Leibniz's explanation of the concord between the real and the mathematical worlds, and his ultimate defense of the applicability of his calculus to the real world, was the unity of the world and God. The laws of reality could therefore not deviate from the ideal laws of mathematics. The universe was the most perfect conceivable, the best of all possible worlds, and rational thinking disclosed its laws.

Though it retained the belief in the mathematical design of nature, the eighteenth century ultimately discarded the philosophical and religious bases for this belief. The core of the entire philosophical position, the doctrine that the universe was designed by God, was gradually undermined by purely mathematico-physical explanations. The religious motivation for mathematical work had begun to lose ground even in the seventeenth century. Galileo had sounded the call for a break. He says in one of his letters, "Yet for my part any discussion of the Sacred Scriptures might have lain dormant forever; no astronomer or scientist who remained within proper bounds has ever got into such things." Then Descartes, by asserting the invariability of the laws of nature, had implicitly restricted God's role. Newton confined the actions of God to keeping the world functioning according to plan. He used the figure of speech of a watchmaker keeping a watch in repair. Thus the role attributed to God became more and more restricted; and, as universal laws embracing heavenly and earthly motions (which Newton himself revealed) began to dominate the intellectual scene and as the continued agreement between predictions and observations bespoke the perfection of these laws, God sank more and more into the background and the mathematical laws of the universe became the focus of attention. Leibniz saw that Newton's *Principia* implied a world functioning according to plan with or without God, and attacked the book as anti-Christian. But the further the development of mathematics proceeded in the eighteenth century the more the religious inspiration and motivation for mathematical work receded.

Unlike Maupertuis and Euler, Lagrange denied any metaphysical implication in the Principle of Least Action. The concern to obtain physically significant results replaced the concern for God's design. As far as mathematical physics was concerned, the complete rejection of God and any metaphysical principles resting on his existence was made by Laplace. There is a well-known story that when Laplace gave Napoleon a copy of his *Mécanique céleste*, Napoleon remarked, "M. Laplace, they tell me you have written this large book on the system of the universe and have never even mentioned its Creator." Laplace is said to have replied, "I have no need of this hypothesis." Toward the end of the century the label "metaphysical" applied to an argument became a term of reproach, though it was often used to condemn what mathematicians could not understand. Thus Monge's theory of characteristics, which was not understood by his contemporaries, was called metaphysical.

5. *The Expansion of Mathematical Activity*

In the eighteenth century the academies of science founded in the middle and at the end of the seventeenth century, rather than the universities, sponsored and supported mathematical research. The academies also supported the journals, which had become the regular outlet for new work. About the only change in the affairs of the academies is that in 1795 the Academy of Sciences of Paris was reconstituted as one of three divisions of the Institute of France.

In Germany, prior to 1800, the universities did no research. They offered a couple of required years of liberal arts followed by specialization in law, theology, or medicine. The great mathematicians were not in the universities but were attached to the Berlin Academy of Science. However, in 1810, Alexander von Humboldt (1769–1859) founded the University of Berlin and introduced the radical idea that professors should lecture on what they wished to and that students could take what they preferred. Hence for the first time professors could lecture on their research interests. Thus Jacobi lectured on his work on elliptic functions at Koenigsberg from 1826 on, though this was still unusual and other teachers had to cover his regular classes. In the nineteenth century many of the German kingdoms, duchies, and free cities founded universities that began to support research professors.

The French universities of the eighteenth century, at least until the Revolution, were no better than those in Germany. However the new government decided to found high-level universities for teaching and research. The organizational work was carried out by Nicolas de Condorcet, who had been active in mathematics. The Ecole Polytechnique, founded in 1794, had Monge and Lagrange as its first mathematics professors. Students competed for admission and were supposed to receive the training that would enable them to become engineers or army officers. Actually the mathematical level of the courses was very high, and the graduates were able to undertake mathematical research. Through this training and through the published lectures, the school exerted a wide and strong influence. In 1808 the French government founded the Ecole Normale Supérieure, which had been preceded by the Ecole Normale; the latter, founded in 1794, lasted only a few months. The new Ecole was devoted to the training of teachers and was divided into two sections, humanities and science. Here too students competed for admission; advanced courses were offered; facilities for study and research were good; and the better students were directed into research.

The nations of Europe differed considerably in mathematical productivity during the eighteenth century. The leading country was France, followed by Switzerland. Germany was relatively inactive, so far as German nationals were concerned, though Euler and Lagrange were supported by the Berlin Academy of Sciences. England too languished. Brook Taylor, Matthew Stewart (1717–85), and Colin Maclaurin were the only prominent mathematicians. England's poor performance in view of its great activity in the seventeenth century may be surprising, but the explanation is readily found. The English mathematicians had not only isolated themselves personally from the Continentals as a consequence of the controversy between Newton and Leibniz, but also suffered by following the geometrical methods of Newton. The English settled down to study Newton instead of nature. Even in their analytical work they used Newton's notation for fluxions and fluents and refused to read anything written in the notation of Leibniz. Moreover, at Oxford and Cambridge, no Jew or Dissenter from the Church of England could even be a student. By 1815 mathematics in England was at its last gasp and astronomy nearly so.

In the first quarter of the nineteenth century the British mathematicians began to take interest in the work on the calculus and its extensions, which had proceeded apace on the Continent. The Analytical Society was formed at Cambridge in 1813 to study this work. George Peacock (1791–1858), John Herschel (1792–1871), Charles Babbage and others undertook to study the principles of "d-ism"—that is, the Leibnizian notation in the calculus, as against those of "dot-age," or the Newtonian notation. Soon the quotient dy/dx replaced $\dot{y}$, and the Continental texts and papers became accessible to English students. Babbage, Peacock, and Herschel translated a one-volume edition of Lacroix's *Traité* and published it in 1816. By 1830 the English were able to join in the work of the Continentals. Analysis in England did prove to be largely mathematical physics, though some entirely new directions of work, algebraic invariant theory and symbolic logic, were also initiated in that country.

6. *A Glance Ahead*

As we know, by the end of the eighteenth century the mathematicians had created a number of new branches of mathematics. But the problems of these branches had become extremely complicated and, with few exceptions, no general methods had been devised to treat them. The mathematicians began to feel blocked. Lagrange wrote to d'Alembert on September 21, 1781, "It appears to me also that the mine [of mathematics] is already very deep and that unless one discovers new veins it will be necessary sooner or later to abandon it. Physics and chemistry now offer the most brilliant riches and easier exploitation; also our century's taste appears to be entirely in this direction and it is not impossible that the chairs of geometry in the Academy will one day become what the chairs of Arabic presently are in the universities."[2] Euler and d'Alembert agreed with Lagrange that mathematics had almost exhausted its ideas, and they saw no new great minds on the horizon. This fear was expressed even as early as 1754 by Diderot in *Thoughts on the Interpretation of Nature:* "I dare say that in less than a century we shall not have three great geometers [mathematicians] left in Europe. This science will very soon come to a standstill where the Bernoullis, Maupertuis, Clairauts, Fontaines, d'Alemberts and Lagranges will have left it. . . . We shall not go beyond this point."

Jean-Baptiste Delambre (1749–1822), who was permanent secretary of the mathematics and physics section of the Institut de France, in a report (*Rapport historique sur le progrès des sciences mathématiques depuis 1789 et sur leur état actuel*, Paris, 1810) said, "It would be difficult and rash to analyze the chances which the future offers to the advancement of mathematics; in almost all its branches one is blocked by insurmountable difficulties; perfection of detail seems to be the only thing which remains to be done. All these difficulties appear to announce that the power of our analysis is practically exhausted. . . ."

The wiser prediction was made in 1781 by Condorcet, who was impressed by Monge's work.

> . . . in spite of so many works often crowned by success, we are far
> from having exhausted all the applications of analysis to geometry,
> and instead of believing that we have approached the end where

2. Lagrange, *Œuvres*, 13, 368.

these sciences must stop because they have reached the limit of the forces of the human spirit, we ought to avow rather we are only at the first steps of an immense career. These new [practical] applications, independently of the utility which they may have in themselves, are necessary to the progress of analysis in general; they give birth to questions which one would not think to propose; they demand that one create new methods. Technical processes are the children of need; one can say the same for the methods of the most abstract sciences. But we owe the latter to needs of a more noble kind, the need to discover the new truths or to know better the laws of nature.

Thus one sees in the sciences many brilliant theories which have remained unapplied for a long time suddenly becoming the foundation of most important applications, and likewise applications very simple in appearance giving birth to ideas of the most abstract theories, for which no one would have felt the need, and directing the work of geometers [mathematicians] to these theories....

Condorcet was of course correct. In fact, mathematics expanded in the nineteenth century even more than in the eighteenth. In 1783, the year in which both Euler and d'Alembert died, Laplace was thirty-four years of age, Legendre thirty-one, Fourier fifteen, and Gauss six.

Just what the new mathematics was we shall see in succeeding chapters. But we shall note here some of the agencies for the propagation of results to which we shall refer later. There was, first of all, a vast expansion in the number of research journals. The *Journal de l'Ecole Polytechnique* was founded at the same time that the school was organized. In 1810 Joseph-Diez Gergonne (1771–1859) started the *Annales de Mathématiques Pures et Appliquées*, which lasted until 1831. This was the first purely mathematical journal. In the meantime August Leopold Crelle (1780–1855), a noteworthy figure because he was a good organizer and helped many young people to obtain university positions, started the *Journal für die reine und angewandte Mathematik* in 1826.[3] By this title Crelle intended to show that he wished to broaden the scope of mathematical interests. Despite Crelle's intentions, the journal was soon filled with specialized mathematical articles and was often referred to humorously as the *Journal für reine unangewandte Mathematik* (journal purely for unapplied mathematics). This journal is also referred to as *Crelle's Journal*, and, from 1855 to 1880, as *Borchardt's Journal*. In 1795 the *Mémoires de l'Académie des Sciences* became the *Mémoires de l'Académie des Sciences de l'Institut de France*. The Academy of Sciences of Paris also started, in 1835, the *Comptes Rendus*, to serve the purpose of giving in four pages or less brief notices of new results. There is an apocryphal story that the restriction to four pages was intended to limit Cauchy, who had been writing prolifically.

In 1836 Liouville founded the French analogue of *Crelle's Journal*, the

---

3. Hereafter referred to as *Jour. für Math.*

Journal de Mathématiques Pures et Appliquées, which is often referred to as *Liouville's Journal*. Louis Pasteur (1822–95) founded the *Annales Scientifiques de l'Ecole Normale Supérieure* in 1864, while Gaston Darboux (1842–1917) launched *Le Bulletin des Sciences Mathématiques* in 1870. Among the numerous other journals, we shall mention the *Mathematische Annalen* (1868), the *Acta Mathematica* (1882), and the *American Journal of Mathematics*, the first mathematical journal in the United States. This was founded in 1878 by J. J. Sylvester when he was a professor at Johns Hopkins University.

Another type of agency has promoted mathematical activity since the nineteenth century. The mathematicians of several countries formed professional societies such as the London Mathematical Society, the first of its kind and organized in 1865, the Société Mathématique de France (1872), the American Mathematical Society (1888), and the Deutsche Mathematiker-Vereinigung (1890). These societies hold regular meetings at which papers are presented; each sponsors one or more journals, in addition to those already mentioned, such as the *Bulletin de la Société Mathématique de France* and the *Proceedings of the London Mathematical Society*.

As the above sketch of newer organizations and publications implies, mathematics expanded enormously in the nineteenth century. In large part this expansion became possible because a small aristocracy of mathematicians was supplanted by a far broader group. The spread of learning made possible the entry of far more scholars from all economic levels. The trend had begun even in the eighteenth century. Euler was the son of a pastor; d'Alembert was an illegitimate child brought up by a poor family; Monge was the son of a peddler; and Laplace was born to a peasant family. The participation of universities in research, the writing of textbooks, and the systematic training of scientists initiated by Napoleon produced a far greater number of mathematicians.

Bibliography

Boutroux, Pierre: *L'Ideal scientifique des mathématiciens*, Libraire Felix Alcan, 1920.

Brunschvicg, Léon: *Les Etapes de la philosophie mathématique*, Presses Universitaires de France, 1947, Chaps. 10–12.

Hankins, Thomas L.: *Jean d'Alembert: Science and Enlightenment*, Oxford University Press, 1970.

Hille, Einar: "Mathematics and Mathematicians from Abel to Zermelo," *Mathematics Magazine*, 26, 1953, 127–46.

Montucla, J. F.: *Histoire des mathématiques*, 2nd ed., 4 vols., 1799–1802, Albert Blanchard (reprint), 1960.

27
Functions of a Complex Variable

The shortest path between two truths in the real domain passes through the complex domain.   JACQUES HADAMARD

1. *Introduction*

From the technical standpoint, the most original creation of the nineteenth century was the theory of functions of a complex variable. The subject is often referred to as the theory of functions, though the abbreviated description implies more than is intended. This new branch of mathematics dominated the nineteenth century almost as much as the direct extensions of the calculus dominated the eighteenth century. The theory of functions, a most fertile branch of mathematics, has been called the mathematical joy of the century. It has also been acclaimed as one of the most harmonious theories in the abstract sciences.

2. *The Beginnings of Complex Function Theory*

As we have seen, complex numbers and even complex functions entered mathematics effectively in connection with integration by partial fractions, the determination of the logarithms of negative and complex numbers, conformal mapping, and the decomposition of a polynomial with real coefficients. Actually the men of the eighteenth century did far more with complex numbers and functions.

In his essay on hydrodynamics, *Essay on a New Theory of the Resistance of Fluids* (1752), d'Alembert considers the motion of a body through a homogeneous, weightless, ideal fluid, and in connection with this study considers the following problem. He seeks to determine two functions p and q whose differentials are

$$(1) \qquad dq = M\,dx + N\,dy, \qquad dp = N\,dx - M\,dy.$$

Since the quantities N and M occur in both dp and dq, it follows that

$$(2) \qquad \frac{\partial p}{\partial x} = \frac{\partial q}{\partial y}, \qquad \frac{\partial p}{\partial y} = -\frac{\partial q}{\partial x}.$$

These equations are now called the Cauchy-Riemann equations. The equations (2) say (Chap. 19, sec. 6) that $q\,dx + p\,dy$ and $p\,dx - q\,dy$ are exact differentials of certain functions. Then the expressions (we shall use i for $\sqrt{-1}$, though this was done only occasionally by Euler and made common practice by Gauss)

$$q\,dx + p\,dy + i(p\,dx - q\,dy) = (q + ip)\left(dx + \frac{dy}{i}\right)$$

and

$$q\,dx + p\,dy - i(p\,dx - q\,dy) = (q - ip)\left(dx - \frac{dy}{i}\right)$$

are also complete differentials, and so $q + ip$ is a function of $x + y/i$ and $q - ip$ is a function of $x - y/i$. D'Alembert sets

$$(3) \qquad q + ip = \xi\left(x + \frac{y}{i}\right) + i\zeta\left(x + \frac{y}{i}\right)$$

$$(4) \qquad q - ip = \xi\left(x - \frac{y}{i}\right) - i\zeta\left(x - \frac{y}{i}\right)$$

where ξ and ζ are functions to be determined and which d'Alembert does determine in special cases. By adding and subtracting (3) and (4) he obtains p and q. The significance of this is that he has shown that p and q are the real and imaginary parts of a complex function.

Euler showed how to use complex functions to evaluate real integrals. He wrote a series of papers from 1776 to the time of his death in 1783 which were published from 1788 on. Two of these were published in 1793 and 1797.[1] Euler remarks that every function of z which for $z = x + iy$ takes the form $M + iN$, where M and N are real functions, also takes on, for $z = x - iy$, the form $M - iN$. This, he states, is the fundamental theorem of complex numbers. He uses this assertion to evaluate real integrals. Suppose

$$(5) \qquad \int Z(z)\,dz = V$$

where z is real. He sets $z = x + iy$ so that V becomes $P + iQ$. Then

$$(6) \qquad P + iQ = \int (M + iN)(dx + i\,dy),$$

where $M + iN$ is now the complex form of $Z(z)$. By his basic assertion,

$$(7) \qquad P - iQ = \int (M - iN)(dx - i\,dy),$$

1. *Nova Acta Acad. Sci. Petrop.*, 7, 1789, 99–133, pub. 1793 = *Opera*, (1), 19, 1–44; *ibid.*, 10, 1792, 3–19, pub. 1797 = *Opera*, (1), 19, 268–86.

so that by separating real and imaginary parts,

$$(8) \qquad P = \int M\,dx + N\,dy, \qquad Q = \int N\,dx + M\,dy.$$

Then $M\,dx - N\,dy$ and $N\,dx + M\,dy$ are exact differentials of P and Q respectively, from which it follows that

$$(9) \qquad \frac{\partial M}{\partial y} = -\frac{\partial N}{\partial x}, \qquad \frac{\partial N}{\partial y} = \frac{\partial M}{\partial x}.$$

Thus by substituting $z = x + iy$ in $Z(z)$, "one obtains two functions M and N which possess the remarkable property that $\partial M/\partial y = -\partial N/\partial x$ and $\partial M/\partial x = \partial N/\partial y$; P and Q have similar properties." Here Euler stresses that M and N, the real and imaginary parts of a complex function, satisfy the Cauchy-Riemann equations. However, his main point is to use the integrals (8) to calculate (5), for P equals the original V. To reduce the integrals in (8) to integrals of functions of one variable, Euler replaces $z = x + iy$ in (5) by $z = r(\cos\theta + i\sin\theta)$ and keeps θ constant. This amounts to integrating along a ray through the origin of the complex plane. He then uses his method to evaluate some integrals.

Laplace too used complex functions to evaluate integrals. In a series of papers that start in 1782 and end with his famous *Théorie analytique des probabilités* (1812), he passes from real to complex integrals, much as Euler did, to evaluate the real integrals. Laplace claimed priority because Euler's papers were published later than his own. However, even the papers of 1793 and 1797 mentioned above were read to the St. Petersburg Academy in March of 1777. Incidentally, in this work Laplace introduced what we now call the Laplace transform method of solving differential equations.

The work of Euler, d'Alembert, and Laplace constituted significant progress in the theory of functions. However, there is an essential limitation in their work. They depended upon separating the real and imaginary parts of $f(x + iy)$ to carry out their analytical work. The complex function was not really the basic entity. It is clear that these men were still very uneasy about the use of complex functions. Laplace, in his 1812 book, remarks, "This transition from real to imaginary can be regarded as a heuristic method, which is like the method of induction long used by mathematicians. However if one uses the method with great care and restraint, one will always be able to prove the results obtained." He does emphasize that the results must be verified.

3. *The Geometrical Representation of Complex Numbers*

A vital step that made the erection of a theory of functions of a complex variable more intuitively reasonable was the geometrical representation of

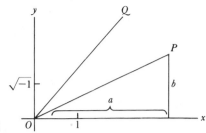

Figure 27.1

complex numbers and of the algebraic operations with these numbers. That many men—Cotes, De Moivre, Euler, and Vandermonde—really thought of complex numbers as points in the plane follows from the fact that all, in attempting to solve $x^n - 1 = 0$, thought of the solutions

$$\cos \frac{2k\pi}{n} + i \sin \frac{2k\pi}{n}$$

as the vertices of a regular polygon. Euler, for example, replaced x and y, which geometrically he visualized as a point in the coordinate plane, by $x + iy$ and then represented the latter by $r(\cos \theta + i \sin \theta)$, which in turn he plots as polar coordinates r and θ. Hence one could say that the plotting of complex numbers as coordinates of points in the plane was known by 1800. However, decisive identification of the two was not made, nor did the algebraic operations with complex numbers have as yet any geometrical meaning. Also missing was the idea of plotting the values of a complex function $u + iv$ of $x + iy$ as points in another plane.

In 1797 the self-taught Norwegian-born surveyor Caspar Wessel (1745–1818) wrote a paper entitled "On the Analytic Representation of Direction; an Attempt," which was published in the memoirs for 1799 of the Royal Academy of Denmark. Wessel sought to represent geometrically directed line segments (vectors) and operations with them. In this paper an axis of imaginaries with $\sqrt{-1}$ (he writes ϵ for $\sqrt{-1}$) as an associated unit is introduced along with the usual x-axis with real unit 1. In Wessel's geometric representation, the vector OP (Fig. 27.1) is the line segment OP drawn from the origin O in the plane of the units $+1$ and $\sqrt{-1}$, and the vector is represented by the complex number, $a + b\sqrt{-1}$. Similarly, the vector OQ is the line segment OQ and is represented by another number, say $c + d\sqrt{-1}$.

Wessel then defines the operations with vectors by defining the operations with complex numbers in geometrical terms. His definitions of the four operations are practically the ones we learn today. Thus the sum of $a + bi$ and $c + di$ is the diagonal of the parallelogram determined by the adjacent sides OP and OQ. The product of $a + bi$ and $c + di$ is a new vector OR, say,

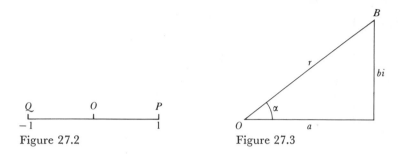

Figure 27.2   Figure 27.3

such that OR is to OQ as OP is to the real unit, and the angle made by OR and the x-axis is the sum of the angles made by OP and OQ. Clearly Wessel thought in terms of vectors rather than associating complex numbers with points in the plane. He applies his geometric representation of vectors to problems of geometry and trigonometry. Despite its great merit, Wessel's paper went unnoticed until 1897, when it was republished in a French translation.

A somewhat different geometric interpretation of complex numbers was given by a Swiss, Jean-Robert Argand (1768–1822). Argand, who was also self-taught and a bookkeeper, published a small book, *Essai sur une manière de représenter les quantités imaginaires dans les constructions géométriques* (1806).[2] He observes that negative numbers are an extension of positive numbers that results from combining direction with magnitude. He then asks, Can we extend the real number system by adding some new concept? Let us consider the sequence 1, x, -1. Can we find an operation that turns 1 into x, and then repeated on x turns x into -1? If we rotate OP (Fig. 27.2) counterclockwise about O through 90° and then repeat this rotation, we do go from P to Q by repeating an operation twice. But, Argand notes, this is precisely what happens if we multiply 1 by $\sqrt{-1}$ and then multiply this product by $\sqrt{-1}$; that is, we obtain -1. Hence we can think of $\sqrt{-1}$ as a rotation through 90° counterclockwise, say, and $-\sqrt{-1}$ as a clockwise rotation through 90°.

To utilize this operational meaning of complex numbers, Argand decided that a typical line segment OB (Fig. 27.3) emanating from the origin, which he calls a directed line, should be represented as $r(\cos \alpha + i \sin \alpha)$, where r is the length. He also regarded the complex number $a + bi$ as symbolizing the geometrical combination OB of a and bi. Argand, like Wessel, showed how complex numbers can be added and multiplied geometrically and applied these geometrical ideas to prove theorems of trigo-

2. A number of papers on Argand's idea and ideas of other writers on the geometrical representation of complex numbers can be found in Volume 4 (1813–14) and Volume 5 (1814–15) of Gergonne's *Annales des Mathématiques*.

nometry, geometry, and algebra. Though Argand's book stirred up some controversy about the geometric interpretation of complex numbers, this was the only contribution to mathematics that he made and his work had little impact. We do, however, still speak of the Argand diagram.

Gauss was more effective in bringing about the acceptance of complex numbers. He used them in his several proofs of the fundamental theorem of algebra (Chap. 25, sec. 2). In the first three proofs (1799, 1815, and 1816) he presupposes the one-to-one correspondence of points of the Cartesian plane and complex numbers. There is no actual plotting of $x + iy$, but rather of x and y as coordinates of a point in the real plane. Moreover, the proofs do not really use complex function theory because he separates the real and imaginary parts of the functions involved. He is more explicit in a letter to Bessel of 1811,[3] wherein he says that $a + ib$ is represented by the point (a, b) and that one can go from one point to another of the complex plane by many paths. There is no doubt, if one judges from the thinking exhibited in the three proofs and in other unpublished works, some of which we shall discuss shortly, that by 1815 Gauss was in full possession of the geometrical theory of complex numbers and functions, though he did say in a letter of 1825 that "the true metaphysics of $\sqrt{-1}$ is elusive."

However, if Gauss still possessed any scruples, by 1831 he had overcome them, and he publicly described the geometric representation of complex numbers. In the second commentary to his paper "Theoria Residuorum Biquadraticorum"[4] and in the "Anzeige" (announcement and brief account) of this paper that he wrote himself for the *Göttingische gelehrte Anzeigen* of April 23, 1831,[5] Gauss is very explicit about the geometrical representation of complex numbers. In Article 38 of the paper he not only gives the representation of $a + bi$ as a point (not a vector, as with Wessel and Argand) in the complex plane, but describes the geometrical addition and multiplication of complex numbers. In the "Anzeige"[6] Gauss says that the transfer of the theory of biquadratic residues into the domain of complex numbers may disturb those who are not familiar with these numbers and may leave them with the impression that the theory of residues is left up in the air. He therefore repeats what he has said in the paper proper about the geometrical representation of complex numbers. He then points out that while fractions, negative numbers, and real numbers were now well understood, complex numbers had merely been tolerated, despite their great value. To many they have appeared to be just a play with symbols. But in this geometrical representation one finds the "*intuitive meaning of complex numbers completely established and more is not needed to admit these quantities*

3. *Werke*, 8, 90–92.
4. *Comm. Soc. Gott.*, 3, 1832 = *Werke*, 2, 95–148; the main content of this paper will be discussed in Chapter 34, sec. 2.
5. *Werke*, 2, 169–78.
6. *Werke*, 2, 174 ff.

into the domain of arithmetic [italics added]." He also says that if the units 1
−1, and $\sqrt{-1}$ had not been given the names positive, negative, and imaginary units but were called direct, inverse, and lateral, people would not
have gotten the impression that there was some dark mystery in these
numbers. The geometrical representation, he says, puts the true metaphysics of imaginary numbers in a new light. He introduced the term
"complex numbers" as opposed to imaginary numbers,[7] and used i for
$\sqrt{-1}$.

4. *The Foundation of Complex Function Theory*

Gauss also introduced some basic ideas about functions of a complex
variable. In the letter to Bessel of 1811,[8] apropos of a paper by Bessel on the
logarithmic integral, $\int dx/x$, Gauss points out the necessity of taking
imaginary (complex) limits into account. Then he asks, "What should one
mean by $\int \phi(x)\, dx$ [Gauss writes $\int \phi x \cdot dx$] when the upper limit is $a + bi$?
Manifestly, if one wants to have clear concepts, he must assume that x takes
on small increments from that value of x for which the integral is 0 to the
value $a + bi$ and then sum all the $\phi(x)\, dx$ .... But the continuous passage
from one value of x to another in the complex plane takes place over a
curve and is therefore possible over many paths. I affirm now that the
integral $\int \phi(x)\, dx$ has only one value even if taken over different paths provided $\phi(x)$ is single-valued and does not become infinite in the space
enclosed by the two paths. This is a very beautiful theorem whose proof is
not difficult and which I shall give on a convenient occasion." Gauss did
not give this proof. He also affirms that if $\phi(x)$ does become infinite, then
$\int \phi(x)\, dx$ can have many values, depending upon whether one chooses a
closed path that encloses the point at which $\phi(x)$ becomes infinite one, two,
or more times.

   Then Gauss returns to the particular case of $\int dx/x$ and says that,
starting from $x = 1$ and going to some value $a + bi$, one obtains a unique
value for the integral if the path does not enclose $x = 0$; but if it does, one
must add $2\pi i$ or $-2\pi i$ to the value obtained by going from $x = 1$ to $x =
a + bi$ without enclosing $x = 0$. Thus there are many logarithms for a given
$a + bi$. Later in the letter, Gauss says that the investigation of the integrals
of functions of complex arguments should lead to most interesting results.
Thus, even before Gauss had published his second and third proofs of the
fundamental theorem—wherein, as in the first proof, he avoided the direct
use of complex numbers and functions except for an occasional writing of
$a + b\sqrt{-1}$—he had very definite ideas about complex functions and about
the integrals of such functions.

7. *Werke*, 2, p. 102.
8. *Werke*, 8, 90–92.

Poisson noted in 1815 and discussed in a paper of 1820[9] the use of integrals of complex functions taken over paths in the complex plane. As an example he gives

(10)
$$\int_{-1}^{1} \frac{dx}{x}.$$

Here he sets $x = e^{i\theta}$, where θ runs from $(2n + 1)\pi$ to 0, and obtains, by treating the integral as a limit of a sum, the value $-(2n + 1)\pi i$.

Then he notes that the value of an integral need not be the same when taken over an imaginary path as when over a real path. He gives the example

(11)
$$\int_{-\infty}^{\infty} \frac{\cos ax}{b^2 + x^2} \, dx,$$

where a and b are positive constants. He lets $x = t + ik$, where k is constant and positive, and obtains the values $\pi(e^{-ab} - e^{ab})/2\pi$ for $k > b$ and πe^{-ab} for $k < b$. The second value is also the correct one for $k = 0$. Thus for two different values of k, which means two different paths, one gets two different results. Poisson was the first man to carry out integrations along a path in the complex plane.

Though these observations of Gauss and Poisson are indeed significant, neither published a major paper on complex function theory. This theory was founded by Augustin-Louis Cauchy. Born in Paris in 1789, in 1805 he entered the Ecole Polytechnique, where he studied engineering. Because he was in poor health he was advised by Lagrange and Laplace to devote himself to mathematics. He held professorships at the Ecole Polytechnique, the Sorbonne, and the Collège de France. Politics had unexpected effects on his career. He was an ardent Royalist and supporter of the Bourbons. When in 1830 a distant branch of the Bourbons took control in France, he refused to swear allegiance to the new monarchy and resigned his professorship at the Ecole Polytechnique. He exiled himself to Turin and taught Latin and Italian for some years. In 1838 he returned to Paris, where he served as professor in several religious institutions, up to the time when the government that took over after the revolution of 1848 did away with oaths of allegiance. In 1848 Cauchy took the chair of mathematical astronomy in the Faculté des Sciences of the Sorbonne. Though Napoleon III restored the oath in 1852, he allowed Cauchy to forgo it. To the condescending gesture of the Emperor he responded by donating his salary to the poor of Sceaux, where he lived. Cauchy, an admirable professor and one of the greatest mathematicians, died in 1857.

Cauchy had universal interests. He knew the poetry of his time and was the author of a work on Hebrew prosody. In mathematics he wrote

9. *Jour. de l'Ecole Poly.*, 11, 1820, 295–341.

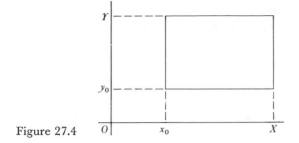

Figure 27.4

over seven hundred papers, second only to Euler in number. His works in a modern edition fill twenty-six volumes and embrace all branches of mathematics. In mechanics he wrote important works on the equilibrium of rods and of elastic membranes and on waves in elastic media. In the theory of light he occupied himself with the theory of waves that Fresnel had started and with the dispersion and polarization of light. He advanced the theory of determinants immensely and contributed basic theorems in ordinary and partial differential equations.

Cauchy's first significant paper in the direction of complex function theory is his "Mémoire sur la théorie des intégrales définies." This paper was read to the Paris Academy in 1814. However, it was not submitted for publication until 1825 and was published in 1827.[10] In the publication Cauchy added two notes that pretty surely reflect developments between 1814 and 1825 and the possible influence of Gauss's work during that period. Let us consider the paper proper for the present. Cauchy says in the preface that he was led to this work by trying to rigorize the passage from the real to the imaginary in processes used by Euler since 1759 and by Laplace since 1782 to evaluate definite integrals, and in fact Cauchy cites Laplace, who observed that the method needed rigorization. But the paper proper does not treat this problem. It treats the question of the interchange of the order of integration in double integrals that arose in hydrodynamical investigations. Euler had said in 1770[11] that this interchange was permissible when the limits for each of the variables under the integral sign were independent of each other, and Laplace apparently agreed because he used this fact repeatedly.

Specifically, Cauchy treats the relation

$$(12) \qquad \int_{x_0}^{X} \int_{y_0}^{Y} f(x, y) \, dy \, dx = \int_{y_0}^{Y} \int_{x_0}^{X} f(x, y) \, dx \, dy$$

wherein x_0, y_0, X, and Y are constants (Fig. 27.4). This interchange of the order of integration holds when $f(x, y)$ is continuous in and on the boundary

10. *Mém. des sav. étrangers*, (2), 1, 1827, 599–799 = *Œuvres*, (1), 1, 319–506.
11. *Novi Comm. Acad. Sci. Petrop.*, 14, 1769, 72–103, pub. 1770 = *Opera*, (1), 17, 289–315.

of the region. Then he introduces two functions $V(x, y)$ and $S(x, y)$ such that

(13) $$\frac{\partial V}{\partial y} = \frac{\partial S}{\partial x} \quad \text{and} \quad \frac{\partial V}{\partial x} = -\frac{\partial S}{\partial y}.$$

Euler had already shown in 1777 how to obtain such functions (see [5], [8], and [9]). Now Cauchy considers an $f(x, y)$ which is given by $\partial V/\partial y = \partial S/\partial x$. In (12) he replaces f on the left side by $\partial V/\partial y$, and f on the right side by $\partial S/\partial x$, so that

(14) $$\int_{x_0}^{X} \int_{y_0}^{Y} \frac{\partial V}{\partial y} \, dy \, dx = \int_{y_0}^{Y} \int_{x_0}^{X} \frac{\partial S}{\partial x} \, dx \, dy,$$

and by using the second equation in (13) he obtains

(15) $$\int_{x_0}^{X} \int_{y_0}^{Y} \frac{\partial S}{\partial y} \, dy \, dx = -\int_{y_0}^{Y} \int_{x_0}^{X} \frac{\partial V}{\partial x} \, dx \, dy.$$

These equalities can be used to evaluate double integrals in either order of integration; however they do not involve complex functions. When Cauchy says in his Introduction [12] that he will "establish rigorously and directly the passage from the real to the imaginary (complex)," it is equations (13) that he has in mind. Cauchy states [13] that these two equations contain the whole theory of the passage from the real to the imaginary.

All of the above is in the body of the 1814 paper, and really gives no explicit indication of how complex function theory is involved. Moreover, though Cauchy used complex functions to evaluate definite real integrals in the same manner as had Euler and Laplace, this use did not involve complex functions as the basic entity. As late as 1821, in his *Cours d'analyse*,[14] he says

$$\cos a + \sqrt{-1} \sin a$$
$$\cos b + \sqrt{-1} \sin b$$
$$\cos (a + b) + \sqrt{-1} \sin (a + b)$$

"are three symbolic expressions which cannot be interpreted according to generally established conventions, and do not represent anything real." The fact that the product of the first and second expressions above equals the third, he says, does not make sense. To make sense of this equation one must equate the real parts and the coefficients of $\sqrt{-1}$. "Every imaginary equation is only the symbolic representation of two equations between real quantities." If we operate on complex expressions according to rules established for *real* quantities, we get exact results that are often important.

12. *Œuvres*, (1), 1, 330.
13. *Œuvres*, (1), 1, 338.
14. *Œuvres*, (2), 3, 154.

In this book he does treat complex numbers and complex variables $u + \sqrt{-1}v$, where u and v are functions of *one* real variable, but always in the sense that the two real components are their significant content. Complex-valued functions of a complex variable were not considered.

In the year 1822, Cauchy took some further steps. From the relations (14) and (15) he had

(16) $$\int_{x_0}^{X} [V(x, Y) - V(x, y_0)] \, dx = \int_{y_0}^{Y} [S(X, y) - S(x_0, y)] \, dy$$

and

(17) $$\int_{x_0}^{X} [S(x, Y) - S(x, y_0)] \, dx = -\int_{y_0}^{Y} [V(X, y) - V(x_0, y)] \, dy.$$

He now had the idea that he could combine these two equations and thus make a statement about $F(z) = F(x + iy) = S + iV$. Thus, by multiplying (16) through by i and adding the two equations, he obtains

$$\int_{x_0}^{X} F(x + iY) \, dx - \int_{x_0}^{X} F(x + iy_0) \, dx$$
$$= \int_{y_0}^{Y} F(X + iy)i \, dy - \int_{y_0}^{Y} F(x_0 + iy)i \, dy,$$

which, by rearranging terms, gives

(18) $$\int_{y_0}^{Y} F(x_0 + iy)i \, dy + \int_{x_0}^{X} F(x + iY) \, dx =$$
$$\int_{x_0}^{X} F(x + iy_0) \, dx + \int_{y_0}^{Y} F(X + iy)i \, dy.$$

This last result is the Cauchy integral theorem for the simple case of complex integration around the boundary of a rectangle (Fig. 27.4). One can express the result thus:

(19) $$\int_{ADC} F(z) \, dz = \int_{ABC} F(z) \, dz.$$

That is, the integral is independent of the path.

The above ideas Cauchy gives in a note of 1822, in his *Résumé des leçons sur le calcul infinitésimal*,[15] and in a footnote to the paper of 1814 which was published in 1827. From these later writings we see how Cauchy did go from real to complex functions.

In 1825 Cauchy wrote another paper, "Mémoire sur les intégrales définies prises entre des limites imaginaires," but this was not published

15. = *Œuvres* (2), 4, 13–256.

until 1874.[16] This paper is considered by many as his most important, and one of the most beautiful in the history of science, though for a long time Cauchy himself did not appreciate its worth.

In this paper he again considers the question of evaluating real integrals by the method of substituting complex values for constants and variables. He treats

$$(20) \qquad \int_{x_0 + iy_0}^{X + iY} f(z)\, dz,$$

where $z = x + iy$, and defines the integral carefully as the limit of the sum

$$\sum_{v=0}^{n-1} f(x_v + iy_v)[(x_{v+1} - x_v) + i(y_{v+1} - y_v)],$$

where $x_0, x_1, \ldots, X$, and $y_0, y_1, \ldots, Y$ are the points of subdivision along a path from (x_0, y_0) to (X, Y). Here $x + iy$ is definitely a point of the complex plane and the integral is over a complex path. He also shows that if we set $x = \phi(t)$, $y = \psi(t)$, where t is real, then the result is independent of the choice of ϕ and ψ, that is, independent of the path, provided that no discontinuity of $f(z)$ lies between two distinct paths. This result generalizes the result for rectangles.

Cauchy formulates his theorem thus: If $f(x + iy)$ is finite and continuous for $x_0 \leq x \leq X$ and $y_0 \leq y \leq Y$, then the value of the integral (20) is independent of the form of the functions $x = \phi(t)$ and $y = \psi(t)$. His proof of this theorem uses the method of the calculus of variations. He considers as an alternative path $\phi(t) + \varepsilon u(t)$, $\psi(t) + \varepsilon v(t)$, and shows that the first variation of the integral with respect to ε vanishes. This proof is not satisfactory. In it Cauchy not only uses the existence of the derivative of $f(z)$ but also the continuity of the derivative, though he does not assume either fact in the statement of the theorem. The explanation is that Cauchy believed a continuous function is always differentiable and that the derivative can be discontinuous only where the function itself is discontinuous. Cauchy's belief was reasonable in that in the earlier years of his work a function meant for him, as for others of the eighteenth and early nineteenth centuries, an analytic expression, and the derivative is given at once by the usual formal rules of differentiation.

In the 1825 paper Cauchy is clearer about a major idea he had already touched on in the paper proper of 1814 and in a footnote to that paper. He considers what happens when $f(z)$ is discontinuous inside or on the boundary

16. *Bull. des Sci. Math.*, 7, 1874, 265–304, and 8, 1875, 43–55, 148–59; this paper is not in Cauchy's *Œuvres*.

of the rectangle (Fig. 27.4). Then the value of the integral along two different paths may be different. If at $z_1 = a + ib$, $f(z)$ is infinite, but the limit

$$F = \lim_{z \to z_1} (z - z_1) f(z)$$

exists, that is, if f has a simple pole at z_1, then the difference in the integrals
is $\pm 2\pi \sqrt{-1} F$. Thus for the function $f(z) = 1/(1 + z^2)$, which is infinite
at $z = \sqrt{-1}$, so that $a = 0$ and $b = 1$,

$$(21) \quad F = \lim_{\substack{x \to 0 \\ y \to 1}} \frac{x + (y - 1)\sqrt{-1}}{[x + (y + 1)\sqrt{-1}][x + (y - 1)\sqrt{-1}]} = \frac{-\sqrt{-1}}{2}.$$

The quantity F itself is what Cauchy called the *résidu intégral* (integral
residue) in his *Exercices de mathématique*.[17] Also, when a function has several
poles in the region bounded by the two paths of integration, Cauchy points
out that one must take the sum of the residues to get the difference in the
integrals over the two paths. In this particular section on residues his two
paths still form a rectangle, but he takes a very large one and lets the sides
become infinite in length to include all the residues.

In the *Exercices*[18] Cauchy points out that the residue of $f(z)$ at z_1 is also
the coefficient of the term $(z - z_1)^{-1}$ in the development of $f(z)$ in powers
of $z - z_1$. Much later, in a paper of 1841,[19] Cauchy gives a new expression
for the residue at a pole, namely,

$$F(z_1) = E[f(z)]_{z_1} = \frac{1}{2\pi i} \int f(z) \, dz,$$

where the integral is taken around a small circle enclosing $z = z_1$. The
concept and development of residues is a major contribution by Cauchy.
His immediate use of all of his results thus far was to evaluate definite
integrals.

It is evident from what Cauchy added in the footnotes to his 1814
paper and from what he wrote in the masterful paper of 1825 that he must
have thought long and hard to realize that some relations between pairs of
real functions achieve their simplest form when complex quantities are
introduced. Just how much he may have learned from the work of Gauss and
Poisson is not known.

During the years 1830 to 1838, when he lived in Turin and in Prague,
his publications were disconnected. He refers to them and repeats most of
the work in his *Exercices d'analyse et de physique mathématique* (4 vols., 1840–47).

17. Four vols., 1826–30 = *Œuvres*, (2), 6–9.
18. Vol. 1, 1826, 23–37 = *Œuvres*, (2), 6, 23–37.
19. *Exercices d'analyse et de physique mathématique*, Vol. 2, 1841, 48–112 = *Œuvres*, (2), 12, 48–112.

In a paper of 1831, published later,[20] he obtained the following theorem: The function $f(z)$ can be expanded according to Maclaurin's formula in a power series that is convergent for all z whose absolute value is smaller than those for which the function or its derivative ceases to be finite or continuous. (The only singularities Cauchy knew at this time were what we now call poles.) He shows that this series is less term for term than a convergent geometric series whose sum is

$$\frac{Z}{Z - z}\overline{f(z)},$$

where Z is the first value for which $f(z)$ is discontinuous and $\overline{f(z)}$ is the largest value of $|f(z)|$ for all z whose absolute value equals $|Z|$. Thus Cauchy gives a powerful and easily applied criterion for the expansibility of a function in a Maclaurin series which uses a comparison series now called a majorant series.

In the proof of the theorem he first shows that

$$f(z) = \frac{1}{2\pi}\int_{-\pi}^{\pi}\frac{\bar{z}f(\bar{z})}{\bar{z} - z}\,d\phi,$$

where $\bar{z} = |Z|e^{i\phi}$. This result is practically what we now call the Cauchy integral formula. He then expands the fraction $\bar{z}/(\bar{z} - z)$ in a geometric series in powers of $z/\bar{z}$ and proves the theorem proper.

In this theorem too Cauchy assumes the existence and continuity of the derivative as necessarily following from the continuity of the function itself. By the time he reproduced this material in his *Exercices*, he had exchanged correspondence with Liouville and Sturm and added to the statement of the above theorem that the region of convergence ends at that value of z for which the function *and* its derivative cease to be finite or continuous. But he was not convinced that one must add conditions on the derivative and in later works he dropped them.

In another major paper on complex function theory, "Sur les intégrales qui s'étendent à tous les points d'une courbe fermée,"[21] Cauchy relates the integral of an [analytic] $f(z) = u + iv$ around a curve bounding a [simply connected] area to an integral over the area. Thus if u and v are functions of x and y,

(22)
$$\iint\left(\frac{\partial u}{\partial x} - \frac{\partial v}{\partial y}\right)dx\,dy = \int u\,dy + \int v\,dx$$

$$\iint\left(\frac{\partial u}{\partial y} + \frac{\partial v}{\partial x}\right)dx\,dy = \int -u\,dx + \int v\,dy,$$

20. *Comp. Rend.*, 4., 1837, 216–218 = *Œuvres*, (1), 4, 38–42; see also *Exercices d'analyse et de physique mathématique*, Vol. 2, 1841, 48–112 = *Œuvres*, (2), 12, 48–112.
21. *Comp. Rend.*, 23, 1846, 251–55 = *Œuvres*, (1), 10, 70–74.

where the double integrals are over the area and the single integrals over the bounding curve. Now, in view of the Cauchy-Riemann equations (cf. [13]), the left sides are 0 and the two right sides are the integrals that appear in

$$\int f(z)\ dz = \int (u + iv)(dx + i\ dy) = \int (u\ dx - v\ dy) + i \int (u\ dy + v\ dx).$$

Hence $\int f(z)\ dz = 0$, and Cauchy has a new proof of the basic theorem on independence of the path. He proves the theorem for a rectangle and then generalizes to a closed curve that does not cut itself. (The theorem was also obtained independently by Weierstrass in 1842). Whether Cauchy got this more fertile formulation of his earlier ideas by learning of Green's work in 1828 (Chap. 28, sec. 4) is not certain, but there are such indications, because Cauchy extended the above result to areas on curved surfaces.

In the above-mentioned paper of 1846 and another of the same year,[22] Cauchy changed his viewpoint toward complex functions as against his work of 1814, 1825, and 1826. Instead of being concerned with definite integrals and their evaluation, he turned to complex function theory proper and to building a base for this theory. In this second paper of 1846, he gives a new statement that involves $\int f(z)\ dz$ around an arbitrary closed curve: If the curve encloses poles, then the value of the integral is $2\pi i$ times the sum of the residues of the function at these poles; that is,

$$(23) \qquad\qquad \int f(z)\ dz = 2\pi iE[f(z)],$$

where $E[f(z)]$ is his notation for the sum of the residues.

He also took up the subject of the integrals of multiple-valued functions.[23] In the first part of the paper, where he treats integrals of single-valued functions, he does not state much more than what Gauss had pointed out in his letter to Bessel apropos of $\int dx/x$ or $\int dx/(1 + x^2)$. The integrals are indeed multiple-valued and their values depend on the path of integration.

But Cauchy goes further to consider multiple-valued functions under the integral sign. Here he says that if the integrand is an expression for the roots of an algebraic or transcendental equation, for example, $\int w^3\ dz$ where $w^3 = z$, and if one integrates over a closed path and returns to the starting point, then the integrand now represents another root. In these cases the value of the integral over the closed path is not independent of the starting

22. "Sur les intégrales dans lesquelles la fonction sous le signe $\int$ change brusquement de valeur," *Comp. Rend.*, 23, 1846, 537 and 557–69 = *Œuvres*, (1), 10, 133–34 and 135–43.
23. "Considérations nouvelles sur les intégrales définies qui s'étendent à tous les points d'une courbe fermée," *Comp. Rend.*, 23, 1846, 689–702 = *Œuvres*, (1), 10, 153–68.

point; and continuation around the path produces different values of the integral. But if one goes around the path enough times so that w returns to its original value, then the values of the integral will repeat and the integral is a periodic function of z. The periodicity modules (*indices de périodicité*) of the integral are not any longer, as in the case of single-valued functions, representable by residues. Cauchy's ideas on the integrals of multiple-valued functions were still vague.

For about twenty-five years, from 1821 on, Cauchy singlehandedly developed complex function theory. In 1843 fellow countrymen began to take up threads of his work. Pierre-Alphonse Laurent (1813–54), who worked alone, published a major result obtained in 1843. He showed[24] that where a function is discontinuous at an isolated point, in place of a Taylor's expansion one must use an expansion in increasing and decreasing powers of the variable. If the function and its derivative are single-valued and continuous in a ring whose center is the isolated point a, then the integral of the function taken over the two circular boundaries of the ring but in opposite directions and properly expanded gives a convergent expansion within the ring itself and with increasing and decreasing powers of z. This Laurent expansion is

$$(24) \qquad f(z) = \sum_{-\infty}^{\infty} a_n (z - a)^n$$

and is an extension of the Taylor expansion. The result was known to Weierstrass in 1841 but he did not publish it.[25]

The subject of multiple-valued functions was taken up by Victor-Alexandre Puiseux. In 1850 Puiseux published a celebrated paper[26] on complex algebraic functions given by $f(u, z) = 0$, f a polynomial in u and z. He first made clear the distinction between poles and branch-points that Cauchy barely perceived and introduced the notion of an essential singular point (a pole of infinite order), to which Weierstrass independently had called attention. Such a point is exemplified by $e^{1/z}$ at $z = 0$. Though Cauchy in the 1846 paper did consider the variation of simple multiple-valued functions along paths that enclosed branch-points, Puiseux clarified this subject too. He shows that if u_1 is one solution of $f(u, z) = 0$ and z varies along some path, the final value of u_1 does not depend upon the path, provided that the path does not enclose any point at which u_1 is infinite or any point where u_1 becomes equal to some other solution (that is, a branch-point).

24. *Comp. Rend.*, 17, 1843, 348–49; the full paper was published in the *Jour. de l'Ecole Poly.*, 23, 1863, 75–204.
25. *Werke*, 1, 51–66.
26. *Jour. de Math.*, 15, 1850, 365–480.

Puiseux also showed that the development of a function of z about a branch-point $z = a$ must involve fractional powers of $z - a$. He then improved on Cauchy's theorem on the expansion of a function in a Maclaurin series. Puiseux obtains an expansion for a solution u of $f(u, z) = 0$ not in powers of z but in powers of $z - c$ and therefore valid in a circle with c as center and not containing any pole or branch-point. Then Puiseux lets c vary along a path so that the circles of convergence overlap and so that the development within one circle can be extended to another. Thus, starting with a value of u at one point, one can follow its variation along any path.

By his significant investigations of many-valued functions and their branch-points in the complex plane, and by his initial work on integrals of such functions, Puiseux brought Cauchy's pioneering work in function theory to the end of what might be called the first stage. The difficulties in the theory of multiple-valued functions and integrals of such functions were still to be overcome. Cauchy did write other papers on the integrals of multiple-valued functions,[27] in which he attempted to follow up on Puiseux's work; and though he introduced the notion of branch-cuts (*lignes d'arrêt*), he was still confused about the distinction between poles and branch-points. This subject of algebraic functions and their integrals was to be pursued by Riemann (sec. 8).

In several other papers in the *Comptes Rendus* for 1851,[28] Cauchy gave some more careful statements about properties of complex functions. In particular, Cauchy affirmed that the continuity of the derivatives, as well as the continuity of the complex function itself, is needed for expansion in a power series. He also pointed out that the derivative at $z = a$ of u as a function of z is independent of the direction in the $x + iy$ plane along which z approaches a and that u satisfies $\partial^2 u/\partial x^2 + \partial^2 u/\partial y^2 = 0$.

In these 1851 papers Cauchy introduced new terms. He used *monotypique* and also *monodrome* when the function was single-valued for each value of z in some domain. A function is *monogen* if for each z it has just one derivative (that is, the derivative is independent of the path). A monodrome, monogenic function that is never infinite he called *synectique*. Later Charles A. A. Briot (1817–82) and Jean-Claude Bouquet (1819–85) introduced "holomorphic" in place of *synectique* and "meromorphic" if the function possessed just poles in that domain.

5. *Weierstrass's Approach to Function Theory*

While Cauchy was developing function theory on the basis of the derivatives and integrals of functions represented by analytic expressions, Karl

27. *Comp. Rend.*, 32, 1851, 68–75 and 162–64 = *Œuvres*, (1), 11, 292–300 and 304–5.
28. *Œuvres*, (1), 11.

Weierstrass developed a new approach. Born in Westphalia in 1815, Weierstrass entered the University of Bonn to study law. After four years in this effort he turned to mathematics in 1838 but did not complete the doctoral work. Instead he secured a state license to become a *gymnasium* (high school) teacher and from 1841 to 1854 he taught youngsters such subjects as writing and gymnastics. During these years he had no contact with the mathematical world, though he worked hard in mathematical research. The few results he published during this period secured for him in 1856 a position teaching technical subject matter in the Industrial Institute in Berlin. In that same year he became a lecturer at the University of Berlin and then professor in 1864. He remained in this post until he died in 1897.

He was a methodical and painstaking man. Unlike Abel, Jacobi, and Riemann he did not have flashes of intuition. In fact he distrusted intuition and tried to put mathematical reasoning on a firm basis. Whereas Cauchy's theory rested on geometrical foundations, Weierstrass turned to constructing the theory of real numbers; when this was done (Chap. 41, sec. 3), about 1841, he built up the theory of analytic functions on the basis of power series, a technique he learned from his teacher Christof Gudermann (1798–1852), and the process of analytic continuation. This work was performed in the 1840s, though he did not publish it then. He contributed to many other topics in the theory of functions and worked on the n-body problem in astronomy and in the theory of light.

It is difficult to date Weierstrass's creations because he did not publish many when he first achieved them. Much of what he had done became known to the mathematical world through his lectures at the University of Berlin. When he published his *Werke* in the 1890s, he did not worry about priority because many of his results had been published by others in the meantime. He was more concerned to present his method of developing function theory.

Power series to represent complex functions already given in analytical form were, of course, known. However, the task, given a power series that defines a function in a restricted domain, to derive other power series that define the same function in other domains on the basis of theorems on power series, was tackled by Weierstrass. A power series in $z - a$ that is convergent in some circle C of radius r about a represents a function that is analytic at each value of z in the circle C. By choosing a point b in the circle and using the values of the function and its derivatives as given by the original series, one may be able to obtain a new power series in $z - b$ whose circle of convergence C' overlaps the first circle. At the points common to the two circles the two series give the same value of the function. However, at the points of C' which are exterior to C, the values of the second series are an analytic continuation of the function defined by the first series. By continuing as far as possible from C' successively to still other circles, one obtains the entire analytic continuation of $f(z)$. The complete $f(z)$ is the collection of values at all points z in all of the circles. Each series is called an element of the function.

It is possible that, during the extension of the domain of the function by the adjunction of more and more circles of convergence, one of the new circles may cover part of a circle not immediately preceding it in the chain and the values of the function in this common part of the new circle and an earlier one may not agree. Then the function is multiple-valued.

The singular points (poles or branch-points) that may arise in this process, which necessarily lie on the boundaries of the circles of convergence of the power series, are included in the function by Weierstrass if the order of a singular point is finite, because at such a point an expansion in powers of $(z - z_0)^{1/n}$ having only a finite number of terms with negative exponents is possible. To obtain expansions about $z = \infty$ Weierstrass uses series in $1/z$. If the function element converges in the entire plane, Weierstrass calls it an entire function, and if it is not a rational integral function, that is, not a polynomial, then it has an essential singularity at ∞ (e.g. $\sin z$).

Weierstrass also gave the first example of a power series whose circle of convergence is its natural boundary; that is, the circle is a curve of singular points, and an example of an analytic expression that can represent different analytic functions in different parts of the plane.

6. *Elliptic Functions*

Paralleling the development of the basic theorems of complex function theory during the first half of the century was a special development dealing with elliptic and later Abelian functions. There is no doubt that Gauss obtained a number of key results in the theory of elliptic functions, because many of these were found after his death in papers he had never published. However, the acknowledged founders of the theory of elliptic functions are Abel and Jacobi.

Niels Henrik Abel (1802–29) was the son of a poor pastor. As a student in Christiania (Oslo), Norway, he had the luck to have Berndt Michael Holmböe (1795–1850) as a teacher. The latter recognized Abel's genius and predicted when Abel was seventeen that he would become the greatest mathematician in the world. After studying at Christiania and at Copenhagen, Abel received a scholarship that permitted him to travel. In Paris he was presented to Legendre, Laplace, Cauchy, and Lacroix, but they ignored him. Having exhausted his funds, he departed for Berlin and spent the years 1825–1827 with Crelle. He returned to Christiania so exhausted that he found it necessary, he wrote, to hold on to the gate of a church. To earn money he gave lessons to young students. He began to receive attention through his published works, and Crelle thought he might be able to secure him a professorship at the University of Berlin. But Abel became ill with tuberculosis and died in 1829.

Abel knew the work of Euler, Lagrange, and Legendre on elliptic integrals and may have gotten suggestions for the work he undertook from remarks made by Gauss, especially in his *Disquisitiones Arithmeticae*. He himself started to write papers in 1825. He presented his major paper on integrals to the Academy of Sciences in Paris on October 30, 1826, for publication in its journal. This paper, "Mémoire sur une propriété générale d'une classe très-étendue de fonctions transcendantes," contained Abel's great theorem (sec. 7). Fourier, the secretary of the Academy at the time, read the introduction to the paper and then referred the paper to Legendre and Cauchy for evaluation, the latter being chiefly responsible. The paper was long and difficult, only because it contained many new ideas. Cauchy laid it aside to favor his own work. Legendre forgot about it. After Abel's death, when his fame was established, the Academy searched for the paper, found it, and published it in 1841.[29] Abel published other papers in *Crelle's Journal* and Gergonne's *Annales* on the theory of equations and elliptic functions. These appeared from 1827 on. Because Abel's main paper of 1826 was not published until 1841, other authors, learning the more limited theorems published in between these dates, obtained independently many of Abel's 1826 results.

The other discoverer of elliptic functions was Carl Gustav Jacob Jacobi (1804–51). Unlike Abel, he lived a quiet life. Born in Potsdam to a Jewish family, he studied at the University of Berlin and in 1827 became a professor at Königsberg. In 1842 he had to give up his post because of ill health. He was given a pension by the Prussian government and retired to Berlin, where he died in 1851. His fame was great even in his own lifetime, and his students spread his ideas to many centers.

Jacobi taught the subject of elliptic functions for many years. His approach to it became the model according to which the theory of functions itself was developed. He also worked in functional determinants (Jacobians), ordinary and partial differential equations, dynamics, celestial mechanics, fluid dynamics, and hyperelliptic integrals and functions. Jacobi has often been labeled a pure mathematician, but, like almost all mathematicians of his own and preceding centuries, he took most seriously the investigation of nature.

While Abel was working on elliptic functions, Jacobi, who had also read Legendre's work on elliptic integrals, started work in 1827 on the corresponding functions. He submitted a paper to the *Astronomische Nachrichten*[30] without proofs. Almost simultaneously, Abel published independently his

29. *Mém. des sav. étrangers*, 7, 1841, 176–264 = *Œuvres*, 145–211.
30. *Astron. Nach.*, 6, 1827, 33–38 = *Werke*, 1, 31–36.

"Recherches sur les fonctions elliptiques."[31] Both had arrived at the key idea of working with inverse functions of the elliptic integrals, an idea Abel had had since 1823. Jacobi next gave proofs of the results he had published in 1827 in several articles of *Crelle's Journal* for the years 1828 to 1830. Thereafter, both published on the subject of elliptic functions; but whereas Abel died in 1829, Jacobi lived until 1851 and was able to publish much more. In particular, Jacobi's *Fundamenta Nova Theoriae Functionum Ellipticarum* of 1829[32] became a key work on elliptic functions.

Through letters from Jacobi, Legendre became familiar with Jacobi's and Abel's work. He wrote to Jacobi on February 9, 1828, "It is a great satisfaction to me to see two young mathematicians so successfully cultivate a branch of analysis which has long been my favorite field, but not at all received as it deserves in my own country." Legendre then published three supplements (1829 and 1832) to his *Traité des fonctions elliptiques* (2 vols., 1825–26), in which he gave accounts of Jacobi's and Abel's work.

The general elliptic integral concerns

$$(25) \qquad u = \int R(x, \sqrt{P(x)}),$$

where $P(x)$ is a third or fourth degree polynomial with distinct roots and $R(x, y)$ is a rational function of x and y. The efforts to deduce some general facts about the function u of x had to fail because the very meaning of the integral was limited for Euler and Legendre. The coefficients of $P(x)$ were real, and the range of x was real and moreover did not contain a root of $P(x) = 0$. With more knowledge of the theory of complex functions some advance might have been made in learning something about u as a function of x, but this knowledge was not available. As it turned out, Abel and Jacobi had a better idea.

To be specific, Legendre had introduced (Chap. 19, sec. 4) the elliptic integrals $F(k, \phi)$, $E(k, \phi)$, and $\pi(n, k, \phi)$. It was Abel who, about 1826, observed that if, to consider $F(k, \phi)$ for example, one studied

$$(26) \qquad u = \int_0^x \frac{dx}{\sqrt{(1 - x^2)(1 - k^2 x^2)}} = \int_0^\phi \frac{d\phi}{\sqrt{1 - k^2 \sin^2 \phi}},$$

where $x = \sin \phi$, then one encountered the same difficulties as when one studied

$$u = \int_0^x \frac{dx}{\sqrt{1 - x^2}} = \text{arc sin } x.$$

The nicer relationships come from studying x as a function of u. Hence Abel proposed to study x as a function of u in the case of the elliptic integrals. Since $x = \sin \phi$, one can also study ϕ as a function of u.

31. *Jour. für Math.*, 2, 1827, 101–81, and 3, 1828, 160–90 = *Œuvres*, 263–388.
32. *Werke*, 1, 49–239.

Jacobi introduced [33] the notation

$$\phi = am\, u$$

for the function ϕ of u defined by (26). He also introduced

$$\cos \phi = \cos am\, u \quad \text{and} \quad \Delta\phi = \Delta am\, u = \sqrt{1 - k^2 \sin^2 \phi}.$$

This notation was abbreviated by Gudermann to

$$x = \sin \phi = \sin am\, u = sn\, u, \qquad \cos \phi = \cos am\, u = cn\, u,$$
$$\Delta\phi = \Delta am\, u = dn\, u.$$

We have at once that

$$sn^2\, u + cn^2\, u = 1, \qquad dn^2\, u + k^2\, sn^2\, u = 1.$$

If ϕ is changed to $-\phi$, then u changes sign. Hence

$$am(-u) = -am\, u, \qquad sn(-u) = -sn\, u, \qquad cn(-u) = cn\, u,$$
$$dn(-u) = dn\, u.$$

The role of π in the trigonometric functions is played here by the quantity K defined by

$$K = \int_0^1 \frac{dx}{\sqrt{(1 - x^2)(1 - k^2 x^2)}} = \int_0^{\pi/2} \frac{d\phi}{\sqrt{1 - k^2 \sin^2 \phi}} = F\!\left(k, \frac{\pi}{2}\right).$$

Associated with K is the transcendental quantity K', which is the same function of k' as K is of k and where k' is defined by $k^2 + k'^2 = 1$, $0 < k < 1$.

What is important (but not proved here) about K and K' is that

$$sn(u \pm 4K) = sn\, u, \qquad cn(u + 4K) = cn\, u, \qquad dn(u \pm 2K) = dn\, u.$$

Hence $4K$ is a period of the elliptic functions $sn\, u$ and $cn\, u$, and $2K$ is a period of $dn\, u$.

The functions $sn\, u$, $cn\, u$, and $dn\, u$ are defined thus far for real x and u. Abel had what he regarded as an element of each function because each was defined just for real values. His next thought was to define the elliptic functions in their totality by introducing complex values of u. As for the knowledge of complex functions, Abel in his visit to Paris had become acquainted with Cauchy's work. He had in fact studied the binomial theorem for complex values of the variable and the exponent. The extension first to pure imaginary values was achieved by what is called Jacobi's imaginary transformation. Abel introduced

$$\sin \theta = i \tan \phi, \qquad \cos \theta = \frac{1}{\cos \phi}, \qquad \Delta(\theta, k) = \frac{\Delta(\phi, k')}{\cos \phi},$$

33. *Fundamenta Nova*, 1829.

where $\theta = am\ i\ u$, so that

$$sn(iu, k) = i\frac{sn(u, k')}{cn(u, k')}, \qquad cn(iu, k) = \frac{1}{cn(u, k')}, \qquad dn(iu, k) = \frac{dn(u, k')}{cn(u, k')}.$$

In addition to allowing his variables to take on pure imaginary values, Abel also developed what is called the addition theorem for elliptic functions. In the case where

$$u = A(x) = \int_0^x \frac{1}{\sqrt{1 - x^2}}\, dx$$

we know that the integral is the multiple-valued function $A(x) = $ arc sin x, and it is true that

(27) $$A(x_1) + A(x_2) = A(x_1 y_2 + x_2 y_1),$$

wherein y_1 and y_2 are the corresponding cosine values; i.e. $y_1 = \sqrt{1 - x_1^2}$. But in this case a great simplification is obtained by introducing the inverse function $x = \sin u$, which is single-valued, and in place of (27) we have the familiar addition theorem for the sine function. Now in the case of

$$u = E(x) = \int_0^x \frac{dx}{\sqrt{R(x)}}$$

where $y^2 = R(x)$ is a polynomial of degree four, Euler had obtained the addition theorem (Chap. 19, sec. 4)

$$E(x_1) + E(x_2) = E(x_3)$$

where x_3 is a known rational function of x_1, x_2, y_1, and y_2, and $y = \sqrt{R(x)}$. Abel thought that for the inverse function $x = \phi(u)$ there might be a simple addition theorem, and this proved to be the case. This result too appeared in his 1827 paper. Thus for real u and v

(28) $$sn(u + v) = \frac{sn\ u\ cn\ v\ dn\ v + sn\ v\ cn\ u\ dn\ u}{1 - k^2 sn^2\ u\ sn^2\ v}$$

with analogous formulas for $cn(u + v)$ and $dn(u + v)$. These are the addition theorems for elliptic functions and the analogues of the addition theorems for elliptic integrals.

Having defined the elliptic functions for real and imaginary values of the arguments, Abel was able with the addition theorems to extend the definitions to complex values. For, if $z = u + iv$, $sn\ z = sn(u + iv)$ now has a meaning in view of the addition theorem, in terms of the sn, cn, and dn of u and of iv separately.

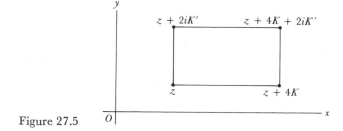

Figure 27.5

It also follows that

$$sn(iu + 2iK', k) = sn(iu, k)$$
(29) $$cn(iu + 4iK', k) = cn(iu, k)$$
$$dn(iu + 4iK', k) = dn(iu, k).$$

Thus the periods (which are not unique) of *sn z* are $4K$ and $2iK'$; those of *cn z* are $4K$ and $2K + 2iK'$; and those of *dn z* are $2K$ and $4iK'$. The important point about the periods is that there are two periods (whose ratio is not real), so that the elliptic functions are doubly periodic. This was one of Abel's great discoveries. The functions are single-valued. They need, therefore, to be studied only in one parallelogram (Fig. 27.5) of the complex plane, for they repeat their behavior in every congruent parallelogram. Elliptic functions besides being single-valued and doubly periodic, have only one essential singularity, at ∞. In fact these properties can be used to define elliptic functions. They do have poles within each period parallelogram.

Though Abel had taken from Legendre what could have been the cream of his life's work by introducing the idea of the inversion of elliptic integrals, which Legendre overlooked and which proved to be the key to exploring them, Legendre praised Abel by saying, "What a head this young Norwegian has." Charles Hermite said Abel left ideas on which mathematicians could work for 150 years.

Many of Abel's results were obtained independently by Jacobi, whose first publication in this field, as already noted, appeared in 1827. Jacobi was conscious of the fact that the basic method he used in his *Fundamenta Nova* was unsatisfactory and in that book to some extent, and in his subsequent lectures, used a different starting point. His lectures were never completely published, but their essential content is fairly well known through letters and notes made by his pupils. In his new approach he built his theory of elliptic functions on the basis of auxiliary functions called theta functions, which are illustrated by

(30) $$\theta(z) = \sum_{n = -\infty}^{\infty} e^{-n^2 t + 2niz},$$

where z and t are complex and $Re(t) > 0$. The series converge absolutely and uniformly in any bounded region of the z-plane. Jacobi introduced four theta functions and then expressed $sn\ u$, $cn\ u$, and $dn\ u$ in terms of these functions. The theta functions are the simplest elements out of which the elliptic functions can be constructed. He also obtained various expressions for the theta functions in the form of infinite series and infinite products. Further study of ideas in Abel's work led Jacobi to relations between the theta functions and the theory of numbers. This connection was taken up later by Hermite, Kronecker, and others. The pursuit of relations among many different forms of theta functions was a major activity for mathematicians of the nineteenth century. It was one of the many fads that sweep through mathematics regularly.

In an important paper of 1835[34] Jacobi showed that a single-valued function of a single variable which for every finite value of the argument has the character of a rational function (that is, is a meromorphic function) cannot have more than two periods, and the ratio of the periods is necessarily a nonreal number. This discovery opened up a new direction of work, namely, the problem of finding all doubly periodic functions. As early as 1844[35] Liouville, in a communication to the French Academy of Sciences, showed how to develop a complete theory of doubly periodic elliptic functions starting from Jacobi's theorem. This theory was a major contribution to elliptic functions. In the double periodicity Liouville had discovered an essential property of the elliptic functions and a unifying point of view for their theory, though the doubly periodic functions are a more general class than those designated by Jacobi as elliptic. However, all the fundamental properties of elliptic functions do hold for the doubly periodic functions.

Weierstrass took up the subject of elliptic functions about 1860. He learned Jacobi's work from Gudermann, and Abel's work through his papers. These impressed him so much that he constantly urged his pupils to read Abel. For his teacher's certificate he took up a problem assigned to him by Gudermann, namely, to represent the elliptic functions as quotients of power series. This he did. In his lectures as a professor he constantly reworked his theory of elliptic functions.

Legendre had reduced the elliptic integrals to three standard forms involving the square root of a fourth degree polynomial. Weierstrass arrived at three different forms involving the square root of a third degree polynomial,[36] namely,

$$\int \frac{dx}{\sqrt{4x^3 - g_2x - g_3}},\ \int \frac{x\ dx}{\sqrt{4x^3 - g_2x - g_3}},\ \int \frac{dx}{(x - a)\sqrt{4x^3 - g_2x - g_3}},$$

34. *Jour. für Math.*, 13, 1835, 55–78 = *Werke*, 2, 23–50.
35. *Comp. Rend.*, 19, 1844, 1261–63, and 32, 1851, 450–52.
36. *Sitzungsber. Akad. Wiss. zu Berlin*, 1882, 443–51 = *Werke*, 2, 245–55; see also *Werke*, 5.

and he introduced as the fundamental elliptic function that which results from "inverting" the first integral. That is, if

$$u = \int_0^x \frac{dx}{\sqrt{4x^3 - g_2 x - g_3}},$$

then the elliptic function x of u is Weierstrass's

$$x = \wp(u) = \wp(u \mid g_2, g_3).$$

In order that $\wp(u)$ should not degenerate into an exponential or trigonometric function, it is necessary that the discriminant $g_2^3 - 27g_3^2 \neq 0$, or, in other words, the three roots of the cubic in x should be unequal. Weierstrass's doubly periodic $\wp(u)$ plays the role of $sn\ u$ in the theory of Jacobi and furnishes the simplest doubly periodic function. He showed that every elliptic function can be expressed very simply in terms of $\wp(u)$ and the derivative of this function. The "trigonometry" of elliptic functions is simpler in Weierstrass's approach, but the functions of Jacobi and the elliptic integrals of Legendre are better for numerical work.

Weierstrass actually started with an element of his $\wp(u)$, namely,

$$\wp(u) = \frac{1}{u^2} + \frac{g_2}{4 \cdot 5} u^2 + \frac{g_3}{4 \cdot 7} u^4 + \cdots, \qquad (g_2, g_3 \text{ complex}),$$

which he obtained by solving the differential equation for dx/du given by the above integral, and then used the addition theorem for $\wp(u)$, in a manner similar to Abel's, to obtain the full function. Weierstrass's work completed, remodeled, and filled with elegance the theory of elliptic functions.

Though we shall not enter into specific details, we cannot leave the subject of elliptic functions without mentioning the work of Charles Hermite (1822–1901), who was a professor at the Sorbonne and at the Ecole Polytechnique. From his student days on, Hermite constantly occupied himself with the subject of elliptic functions. In 1892 he wrote, "I cannot leave the elliptic domain. Where the goat is attached she must graze." He produced basic results in the theory proper and studied the connection with the theory of numbers. He applied elliptic functions to the solution of the fifth degree polynomial equation and treated problems of mechanics involving the functions. He is known also for his proof of the transcendence of e and his introduction of the Hermite polynomials.

7. Hyperelliptic Integrals and Abel's Theorem

The successes achieved in the study of the elliptic integrals (25) and the corresponding functions encouraged the mathematicians to tackle a more difficult type, hyperelliptic integrals.

The hyperelliptic integrals are of the form

(31) $$\int R(x, y)\, dx,$$

where $R(x, y)$ is a rational function of x and y, $y^2 = P(x)$, and the degree of $P(x)$ is at least five. When $P(x)$ is of degree five or six, the integrals were called ultraelliptic in the mid-nineteenth century. To emphasize complex values it is common to write

(32) $$\int R(u, z)\, dz;$$

and $P(z)$ is often written as

(33) $$u^2 \equiv P(z) = A(z - e_1) \cdots (z - e_n).$$

Of course u is a multiple-valued function of z.

Among the integrals of the form (32) there are some that are everywhere finite. These basic ones are

(34) $$u_1 = \int \frac{dz}{u}, u_2 = \int \frac{z\, dz}{u}, \cdots, u_p = \int \frac{z^{p-1}\, dz}{u},$$

where u is given by (33) and $p = (n - 2)/2$ or $(n - 1)/2$ according as n is even or odd. For $n = 6$ (and so $p = 2$), there are two such integrals. The general integrals (32) have at most poles and logarithmically singular points, that is, singular points that behave like $\log z$ at $z = 0$. Those integrals of the first class, that is, those which are everywhere finite and so have no singular points, can always be expressed in terms of the p integrals (34), which are linearly independent.

For the case $n = 6$ (and $p = 2$) integrals of the second class are exemplified by

(35) $$\int \frac{z^2\, dz}{\sqrt{P(z)}}, \quad \int \frac{z^3\, dz}{\sqrt{P(z)}},$$

where $P(z)$ is a polynomial of the sixth degree. The integrals of the first and second kind for $n = 6$ have four periods each.

The hyperelliptic integrals are functions of the upper limit z if the lower limit is fixed. Suppose we denote one such function by w. Then, as in the case of the elliptic integrals, one can raise the question as to what is the inverse function z of w. This problem was tackled by Abel, but he did not solve it. Jacobi then tackled it.[37] Let us consider with Jacobi the particular hyperelliptic integrals

(36) $$w = \int_0^z \frac{dz}{\sqrt{P(z)}}, \quad \text{and} \quad w = \int_0^z \frac{z\, dz}{\sqrt{P(z)}},$$

37. *Jour. für Math.*, 9, 1832, 394–403 = *Werke*, 2, 7–16, and *Jour. für Math.*, 13, 1835, 55–78 = *Werke*, 2, 25–50 and 516–21.

where $P(z)$ is a fifth or sixth degree polynomial. Here the determination of z as a single-valued function of w proved to be hopeless. Jacobi showed in fact that the mere inversion of such integrals with $P(z)$ of degree five did not lead to a monogenic function. The inverse functions appeared unreasonable to Jacobi because in each case z as a function of w is infinitely valued; such functions were not well understood at the time.

Jacobi decided to consider combinations of such integrals. Guided by Abel's theorem (see below), whose statement at least he knew because that much had been published by this time, Jacobi did the following. Consider the equations

$$(37) \qquad \int_0^{z_1} \frac{dz}{\sqrt{P(z)}} + \int_0^{z_2} \frac{dz}{\sqrt{P(z)}} = w_1$$

$$(38) \qquad \int_0^{z_1} \frac{z\,dz}{\sqrt{P(z)}} + \int_0^{z_2} \frac{z\,dz}{\sqrt{P(z)}} = w_2.$$

Jacobi succeeded in showing that the symmetric functions $z_1 + z_2$ and $z_1 z_2$ are each single-valued functions of w_1 and w_2 with a system of *four* periods. The functions z_1 and z_2 of the two variables w_1 and w_2 can then be obtained. He also gave an addition theorem for these functions. Jacobi left many points incomplete. "For Gaussian rigor," he said, "we have no time."

The study of a generalization of the elliptic and hyperelliptic integrals was begun by Galois, but the more significant initial steps were taken by Abel in his 1826 paper. He considered (32), that is,

$$(39) \qquad \int R(u, z)\,dz,$$

but in place of (33), where u and z are connected merely by a polynomial as in $u^2 = P(z)$, Abel considered a general algebraic equation in z and u,

$$(40) \qquad f(u, z) = 0.$$

Equations (39) and (40) define what is called an *Abelian integral*, which then includes as special cases the elliptic and hyperelliptic integrals.

Though Abel did not carry the study of Abelian integrals very far, he proved a key theorem in the subject. Abel's basic theorem is a very broad generalization of the addition theorem for elliptic integrals (Chap. 19, sec. 4). The theorem and proof is in his Paris paper of 1826 and the statement of it is in *Crelle's Journal* for 1829.[38] Let us consider the integral

$$(41) \qquad \int R(x, y)\,dx,$$

38. *Jour. für Math.*, 4, 212–15 = *Œuvres*, 515–17.

where x and y are related by $f(x, y) = 0$, f being a polynomial in x and y. Abel writes as though x and y are real variables, though occasionally complex numbers appear. Loosely stated, Abel's theorem is this: A sum of integrals of the form (41) can be expressed in terms of p such integrals plus algebraic and logarithmic terms. Moreover, the number p depends only on the equation $f(x, y) = 0$ and is in fact the genus of the equation.

To obtain a more precise statement, let y be the algebraic function of x defined by the equation

$$(42) \qquad f(x, y) = y^n + A_1 y^{n-1} + \cdots + A_n = 0,$$

where the A_i are polynomials in x and the polynomial (42) is irreducible into factors of the same form. Let $R(x, y)$ be any rational function of x and y. Then the sum of any number m of similar integrals

$$(43) \qquad \int^{(x_1, y_1)} R(x, y)\, dx + \cdots + \int^{(x_m, y_m)} R(x, y)\, dx$$

with fixed (but arbitrary) lower limits is expressible by rational functions of $x_1, y_1, \ldots, x_m$, and y_m and logarithms of such rational functions, with the addition of a sum of a certain number p of integrals

$$(44) \qquad \int^{(z_1, s_1)} R(x, y)\, dx, \cdots, \int^{(z_p, s_p)} R(x, y)\, dx,$$

wherein $z_1, \cdots, z_p$ are values of x determinable from $x_1, y_1, \cdots, x_m$, and y_m as the roots of an algebraic equation whose coefficients are rational functions of $x_1, y_1, \cdots, x_m$, and y_m, and $s_1, \cdots$, and s_p are the corresponding values of y determined by (42) with any s_i determinable as a rational function of the z_i and the $x_1, y_1, \cdots, x_m$, and y_m. The relations thus determining $(z_1, s_1), \cdots$, (z_p, s_p) in terms of $(x_1, y_1), \cdots, (x_m, y_m)$ must be supposed to hold at all stages of the integration; in particular these relations determine the lower limits of the last p integrals in terms of the lower limits of the first m integrals. The number p does not depend on m nor on the form of the rational function $R(x, y)$, nor on the values $x_1, y_1, \cdots, x_m$, and y_m, but does depend on the fundamental equation (42) which relates y and x.

In the case of those hyperelliptic integrals where $f = y^2 - P(x)$ and $P(x)$ is a sixth degree polynomial and where p, which is $(n - 2)/2$, is 2, the main part of Abel's theorem says that

$$(45) \quad \int_0^{x_1} R(x, y)\, dx + \cdots + \int_0^{x_m} R(x, y)\, dx$$

$$= \int_0^A R(x, y)\, dx + \int_0^B R(x, y)\, dx + R_1(x_1, y_1, \cdots, x_m, y_m, A, y(A), B, y(B))$$

$$+ \sum \text{const.} \log R_2(x_1, y_1, \cdots, x_m, y_m, A, y(A), B, y(B)),$$

where R_1 and R_2 are rational functions of their variables.

Abel actually calculated the number p for a few cases of general $f(x, y) = 0$. Though he did not see the full significance of his result, he certainly recognized the notion of genus before Riemann and founded the subject of Abelian integrals. His paper was very difficult to understand, partly because he tried to prove what we would today call an existence theorem by actually computing the result. Later proofs simplified Abel's considerably (see also Chap. 39, sec. 4). Abel did not consider the inversion problem. All of the work on the inversion of hyperelliptic and Abelian integrals up to the entry of Riemann on the scene was hampered by the limited methods of handling multiple-valued functions.

8. *Riemann and Multiple-Valued Functions*

About 1850 one period of achievement in the theory of functions came to an end. Rigorous methods, such as Weierstrass provided, sharp delineation of results, and unquestionable existence proofs denote in any mathematical discipline an important but also the last stage of a development. Further development must be preceded by a period of free, numerous, disconnected, often accidentally discovered, and perhaps disordered creations. Abel's theorem was one such step. A new period of discovery in the theory of algebraic functions, their integrals, and the inverse functions is due to Riemann. Riemann actually offered a much broader theory, namely, the treatment of multiple-valued functions, thus far only touched upon by Cauchy and Puiseux, and thereby paved the way for a number of different advances.

Georg Friedrich Bernhard Riemann (1826–66) was a student of Gauss and Wilhelm Weber. He came to Göttingen in 1846 to study theology but soon turned to mathematics. His doctoral thesis of 1851, written under Gauss, and entitled "Grundlagen für eine allgemeine Theorie der Functionen einer veränderlichen complexen Grösse"[39] is a basic paper in complex function theory. Three years later he became a *Privatdozent* at Göttingen, that is, he was privileged to give lectures and charge students a fee. To qualify as *Privatdozent* he wrote the *Habilitationsschrift*, "Über die Darstellbarkeit einer Function durch eine trigonometrische Reihe," and gave a qualifying lecture, the *Habilitationsvortrag*, "Über die Hypothesen welche der Geometrie zu Grunde liegen." These were followed by a number of other famous papers. Riemann succeeded Dirichlet as a professor of mathematics at Göttingen in 1859. He died of tuberculosis.

Riemann is often described as a pure mathematician, but this is far from correct. Though he made numerous contributions to mathematics proper, he was deeply concerned with physics and the relationship of

39. *Werke*, 3–43.

mathematics to the physical world. He wrote papers on heat, light, the theory of gases, magnetism, fluid dynamics, and acoustics. He attempted to unify gravitation and light and investigated the mechanism of the human ear. His work on the foundations of geometry sought to ascertain what is absolutely reliable about our knowledge of physical space (Chap. 37). He himself states that his work on physical laws was his chief interest. As a mathematician, he used geometrical intuition and physical arguments freely. It seems very likely, on the basis of evidence given by Felix Klein, that Riemann's ideas on complex functions were suggested to him by his studies on the flow of electrical currents along a plane. The potential equation is central in that subject and became so in Riemann's approach to complex functions.

The key idea in Riemann's approach to multiple-valued functions is the notion of a Riemann surface. The function $w^2 = z$ is multiple-valued and in fact there are two values of w for each value of z. To work with this function and keep the two sets of values $\sqrt{z}$ and $-\sqrt{z}$ apart, that is, to separate the branches, Riemann introduced one plane of z-values for each branch. Incidentally he also introduced a point on each plane corresponding to $z = \infty$. The two planes are regarded as lying one above the other and joined, first of all, at those values of z where the branches give the same w-values. Thus for $w^2 = z$ the two planes, or sheets as they are called, are joined at $z = 0$ and $z = \infty$.

Now $w = +\sqrt{z}$ is represented by the z-values only on the upper sheet and $w = -\sqrt{z}$ by z-values on the lower sheet. As long as one considers z values on the upper sheet, one understands that one must compute $w_1 = +\sqrt{z}$. However, as z moves in a circle around the origin on that sheet (Fig. 27.6) so that the θ in $z = \rho(\cos\theta + i\sin\theta)$ varies from 0 to 2π, $\sqrt{z}$ covers one half of a circle in the complex plane in which the w-values are mapped. Now let z move into the second sheet as it crosses, say, the positive x-axis. As z moves on this second sheet, we take for w the values given by $w_2 = -\sqrt{z}$. As z makes another circuit of the origin, so that θ ranges from 2π to 4π while in the second sheet, we obtain the range of values of $w_2 = -\sqrt{z}$ for this path, and the polar angle of these w-values ranges from π to 2π. As z crosses the positive x-axis again, we regard it as traveling over the first sheet. Thus by two circuits of z-values around the origin, one on each sheet, we get the full range of w-values for the function $w^2 = z$. Moreover, and this is essential, w becomes a single-valued function of z if z ranges over the Riemann surface, which is the aggregate of the two sheets.

To distinguish paths on one sheet from those on the other we agree in the case of $w^2 = z$ to regard the positive x-axis as a branch-cut. This joins the points $z = 0$ and $z = \infty$. That is, whenever z crosses this cut, that branch of w must be taken which belongs to the sheet into which z passes.

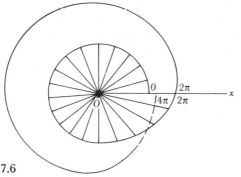

Figure 27.6

The branch-cut need not be the positive x-axis but it must, in the present case, join 0 and ∞. The points 0 and ∞ are called branch-points because the branches of $w^2 = z$ are interchanged when z describes a closed path about each.

The function $w^2 = z$ and, consequently, its associated Riemann surface, are especially simple. Let us consider the function $w^2 = z^3 - z$. This function also has two branches that become equal at $z = 0$, $z = 1$, $z = -1$, and $z = \infty$. Moreover, though we shall not present the full argument, all four of these points are branch-points, because if z makes a circuit around any one of them the value of w changes from that of one branch to that of another. The branch-cuts can be taken to be the line segments from 0 to 1, 0 to -1, 1 to ∞, and -1 to ∞. As z crosses any one of these cuts, the value of w changes from those it takes on one branch to those of the second.

For more complicated multiple-valued functions the Riemann surface is more complicated. An n-valued function requires an n-sheeted Riemann surface. There may be many branch-points, and one must introduce branch-cuts joining each two. Moreover the sheets that come together at one branch-point need not be the same as those that come together at another. If k sheets coincide at a branch-point, the order of the point is said to be $k - 1$. However, two sheets of a Riemann surface may touch at a point but the branches of the function may not change as z goes completely around the point. Then the point is not a branch-point.

It is not possible to represent Riemann surfaces accurately in three-dimensional space. For example, the two sheets for $w^2 = z$ must intersect along the positive x-axis if represented in three dimensions, so that one must be cut along the positive x-axis, whereas the mathematics requires smooth passage from the first sheet into the second and then, after a circuit around $z = 0$, back to the first sheet again.

Riemann surfaces are not merely a way of portraying multiple-valued functions but, in effect, make such functions single-valued on the surface as opposed to the z-plane. Then theorems about single-valued functions can be extended to multiple-valued functions. For example, Cauchy's theorem about integrals of *single-valued* functions being 0 over a curve bounding a domain (in which the function is analytic) was extended by Riemann to multiple-valued functions. The domain of analyticity must be simply connected (contractable to a point) on the *surface*.

Riemann thought of his surface as an *n*-sheeted duplication of the plane, each replica completed by a point at infinity. However, it is difficult to follow all of the arguments involving such a surface by visualizing in terms of the *n* interconnected planes. Hence mathematicians since Riemann's time have suggested equivalent models that are easier to contemplate. It is known that a plane may be transformed to a sphere by stereographic projection (Chap. 7, sec. 5). Hence we can construct a model of the Riemann surface by considering *n* concentric spheres of about the same radius. The sequence of spheres is to be the same as the sequence of plane sheets. The branch-points of the planes and the branch-cuts are likewise transformed to the spheres, so that these spheres wind into each other along branch-cuts. We now think of the set of spheres as the domain of z, and the multiple-valued function w of z is single-valued on this set of spherical sheets.

In our exposition thus far of Riemann's ideas, we have started with a function $f(w, z) = 0$, which is an irreducible polynomial in w and z, and have pointed out what its Riemann surface is. This was not Riemann's approach. He started with a Riemann surface and proposed to show that there is an equation $f(w, z) = 0$ that belongs to it and, further, that there are other single-valued and multiple-valued functions defined on this Riemann surface.

Riemann's definition of a single-valued analytic function $f(z) = u + iv$ is that the function is analytic at a point and in its neighborhood if it is continuous and differentiable and satisfies what we now call the Cauchy-Riemann equations

$$(46) \qquad \frac{\partial u}{\partial x} = \frac{\partial v}{\partial y} \quad \text{and} \quad \frac{\partial u}{\partial y} = -\frac{\partial v}{\partial x}.$$

These equations, as we know, appeared in the work of d'Alembert, Euler, and Cauchy. Incidentally, Riemann was the first to require that the existence of the derivative dw/dz mean that the limit of $\Delta w/\Delta z$ must be the same for *every* approach of $z + \Delta z$ to z. (This condition distinguishes complex functions, for in the case of a real function $u(x, y)$, the existence of the first derivatives of u for all directions of approach to some (x_0, y_0) does not guarantee analyticity.) He then sought what we might describe as the minimum conditions under which a function of $x + iy$ could be determined as a whole in whatever domain it existed. It is apparent from the Cauchy-

Riemann equations that u and v satisfy the two-dimensional potential equation

(47)
$$\frac{\partial^2 w}{\partial x^2} + \frac{\partial^2 w}{\partial y^2} = 0.$$

Riemann had the idea that the complex function could be determined at once in the totality of its domain of existence by using the fact that u satisfies the potential equation.

Specifically, Riemann supposes that a function w of position on the Riemann surface is determined, except as to an additive constant, by the real function $u(x, y)$ if u is subject to the conditions:

(1) It satisfies the potential equation at all points on the surface where its derivatives are not infinite.

(2) If u should be multiple-valued, its values at any point on the surface differ by linear combinations of integral multiples of real constants. (These real constants are the real parts of the periodicity moduli of w, which we shall discuss later.)

(3) u may have specified infinities (poles) of given form at assigned points on the surface. These infinities should belong to the real parts of the terms that give the infinities of w. He does assume further as a subsidiary condition that u may have finite values along a closed curve bounding a portion of the surface or that a relation between the boundary values of u and v may exist. Riemann is vague on how general such a relation may be.

These conditions should determine u. Once u is determined, then, in view of the Cauchy-Riemann equations,

(48)
$$v = \int \left(-\frac{\partial u}{\partial y}\, dx + \frac{\partial u}{\partial x}\, dy \right).$$

Thus v is also determined, and so is w. It is important to note that for Riemann the domain of u was any part of a Riemann surface, including possibly the whole surface. In his doctoral thesis he considered surfaces with boundaries and only later used closed surfaces, that is, surfaces without boundaries, such as a torus.

To determine u, Riemann's essential tool was what he called the Dirichlet principle, because he learned it from Dirichlet; but he extended it to domains on Riemann surfaces and, moreover, prescribed singularities for u in the domain and prescribed jumps (conditions 2 and 3 above). The Dirichlet principle says that the function u that minimizes the Dirichlet integral

$$\iint \left\{ \left(\frac{\partial u}{\partial x}\right)^2 + \left(\frac{\partial u}{\partial y}\right)^2 \right\} dx\, dy$$

satisfies the potential equation. The latter is in fact the necessary condition that the first variation of the Dirichlet integral vanish (see also Chap. 28, sec. 4). Since the integrand in the Dirichlet integral is positive and so has a lower bound which is greater than or at worst zero, Riemann concluded that there must be a function u that minimizes the integral and so satisfies the potential equation. Thus the existence of the function u and therefore by (48) of $f(z)$ which belongs to the Riemann surface and may even have prescribed singularities and complex jumps (periodicity modules) was assured as far as Riemann was concerned.

Once the existence of functions on a given Riemann surface as their domain has been established, one can show that there is a fundamental equation that can be associated with the given surface; that is, there is an $f(w, z) = 0$ that has the given surface as its surface. Just how the surface corresponds to the relationship between w and z Riemann does not state. Actually this $f(w, z) = 0$ is not unique. As a matter of fact, from every rational function w_1 of w and z on the surface one can obtain through $f(w, z) = 0$ another equation $f_1(w_1, z) = 0$ which, if irreducible, has the same Riemann surface. This is a feature of Riemann's method.

To investigate further the kinds of functions that can exist on a Riemann surface, it is necessary to become acquainted with Riemann's notion of the connectivity of a Riemann surface. A Riemann surface may have boundary curves or be closed like a sphere or a torus. If it is the Riemann surface of an algebraic function, that is, if $f(w, z) = 0$ defines w as a function of z and f is a polynomial in w and z, then the surface is closed. If f is irreducible, that is, cannot be expressed as a product of such polynomials, then the surface consists of one piece or is said to be connected.

A plane or a sphere is a surface such that any closed curve divides it into two parts so that it is not possible to pass continuously from a point in one part to a point in the second without crossing the closed curve. Such a surface is said to be *simply* connected. If, however, it is possible to draw some closed curve on a surface and the curve does not disconnect the surface, then the surface is not simply connected. For example one may draw two different closed curves on the torus (Fig. 27.7), and even the presence of both does not disconnect the surface.

Riemann wished to assign a number that indicated the connectivity of his surface. He regarded the poles and branch-points as part of the surface and because he had algebraic functions in mind, his surfaces were closed. By removing a small portion of one sheet the surface had a boundary curve C. He then thought of the surface as cut by a non-self-intersecting curve which runs from the boundary C to another point of the boundary C. Such a curve is called a cross-cut (*Querschnitt*). This cross-cut and C are regarded as a new boundary and a second cross-cut can be introduced that starts at one point of the (new) boundary and ends at another and does not cross the (new) boundary.

A sufficient number of these cross-cuts is introduced so as to cut up a

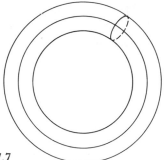

Figure 27.7

Riemann surface that may be multiply connected into one simply connected surface. Thus, if a surface is simply connected, no cross-cut is necessary and the surface has connectivity (*Grundzahl*) 1. A surface is called doubly connected if by one appropriate cross-cut it is changed into a single simply connected surface. Then the connectivity is 2. A plane ring and a spherical surface with two holes are examples. A surface is called triply connected when by two appropriate cross-cuts it is converted into a single simply connected surface. Then the connectivity is 3. An example is the (surface of a) torus with a hole in it. In general a surface will be said to be N-ply connected or has connectivity N if by $N - 1$ appropriate cross-cuts it can be changed into a single surface that is simply connected. A spherical surface with n holes in it has connectivity N.

It is now possible to relate the connectivity of a Riemann surface (with one boundary) and the number of branch-points. Each branch-point, r_i, say, is to be counted according to the multiplicity of the number of branches of the function which interchange at that point. If the number is w_i, $i = 1, 2, \ldots, r$, then the multiplicity of r_i is $w_i - 1$. Suppose the surface has q sheets. Then the connectivity N is given by

$$N = \sum_i w_i - 2q + 3.$$

One can show that the connectivity N of a closed surface with a single boundary is $2p + 1$. Hence

$$2p = \sum w_i - 2q + 2.$$

The integer p is called the genus of the Riemann surface and of the associated equation $f(w, z) = 0$. This relation was established by Riemann.

A special case of considerable importance is the surface of

$$w^2 = (z - a_1)(z - a_2) \cdots (z - a_n),$$

which has a two-leaved Riemann surface. There are n finite branch-points, and $z = \infty$ is a branch-point when n is odd. Then $\sum w_i = n$ or $n + 1$ and $2q = 4$. The genus p of the surface is given by

$$
p = \begin{cases} \dfrac{n-2}{2} & \text{when } n \text{ is even} \\[2ex] \dfrac{n-1}{2} & \text{when } n \text{ is odd.} \end{cases}
$$

Given a Riemann surface determined by $f(w, z) = 0$, we know that w is a single-valued function of the points on the surface. Then every rational function of w and z is also a single-valued function of position on the surface (because we can replace w in the rational function by its value in terms of z). Also the branch-points of this rational function, though not its poles, are the same as those of f. Conversely, one can prove that every single-valued function of position on the surface having poles of finite order is a rational function of w and z.

Even in the case of simple single-valued functions defined on the ordinary plane, the integrals of such functions can be multiple-valued. Thus

$$
\int_0^z \frac{dz}{1 + z^2} = w + n\pi
$$

where w is the value of the integral along, say, a straight line path from 0 to z, and n depends upon how the path from 0 to z circles $\pm i$. Likewise, when considering functions single-valued on a Riemann surface, say a rational function of w and z on the surface, the integral of such a function may be multiple-valued. This does indeed occur. If one then introduces cross-cuts to make the surface simply connected, and if the integral takes a path from z_1 to z_2, each time the path crosses a cross-cut a constant value I is added to the basic value U of the integral for a path on a simply connected portion of the surface. If the path should cross the cross-cut m times in the same direction, then the value mI is added to U. The constant I is called a periodicity modulus. Each cross-cut introduces its own periodicity modulus, and if the connectivity of the surface is $N + 1$, there are N linearly independent periodicity moduli. Let these be $I_1, I_2, \ldots, I_n$. Then the value of the integral of the original single-valued function taken over its original path is

$$
U + m_1 I_1 + m_2 I_2 + \cdots + m_n I_n,
$$

where $m_1, m_2, \ldots, m_n$ are integers. The I_i are generally complex numbers.

9. *Abelian Integrals and Functions*

While four important papers Riemann published in the *Journal für Mathematik*[40] repeat many of the ideas in his dissertation, they are primarily devoted to Abelian integrals and functions. The fourth paper is the one that gave the subject its major development. All four were difficult to understand; "they were a book with seven seals." Fortunately many fine mathematicians later elaborated on and explained the material. Riemann pulled together the work of Abel and Jacobi, which stemmed largely from real functions, and Weierstrass's treatment, which used complex functions.

Since Riemann had clarified the concept of multiple-value functions, he could be clearer about Abelian integrals. Let $f(w, z) = 0$ be the equation of a Riemann surface and let $\int R(w, z)\, dz$ be an integral of a rational function of w and z on this Riemann surface. Riemann classified Abelian integrals as follows. Among the *integrals* of rational functions of w and z on a Riemann surface determined by an equation $f(w, z) = 0$, there are some which, though multiple-valued functions on the uncut surface, are everywhere finite. These are called integrals of the first kind. The number of such linearly independent integrals is equal to the genus p of the surface if the connectivity is $2p + 1$. If the $2p$ cross-cuts are introduced, each integral is a single-valued function for a path in a region bounded by cross-cuts. If the path should cross a cross-cut, then the periodicity moduli we discussed in the preceding section must be brought in and the value of the integral is this: If W is its value from a fixed point to z, then all possible values are

$$W + \sum_{r=1}^{2p} m_r \omega_r,$$

where the m_r are integers and the ω_r are periodicity moduli for this integral.

The integrals of the second kind have algebraic but not logarithmic infinities. An elementary integral of the second kind has an infinity of first order at one point on the Riemann surface. If $E(z)$ is a value of the integral at one point on the surface (the upper limit of the integral), then all the values of the integral are included in

$$E(z) + \sum_{r=1}^{2p} n_r \varepsilon_r,$$

where the n_r are integers and the ε_r are the periodicity moduli for this integral. Two elementary integrals with an infinity at a common point on the Riemann surface differ by an integral of the first kind. From this we can infer that there are $p + 1$ linearly independent elementary integrals of the second kind with an infinity at the same point on the Riemann surface.

40. Vol. 54, 1857, 115–55 = *Werke*, 88–144.

Integrals having logarithmic infinities are called integrals of the third kind. It turns out that each must have two logarithmic infinities. If such an integral has no algebraic infinities, that is, no algebraic terms in the expansion of the integral in the vicinity of either of the points at which it has logarithmic infinities, then the integral is called an *elementary* integral of the third kind. There are $p + 1$ linearly independent elementary integrals of the third kind having their logarithmic infinities at the same two points on the Riemann surface. Every Abelian integral is a sum of integrals of the three kinds.

The analysis of Abelian integrals sheds light on what kinds of functions can exist on a Riemann surface. Riemann treats two classes of functions; the first consists of single-valued functions on the surface whose singularities are poles. The second class consists of functions that are one-valued on the surface provided with cross-cuts but discontinuous along each cross-cut. Indeed, such a function differs by a complex constant h_ν on one side of the νth cross-cut from its value on the other side. This second type of function may also have poles and logarithmic infinities. Riemann shows that the functions of the first class are algebraic and those of the second class are integrals of algebraic functions.

There are also everywhere finite functions on the surface. One can represent such a function by means of functions of the first of the above classes. One can also build up algebraic functions on a surface by combining integrals of the second and third kind. Thus Riemann showed that algebraic functions can be represented as a sum of transcendental functions. Also single-valued functions algebraically infinite in a given number of points can be represented by rational functions. A function that is one-valued on the entire surface is the integrand of an everywhere finite integral. The function can then be represented as a rational function in w and z and can have the form $\phi(w, z)/\partial f/\partial w$ where $f(w, z) = 0$ is the equation of the surface. The function ϕ, which enters here and into the construction of integrals of the first kind, is called the adjoint polynomial of $f(w, z) = 0$. Its degree is generally $n - 3$ when the degree of f is n.

The importance of rational functions on a Riemann surface derives from the fact, just mentioned, that every function single-valued on the surface and having no essential singularities is a rational function. Such a function has as many zeros as poles and takes on every value the same number of times. Moreover, once the equation $f(w, z) = 0$, which defines the surface, is fixed, all other functions of position on the surface are co-extensive in their totality with rational functions of w and z and integrals of such functions.

Weierstrass also worked on Abelian integrals during the 1860s. But he and the other successors of Riemann in this field set up transcendental functions from the algebraic functions, the reverse of Riemann's procedure.

They did so because they had reason to distrust the Dirichlet principle. Weierstrass, in a paper read in 1870,[41] pointed out that the existence of a function that minimizes the Dirichlet integral was not established. Riemann himself was of another mind. He recognized the problem of establishing the existence of a minimizing function for the Dirichlet integral before Weierstrass made his statement, but declared that the Dirichlet principle was just a convenient tool that happened to be available; the existence of the function u, he said, was nevertheless correct. Helmholtz's remark on this point is also interesting: "... for us physicists the [use of the] Dirichlet principle remains a proof."[42]

Another of the new investigations in complex function theory that Riemann launched is the inversion of Abelian integrals, that is, to determine the function z of u when

$$u = \int_0^z R(z, w) \, dz$$

and of course w and z are related by an algebraic equation. The function z of u is not only multiple-valued but is not clearly definable. As in the case of hyperelliptic integrals, Riemann took sums of p Abelian integrals and defined new *Abelian functions* of p variables that are one-valued and $2p$-tuply periodic. By a $2p$-tuply periodic function of p-variables is meant that there exist $2p$ sets of quantities $\omega_{1k}, \omega_{2k}, \ldots, \omega_{pk}, k = 1, 2, \ldots, 2p$, each set containing a period of each of the p-variables. Riemann proved that a single-valued function cannot have more than $2p$ sets of simultaneous periods. The Abelian functions, expressed in terms of theta functions in p-variables, are generalizations of the elliptic functions.

One of the noteworthy results on functions on a Riemann surface of genus p is now known as the Riemann-Roch theorem. The work on this result was begun by Riemann and completed by Gustav Roch (1839–66).[43] Essentially, the theorem determines the number of linearly independent meromorphic functions on the surface that have at most a specified finite set of poles. More specifically, suppose w is a single-valued function on the surface and has poles of first order at the points $c_1, \ldots, c_m$, but not elsewhere. The positions c_i need not be independent. If q linearly independent functions (adjoint functions) vanish on them, then w contains $m - p + q + 1$ arbitrary constants. It is a linear combination of arbitrary multiples of $m - p + q$ functions, each having $p - q + 1$ poles of the first order, $p - q$ of which are common to all the functions in the combination.

41. *Werke*, 2, 49–54.
42. For the subsequent history of the Dirichlet problem and Dirichlet principle, see Chap. 28, secs. 4 and 8.
43. *Jour. für Math.*, 64, 1864, 372–76.

10. Conformal Mapping

To complete the theory of his doctoral thesis, Riemann closes with some applications of the theory of functions to conformal mapping. The general problem of conformal mapping of a plane into a plane (which comes from map-making) was solved by Gauss in 1825. His result amounts to the fact that such a conformal map is set up by any analytic $f(z)$—though Gauss did not use complex function theory. Riemann knew that an analytic function establishes a conformal mapping from the z- to the w-plane, but he was concerned to extend this to Riemann surfaces. Thus a new chapter in conformal mapping was opened up.

At the close of his thesis Riemann gives the following theorem: Two given simply connected plane surfaces (he includes simply connected domains on Riemann surfaces) can be mapped one-to-one and conformally on each other, and one inner point and one boundary point on one surface can be assigned to arbitrarily chosen inner and boundary points on the other. Thereby the entire mapping is determined. This theorem contains as a special case the basic result that, given any simply connected domain D with a boundary that contains more than one point and given a point A of this domain and a direction T at this point, there exists a function $w = f(z)$ that is analytic in D and maps D conformally and biuniquely into a circle of radius 1 centered at the origin of the w-plane. Under this mapping A goes into the origin and T is sent into the direction of the positive real axis. This latter statement is usually described as the Riemann mapping theorem.

Riemann proved his theorem by using the Dirichlet principle, but since this principle was found faulty at the time, the mathematicians sought a sound proof. Carl Gottfried Neumann and Hermann Amandus Schwarz did prove (1870) that one could map a simply connected plane region onto a circle. However, they could not handle multileaved simply connected domains.

Incidentally, the emphasis on mapping a simply connected region conformally on a circle is explained by the fact that, to map one simply connected region on another conformally, it is sufficient to map each on a circle and then a product of two conformal mappings will do the trick.

While the proof of the Riemann mapping theorem remained open, a number of special results on conformal mapping were obtained. Of these a most useful one for the solution of partial differential equations was given by Schwarz[44] and Elwin Bruno Christoffel.[45] Their theorem shows how to map a polygon and its interior (Fig. 27.8) in the z-plane conformally into

44. *Jour. für Math.*, 70, 1869, 105–20 = *Ges. Abh.*, 2, 65–83.
45. *Annali di Mat.*, (2), 1, 1867, 95–103, and (2), 4, 1871, 1–9 = *Ges. Abh.*, 2, 56 ff.

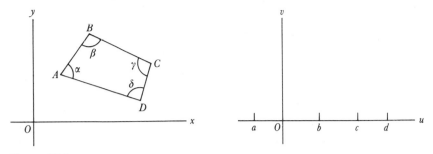

Figure 27.8

the upper half of the w-plane. The mapping is given by

$$z = c \int_a^z (w - a)^{(\alpha/\pi) - 1}(w - b)^{(\beta/\pi) - 1} \cdots dw + c',$$

where c and c' are determinable from the position of the polygon and where $a, b, c, \ldots$, correspond to $A, B, C, \ldots$. The mapping has proved to be very useful in solving the potential (Laplace) equation.

11. *The Representation of Functions and Exceptional Values*

The development of complex function theory proceeded apace in the latter part of the nineteenth century, and we shall have occasion in later chapters to consider some of these developments. However, a few of the many creations that bear primarily on complex functions themselves will be noted here.

Among single-valued complex functions, the entire functions, that is, those having no singularities in the finite part of the plane, which includes polynomials, e^z, sin z, and cos z, proved to be of considerable interest because they are roughly the analogues of the elementary real functions. For such functions Liouville's theorem states that every bounded entire function is a constant.[46] Weierstrass extended to entire functions the theorem on the decomposition of real polynomials into linear factors. The theorem Weierstrass established,[47] which probably dates from the 1840s, is called the factorization theorem, and states: If $G(z)$ is an entire function that does not vanish identically but has an infinite number of roots (that is, is not a polynomial), then $G(z)$ can be written as the infinite product

$$G(z) = \Gamma(z) z^m \prod_{n=1}^{\infty} \left(1 - \frac{z}{a_n}\right) e^{g_n(z)},$$

46. The theorem is due to Cauchy (*Comp. Rend.*, 19, 1844, 1377–81 = *Œuvres*, (1), 8, 378–85). C. W. Borchardt heard it in Liouville's lectures of 1847 and attributed the theorem to him.
47. *Abh. König. Akad. der Wiss.*, Berlin, 1876, 11–60 = *Werke*, 2, 77–124.

where

$$g_n(z) = \frac{z}{a_n} + \frac{1}{2}\left(\frac{z}{a_n}\right)^2 + \cdots + \frac{1}{m_n}\left(\frac{z}{a_n}\right)^{m_n}.$$

$\Gamma(z)$ is an entire function having no zeros; the a_n are the zeros of $G(z)$; and z^m represents the zero at $z = 0$ of multiplicity m if $G(z)$ has such a zero. The individual factors of the products are called the prime factors of $G(z)$.

Next to entire functions in order of complexity are the meromorphic functions, which can have only poles in the finite region of the complex plane. In his paper of 1876[48] Weierstrass showed that a meromorphic function can be expressed as the quotient of two entire functions. The theorem was extended by Gösta Mittag-Leffler (1846–1927) in a paper of 1877.[49] A function meromorphic in an arbitrary region can be expressed as the quotient of two functions, each analytic in the region. In both Weierstrass's and Mittag-Leffler's theorems the numerator and denominator do not vanish at the same point in the region.

Another topic that has engaged the attention of numerous mathematicians is the range of values that various types of complex functions can take on. A series of results was obtained by (Charles) Emile Picard (1856–1941), a professor of higher analysis at the Sorbonne and permanent secretary of the Paris Academy of Sciences. Picard in 1879[50] showed that an entire function can omit at most one finite value without reducing to a constant, and if there exist at least two values, each of which is taken on only a finite number of times, then the function is a polynomial. Otherwise the function takes on every value, other than the exceptional one, an infinite number of times. If the function is meromorphic, infinity being an admissible value, at most two values can be omitted without the function reducing to a constant.

In this same paper he extended a result of Julian W. Sochozki (1842–1927) and Weierstrass and proved that in any neighborhood of an isolated essential singular point a function takes on all values, with the possible exception of at most one (finite) value. The result is deep and has a multitude of consequences. Indeed a number of other results and alternative proofs were created, which carried the topic well into the twentieth century.

In the subject of complex functions, the nineteenth century ended with a return to fundamentals. The proofs of the Cauchy integral theorem in the nineteenth century used the fact that df/dz is continuous. Edouard Goursat (1858–1936) proved[51] Cauchy's theorem, $\int f(z)\,dz = 0$ around a closed

48. *Abh. König. Akad. der Wiss.*, Berlin, 1876, 11–60 = *Werke*, 2, 77–124.
49. *Öfversigt af Kongliga Vetenskops-Akademiens Förhandlingar*, 34, 1877, #1, 17–43; see also *Acta Math.*, 4, 1884, 1–79.
50. *Ann. de l'Ecole Norm. Sup.*, (2), 9, 1880, 145–66.
51. *Amer. Math. Soc., Trans.*, 1, 1900, 14–16.

curve C, without assuming the continuity of the derivative $f'(z)$ in the closed region bounded by the curve C. The existence of $f'(z)$ was sufficient. Goursat pointed out that the continuity of $f(z)$ and existence of the derivative are sufficient to characterize analyticity.

As our sketch of the rise of the theory of complex functions has shown, Cauchy, Riemann, and Weierstrass are the three major founders of the theory of functions. For a long time their respective ideas and methods were pursued independently by their followers. Then Cauchy's and Riemann's ideas were fused and Weierstrass's ideas were gradually deduced from the Cauchy-Riemann view, so that the idea of starting from the power series is no longer emphasized. Moreover, the rigor of the Cauchy-Riemann view was improved, so that from this standpoint also Weierstrass's approach is not essential. Full unification took place only at the beginning of the twentieth century.

Bibliography

Abel, N. H.: *Œuvres complètes*, 2 vols., 1881, Johnson Reprint Corp., 1964.

————: *Mémorial publié à l'occasion du centénaire de sa naissance*, Jacob Dybwad, Kristiania, 1902.

Brill, A., and M. Noether: "Die Entwicklung der Theorie der algebraischen Functionen in älterer und neuerer Zeit," *Jahres. der Deut. Math.-Verein.*, 3, 1892/3, 109–556, 155–86 in particular.

Brun, Viggo: "Niels Henrik Abel. Neue biographische Funde," *Jour. für Math.*, 193, 1954, 239–49.

Cauchy, A. L.: *Œuvres complètes*, 26 vols., Gauthier-Villars, 1882–1938, relevant papers.

Crowe, Michael J.: *A History of Vector Analysis*, University of Notre Dame Press, 1967, Chap. 1.

Enneper, A.: *Elliptische Funktionen: Theorie und Geschichte*, 2nd ed., L. Nebert, 1890.

Hadamard, Jacques: *Notice sur les travaux scientifiques de M. Jacques Hadamard*, Gauthier-Villars, 1901.

Jacobi, C. G. J.: *Gesammelte Werke*, 7 vols. and Supplement, G. Reimer, 1881–91; Chelsea reprint, 1968.

Jourdain, Philip E. B.: "The Theory of Functions with Cauchy and Gauss," *Bibliotecha Mathematica*, (3), 6, 1905, 190–207.

Klein, Felix: *Vorlesungen über die Entwicklung der Mathematik im 19. Jahrhundert*, 2 vols., Chelsea (reprint), 1950.

Lévy, Paul, et al.: "La Vie et l'œuvre de J. Hadamard," *L'Enseignement Mathématique*, (2), 13, 1967, 1–72.

Markuschewitsch, A. I.: *Skizzen zur Geschichte der analytischen Funktionen*, V E B Deutscher Verlag der Wissenschaften, 1955.

Mittag-Leffler, G.: "An Introduction to the Theory of Elliptic Functions," *Annals of Math.*, 24, 1922/3, 271–351.

————: "Die ersten 40 Jahre des Lebens von Weierstrass," *Acta Math.*, 39, 1923, 1–57.

Ore, O.: *Niels Henrik Abel, Mathematician Extraordinary*, University of Minnesota Press, 1957.

Osgood, W. F.: "Allgemeine Theorie der analytischen Funktionen," *Encyk. der Math. Wiss.*, B. G. Teubner, 1901–21, II B1, 1–114.

Reichardt, Hans, ed.: *Gauss: Leben und Werk*, Haude und Spenersche Verlagsbuchhandlung, 1960; B. G. Teubner, 1957, 151–82.

Riemann, Bernhard: *Gesammelte mathematische Werke*, 2nd ed., Dover (reprint), 1953.

Schlesinger, L.: "Über Gauss' Arbeiten zu Funktionenlehre," *Nachrichten König. Ges. der Wiss. zu Gött.*, 1912, Beiheft, 1–143; also in Gauss's *Werke*, 10_2, 77 ff.

Smith, David Eugene: *A Source Book in Mathematics*, 2 vols., Dover (reprint), 1959, pp. 55–66, 404–10.

Staeckel, Paul: "Integration durch imaginäres Gebiet," *Bibliotecha Mathematica*, (3), 1, 1900, 109–28.

Valson, C. A.: *La Vie et les travaux du baron Cauchy*, 2 vols., Gauthier-Villars, 1868.

Weierstrass, Karl: *Mathematische Werke*, 7 vols., Mayer und Müller, 1895–1924.

28
Partial Differential Equations in the Nineteenth Century

> The profound study of nature is the most fertile source of mathematical discoveries.   JOSEPH FOURIER

1. *Introduction*

The subject of partial differential equations, which had its beginnings in the eighteenth century, burgeoned in the nineteenth. As physical science expanded, in both the variety and depth of the phenomena investigated, the number of new types of differential equations increased; and even the types already known, the wave equation and the potential equation, were applied to new areas of physics. Partial differential equations became and remain the heart of mathematics. Their importance for physical science is only one of the reasons for assigning them this central place. From the standpoint of mathematics itself, the solution of partial differential equations created the need for mathematical developments in the theory of functions, the calculus of variations, series expansions, ordinary differential equations, algebra, and differential geometry. The subject has become so extensive that in this chapter we can give only a few of the major results.

We are accustomed today to classifying partial differential equations according to types. At the beginning of the nineteenth century, so little was known about the subject that the idea of distinguishing the various types could not have occurred. The physical problems dictated which equations were to be pursued and the mathematicians passed freely from one type to another without recognizing some differences among them that we now consider fundamental. The physical world was and still is indifferent to the mathematicians' classification.

2. *The Heat Equation and Fourier Series*

The first big nineteenth-century step, and indeed one of enormous importance, was made by Joseph Fourier (1768–1830). Fourier did very well

671

as a young student of mathematics but had set his heart on becoming an army officer. Denied a commission because he was the son of a tailor, he turned to the priesthood. When he was offered a professorship at the military school he had attended he accepted and mathematics became his life interest.

Like other scientists of his time, Fourier took up the flow of heat. The flow was of interest as a practical problem in the handling of metals in industry and as a scientific problem in attempts to determine the temperature in the interior of the earth, the variation of that temperature with time, and other such questions. He submitted a basic paper on heat conduction to the Academy of Sciences of Paris in 1807.[1] The paper was judged by Lagrange, Laplace, and Legendre and was rejected. But the Academy did wish to encourage Fourier to develop his ideas, and so made the problem of the propagation of heat the subject of a grand prize to be awarded in 1812. Fourier submitted a revised paper in 1811, which was judged by the men already mentioned and others. It won the prize but was criticized for its lack of rigor and so not published at that time in the *Mémoires* of the Academy. Fourier resented the treatment he received. He continued to work on the subject of heat and, in 1822, published one of the classics of mathematics, *Théorie analytique de la chaleur*.[2] It incorporated the first part of his 1811 paper practically without change. This book is the main source for Fourier's ideas. Two years later he became secretary of the Academy and was able to have his 1811 paper published in its original form in the *Mémoires*.[3]

In the interior of a body that is gaining or losing heat, the temperature is generally distributed nonuniformly and changes at any one place with time. Thus the temperature T is a function of space and time. The precise form of the function depends upon the shape of the body, the density, the specific heat of the material, the initial distribution of T, that is, the distribution at time $t = 0$, and the conditions maintained at the surface of the body. The first major problem Fourier considered in his book was the determination of the temperature T in a homogeneous and isotropic body as a function of x, y, z, and t. He proved on the basis of physical principles that T must satisfy the partial differential equation, called the heat equation in three space dimensions,

$$(1) \qquad \left(\frac{\partial^2 T}{\partial x^2} + \frac{\partial^2 T}{\partial y^2} + \frac{\partial^2 T}{\partial z^2} \right) = k^2 \frac{\partial T}{\partial t},$$

where k^2 is a constant whose value depends on the material of the body.

1. The manuscript is in the library of the *Ecole des Ponts et Chaussées*.
2. *Œuvres*, 1.
3. *Mém. de l'Acad. des Sci.*, *Paris*, (2), 4, 1819/20, 185–555, pub. 1824, and 5, 1821/22, 153–246, pub. 1826; only the second part is reproduced in Fourier's *Œuvres*, 2, 3–94.

Fourier then solved specific heat conduction problems. We shall consider a case that is typical of his method, the problem of solving equation (1) for the cylindrical rod whose ends are kept at 0° temperature and whose lateral surface is insulated so that no heat flows through it. Since this rod involves only one space dimension, (1) becomes

$$(2) \qquad \frac{\partial^2 T}{\partial x^2} = k^2 \frac{\partial T}{\partial t}$$

subject to the boundary conditions

$$(3) \qquad T(0, t) = 0, \qquad T(l, t) = 0, \qquad \text{for } t > 0,$$

and the initial condition

$$(4) \qquad T(x, 0) = f(x) \qquad \text{for } 0 < x < l.$$

To solve this problem Fourier used the method of separation of variables. He let

$$(5) \qquad T(x, t) = \phi(x)\psi(t).$$

On substitution in the differential equation, he obtained

$$\frac{\phi''(x)}{k^2\phi(x)} = \frac{\psi'(t)}{\psi(t)}.$$

He then argued (cf. [30] of Chap. 22) that each of these ratios must be a constant, $-\lambda$ say, so that

$$(6) \qquad \phi''(x) + \lambda k^2 \phi(x) = 0$$

and

$$(7) \qquad \psi'(t) + \lambda\psi(t) = 0.$$

However the boundary conditions (3), in view of (5), imply that

$$(8) \qquad \phi(0) = 0 \quad \text{and} \quad \phi(l) = 0.$$

The general solution of (6) is

$$\phi(x) = b \sin\left(\sqrt{\lambda}kx + c\right).$$

The condition that $\phi(0) = 0$ implies that $c = 0$. The condition $\phi(l) = 0$ imposes a limitation on λ, namely that $\sqrt{\lambda}$ must be an integral multiple of π/kl. Hence there are an infinite number of admissible values λ_ν of λ or

$$(9) \qquad \lambda_\nu = \left(\frac{\nu\pi}{kl}\right)^2, \qquad \nu \text{ integral.}$$

These λ_ν are what we now call the eigenvalues or characteristic values.

Since the general solution of (7) is an exponential function but λ is now limited to the λ_ν, then, in view of (5), Fourier had, so far, that

$$T_\nu(x, t) = b_\nu e^{-(\nu^2\pi^2/k^2l^2)t} \sin \frac{\nu\pi x}{l},$$

where b_ν now denotes the constant in place of b and $\nu = 1, 2, 3, \ldots$. However the equation (2) is linear, so that a sum of solutions is a solution. Hence one can assert that

$$(10) \qquad T(x, t) = \sum_{\nu=1}^{\infty} b_\nu e^{-(\nu^2\pi^2/k^2l^2)t} \sin \frac{\nu\pi x}{l}.$$

To satisfy the initial condition (4), one must have for $t = 0$

$$(11) \qquad f(x) = \sum_{\nu=1}^{\infty} b_\nu \sin \frac{\nu\pi x}{l}.$$

Fourier then faced the question, Can $f(x)$ be represented as a trigonometric series? In particular, can the b_ν be determined?

Fourier proceeded to answer these questions. Though by this time he was somewhat conscious of the problem of rigor, he proceeded formally in the eighteenth-century spirit. To follow Fourier's work we shall, for simplicity, let $l = \pi$. Thus we consider

$$(12) \qquad f(x) = \sum_{\nu=1}^{\infty} b_\nu \sin \nu x, \qquad \text{for } 0 < x < \pi.$$

Fourier takes each sine function and expands it by Maclaurin's theorem into a power series; that is, he uses

$$(13) \qquad \sin \nu x = \sum_{n=1}^{\infty} \frac{(-1)^{n-1}\nu^{2n-1}}{(2n-1)!} x^{2n-1}$$

to replace $\sin \nu x$ in (12). Then, by interchanging the order of the summations, an operation unquestioned at the time, he obtains

$$(14) \qquad f(x) = \sum_{n=1}^{\infty} \frac{(-1)^{n-1}}{(2n-1)!} \left(\sum_{\nu=1}^{\infty} \nu^{2n-1} b_\nu \right) x^{2n-1}.$$

Thus $f(x)$ is expressed as a power series in x, which implies a strong restriction on the admissible $f(x)$ that was not presupposed for the $f(x)$ Fourier treats. This power series must be the Maclaurin series for $f(x)$, so that

$$(15) \qquad f(x) = \sum_{k=0}^{\infty} \frac{1}{k!} f^{(k)}(0) x^k.$$

By equating coefficients of like powers of x in (14) and (15), Fourier finds that $f^{(k)}(0) = 0$ for even k and beyond that

$$\sum_{\nu=1}^{\infty} \nu^{2n-1} b_\nu = (-1)^{n-1} f^{(2n-1)}(0), \qquad n = 1, 2, 3, \cdots.$$

Now the derivatives of $f(x)$ are known, because $f(x)$ is a given initial condition. Hence the b_ν are an infinite set of unknowns in an infinite system of linear algebraic equations.

In a previous problem, wherein he faced this same kind of system, Fourier took the first k terms and the right-hand constant of the first k equations, solved these, and by obtaining a general expression for $b_{\nu,k}$, which denotes the approximate value of b_ν obtained from the first k equations, he boldly concluded that $b_\nu = \lim_{k \to \infty} b_{\nu,k}$. However, this time he had much difficulty in determining the b_ν. He took several different $f(x)$'s and showed how to determine the b_ν by very complicated procedures that involved divergent expressions. Using these special cases as a guide he obtained an expression for b_ν involving infinite products and infinite sums. Fourier realized that this expression was rather useless, and by further bold and ingenious, though again often questionable, steps finally arrived at the formula

$$(16) \qquad\qquad b_\nu = \frac{2}{\pi} \int_0^\pi f(s) \sin \nu s \, ds.$$

The conclusion was to an extent not new. We have already related (Chap. 20, sec. 5) how Clairaut and Euler had expanded some functions in Fourier series and had obtained the formulas

$$(17) \qquad a_n = \frac{1}{\pi} \int_{-\pi}^{\pi} f(x) \cos nx \, dx, \qquad b_n = \frac{1}{\pi} \int_{-\pi}^{\pi} f(x) \sin nx \, dx, \qquad n \geq 1.$$

Moreover, Fourier's results as derived thus far were limited, because he assumed his $f(x)$ had a Maclaurin expansion, which means an infinite number of derivatives. Finally, Fourier's method was certainly not rigorous and was more complicated than Euler's. Whereas Fourier had to use an infinite system of linear algebraic equations, Euler proceeded more simply by using the properties of trigonometric functions.

But Fourier now made some remarkable observations. He noted that each b_ν can be interpreted as the area under the curve of $y = (2/\pi) f(x) \cdot \sin \nu x$ for x between 0 and π. Such an area makes sense even for very arbitrary functions. The functions need not be continuous or could be known only graphically. Hence Fourier concluded that *every* function $f(x)$ could be represented as

$$(18) \qquad f(x) = \sum_{v=1}^{\infty} b_v \sin vx, \qquad \text{for } 0 < x < \pi.$$

This possibility had, of course, been rejected by the eighteenth-century masters except for Daniel Bernoulli.

How much Fourier knew of the work of his predecessors is not clear. In a paper of 1825 he says that Lacroix had informed him of Euler's work but he does not say when this happened. In any case Fourier was not deterred by the opinions of his predecessors. He took a great variety of functions $f(x)$, calculated the first few b_v for each function, and plotted the sum of the first few terms of the sine series (18) for each one. From this graphical evidence he concluded that the series always represents $f(x)$ over $0 < x < \pi$, whether or not the representation holds outside this interval. He points out in his book (p. 198) that two functions may agree in a given interval but not necessarily outside that interval. The failure to see this explains why earlier mathematicians could not accept that an arbitrary function can be expanded in a trigonometric series. What the series does give is the function in the domain 0 to π, in the present case, and periodic repetitions of it outside.

Once Fourier obtained the above simple result for the b_v, he, like Euler, realized that each b_v can be obtained by multiplying the series (18) by $\sin vx$ and integrating from 0 to π. He also points out that this procedure is applicable to the representation

$$(19) \qquad f(x) = \frac{a_0}{2} + \sum_{v=1}^{\infty} a_v \cos vx.$$

He considers next the representation of any $f(x)$ in the interval $(-\pi, \pi)$. The series (18) represents an odd function $[f(x) = -f(-x)]$ and the series (19) an even function $[f(x) = f(-x)]$. But any function can be represented as the sum of an odd function $f_0(x)$ and an even function $f_e(x)$ where

$$f_0(x) = \frac{1}{2}[f(x) - f(-x)], \qquad f_e = \frac{1}{2}[f(x) + f(-x)].$$

Then any $f(x)$ can be represented in the interval $(-\pi, \pi)$ by

$$(20) \qquad f(x) = \frac{a_0}{2} + \sum_{v=1}^{\infty} (a_v \cos vx + b_v \sin vx)$$

and the coefficients can be determined by multiplying through by $\cos vx$ or $\sin vx$ and integrating from $-\pi$ to π, which yields (17).

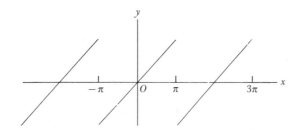

Figure 28.1

Fourier never gave any complete proof that an "arbitrary" function can be represented by a series such as (20). In the book he gives some loose arguments, and in his final discussion of this point (paragraphs 415, 416, and 423) he gives a sketch of a proof. But even there Fourier does not state the conditions that a function must satisfy to be expansible in a trigonometric series. Nevertheless Fourier's conviction that this was possible is expressed throughout the book. He also says[4] that his series are convergent no matter what $f(x)$ may be, whether or not one can assign an analytic expression to $f(x)$ and whether or not the function follows any regular law. Fourier's conviction that any function can be expanded in a Fourier series rested on the geometrical evidence described above. About this he says in his book (p. 206), "Nothing has appeared to us more suitable than geometrical constructions to demonstrate the truth of the new results and to render intelligible the forms which analysis employs for their expressions."

Fourier's work incorporated several major advances. Beyond furthering the theory of partial differential equations, he forced a revision in the notion of function. Suppose the function $y = x$ is represented by a Fourier series (20) in the interval $(-\pi, \pi)$. The series repeats its behavior in each interval of length 2π. Hence the function given by the series looks as shown in Figure 28.1. Such functions cannot be represented by a single (finite) analytic expression, whereas Fourier's predecessors had insisted that a function must be representable by a single expression. Since the entire function $y = x$ for all x is not represented by the series, they could not see how an arbitrary function, which is not periodic, could be represented by the series, though Euler and Lagrange had actually done so for particular nonperiodic functions. Fourier is explicit that his series can represent functions that also have different analytical expressions in different parts of the interval $(0, \pi)$ or $(-\pi, \pi)$, whether or not the expressions join one another continuously. He points out, finally, that his work settles the arguments on solutions of the vibrating-string problem in favor of Daniel Bernoulli. Fourier's work marked the break from analytic functions or functions developable in Taylor's series. It is also significant that a Fourier series

4. Page 196 = *Œuvres*, 1, 210.

represents a function over an entire interval, whereas a Taylor series represents a function only in a neighborhood of a point at which a function is analytic, though in special cases the radius of convergence may be infinite.

We have already noted that Fourier's paper of 1807, in which he had maintained that an arbitrary function can be expanded in a trigonometric series, was not well received by the Academy of Sciences of Paris. Lagrange in particular denied firmly the possibility of such expansions. Though he criticized only the lack of rigor in the paper, he was certainly disturbed by the generality of the functions that Fourier entertained, because Lagrange still believed that a function was determined by its values in an arbitrarily small interval, which is true of analytic functions. In fact Lagrange returned to the vibrating-string problem and, with no better insight than he had shown in earlier work, insisted on defending Euler's contention that an arbitrary function cannot be represented by a trigonometric series. Siméon-Denis Poisson (1781–1840) later asserted that Lagrange had shown that an arbitrary function can be represented by a Fourier series but Poisson, who was envious of Fourier, said this to rob Fourier of the credit.

Fourier's work made explicit another fact that was also implicit in the eighteenth-century work of Euler and Laplace. These men had expanded functions in series of Bessel functions and Legendre polynomials in order to solve specific problems. The general fact that a function might be expanded in a series of functions such as the trigonometric functions, Bessel functions, and Legendre polynomials was thrust into the light by Fourier's work. He showed, further, how the initial condition imposed on the solution of a partial differential equation could be met, and so advanced the technique of solving such equations. Fourier's paper of 1811, though not published until 1824–26, was accessible to other men in the meantime, and his ideas, at first grudgingly accepted, finally won favor.

Fourier's method was taken up immediately by the formerly critical Poisson, one of the greatest of nineteenth-century analysts and a first-class mathematical physicist. Though his father had wanted him to study medicine, he became a student and then professor at the fountainhead of nineteenth-century French mathematics, the Ecole Polytechnique. He worked in the theory of heat, was one of the founders of the mathematical theory of elasticity, and was one of the first to suggest that the theory of the gravitational potential be carried over to static electricity and magnetism.

Poisson was so much impressed with Fourier's evidence that arbitrary functions can be expanded in a series of functions that he believed all partial differential equations could be solved by series expansions; each term of the series would itself be a product of functions (cf. [10]), one for each independent variable. These expansions, he thought, embraced the most general solutions. He also believed that if an expansion diverged, this meant that one should seek an expansion in terms of other functions. Of course Poisson was far too optimistic.

From about 1815 on he himself solved a number of heat conduction problems and used expansions in trigonometric functions, Legendre polynomials, and Laplace surface harmonics. We shall encounter some of this work later. Much of Poisson's work on heat conduction was presented in his *Théorie mathématique de la chaleur* (1835).

3. Closed Solutions; the Fourier Integral

Despite the success and impact of Fourier's series solutions of partial differential equations, one of the major efforts during the nineteenth century was to find solutions in closed form, that is, in terms of elementary functions and integrals of such functions. Such solutions, at least of the kind known in the eighteenth and early nineteenth centuries, were more manageable, more perspicuous, and more readily used for calculation.

The most significant method of solving partial differential equations in closed form, which arose from work initiated by Laplace, was the Fourier integral. The idea is due to Fourier, Cauchy, and Poisson. It is impossible to assign priority for this important discovery, because all three presented papers orally to the Academy of Sciences that were not published until some time afterward. But each heard the others' papers, and one cannot tell from the publications what each may have taken from the verbal accounts.

In the last section of his prize paper of 1811, Fourier treated the propagation of heat in domains that extend to infinity in one direction. To obtain an answer for such problems, he starts with the general form of the solution of the heat equation for a bounded domain, namely (cf. [10]),

$$(21) \qquad u = \sum_{n=1}^{\infty} a_n e^{-kq_n^2 t} \cos q_n x,$$

where the q_n are determined by the boundary conditions and the a_n are determined by the initial conditions. Fourier now regards the q_n as abscissas of a curve and the a_n as the ordinates of that curve. Then $a_n = Q(q_n)$ where Q is some function of q. He then replaces (21) by

$$(22) \qquad u = \int_0^{\infty} Q(q) e^{-kq^2 t} \cos qx \, dq,$$

and seeks to determine Q. He goes back to the formula for the coefficients

$$a_n = \frac{2}{\pi} \int_0^{\pi} \phi(x) \cos nx \, dx,$$

where $\phi(x)$ would usually be the initial function. By a "limiting process" that replaces a_n by Q and n by q, he obtains

$$(23) \qquad Q = \frac{2}{\pi} \int_0^\infty F(x) \cos qx \, dx,$$

where $F(x)$, an even function, is the given initial temperature in the infinite domain. Then by using (23) in (22) and by an interchange of the order of integration, which Fourier does not bother to question, he has

$$u = \frac{2}{\pi} \int_0^\infty F(\alpha) \, d\alpha \int_0^\infty e^{-kq^2t} \cos qx \cos q\alpha \, dq.$$

Fourier then does the analogous thing for an odd $F(x)$, and so finally obtains

$$(24) \qquad u = \frac{1}{\pi} \int_{-\infty}^\infty F(\alpha) \, d\alpha \int_0^\infty e^{-kq^2t} \cos q(x - \alpha) \, dq.$$

Thus the solution is expressed in closed form. Now for $t = 0$, u is $F(x)$, which could be any given function. Hence Fourier asserts that, for an *arbitrary $F(x)$*,

$$(25) \qquad F(x) = \frac{1}{\pi} \int_{-\infty}^\infty F(\alpha) \, d\alpha \int_0^\infty \cos q(x - \alpha) \, dq,$$

and this is one form of the Fourier double-integral representation of an arbitrary function. In his book Fourier showed how to solve many types of differential equations with this integral. One use lies in the fact that if (24) is obtained by any process, then (25) shows that u satisfies the initial condition at $t = 0$. Another use is more evident if one writes the Fourier integral in exponential form, using the Euler relation, $e^{ix} = \cos x + i \sin x$. Then (25) becomes

$$F(x) = \frac{1}{2\pi} \int_{-\infty}^\infty e^{iqx} \, dq \int_{-\infty}^\infty F(\alpha) e^{-iq\alpha} \, d\alpha.$$

This form shows that $F(x)$ can be resolved into an infinite number of harmonic components with continuously varying frequency $q/2\pi$ and with amplitude $(1/2\pi) \int_{-\infty}^\infty F(\alpha) e^{-iq\alpha} \, d\alpha$, whereas the ordinary Fourier series resolves a given function into an infinite but discrete set of harmonic components.

Cauchy's derivation of the Fourier integral is somewhat similar. The paper in which it appeared, "Théorie de la propagation des ondes," received the prize of the Paris Academy for 1816.[5] This paper is the first large investigation of waves on the surface of a fluid, a subject initiated by

5. *Mém. divers savans*, 1, 1827, 3–312 = *Œuvres*, (1), 1, 5–318; see also Cauchy, *Nouv. Bull. de la Soc. Phil.*, 1817, 121–24 = *Œuvres*, (2), 2, 223–27.

Laplace in 1778. Though Cauchy sets up the general hydrodynamical equations he limits himself almost at once to special cases. In particular he considers the equation

$$\frac{\partial^2 q}{\partial x^2} + \frac{\partial^2 q}{\partial y^2} = 0$$

in which q is what was later called a velocity potential and x and y are spatial coordinates. He writes down without explanation the solution (cf. [22])

(26) $$q = \int_0^\infty \cos mx \, e^{-ym} f(m) \, dm,$$

wherein $f(m)$ is arbitrary thus far. Since $y = 0$ on the surface, q reduces to a given $F(x)$,

(27) $$F(x) = \int_0^\infty \cos mx \, f(m) \, dm.$$

Then Cauchy shows that

(28) $$f(m) = \frac{2}{\pi} \int_0^\infty \cos mu \, F(u) \, du.$$

With this value of $f(m)$

(29) $$F(x) = \frac{2}{\pi} \int_0^\infty \int_0^\infty \cos mx \cos mu \, F(u) \, du \, dm.$$

Cauchy thus obtains not only the Fourier double-integral representation of $F(x)$, but he also has the Fourier transform from $f(m)$ to $F(x)$ and the inverse transform. Given $F(x)$, $f(m)$ is determined by (28) and can be used in (26).

Shortly after Cauchy turned in his prize paper, Poisson, who could not compete for the prize because he was a member of the Academy, published a major work on water waves, "Mémoire sur la théorie des ondes."[6] In this work he derives the Fourier integral in about the same manner as Cauchy.

4. The Potential Equation and Green's Theorem

The next significant development centered about the potential equation, though the principal result, Green's theorem, has application to many other types of differential equations. The potential equation had figured in the eighteenth-century work on gravitation and had also appeared in the nineteenth-century work on heat conduction, for when the temperature

6. *Mém. de l'Acad. des Sci., Paris*, (2), 1, 1816, 71–186.

distribution in a body, though varying from point to point, remains the same as time varies, or is in the steady state, then T in (1) is independent of time and the heat equation reduces to the potential equation. The emphasis on the potential equation for the calculation of gravitational attraction continued in the early nineteenth century but was accentuated by a new class of applications to electrostatics and magnetostatics. Here too the attraction of ellipsoids was a key problem.

One correction in the theory of gravitational attraction as expressed by the potential equation was made by Poisson.[7] Laplace (Chap. 22, sec. 4) had assumed that the potential equation

$$(30) \qquad \frac{\partial^2 V}{\partial x^2} + \frac{\partial^2 V}{\partial y^2} + \frac{\partial^2 V}{\partial z^2} = 0,$$

wherein V is a function of x, y, and z, holds at any point (x, y, z) whether inside or outside of the body that exerts the gravitational attraction. Poisson showed that if (x, y, z) lies inside the attracting body, then V satisfies

$$(31) \qquad \frac{\partial^2 V}{\partial x^2} + \frac{\partial^2 V}{\partial y^2} + \frac{\partial^2 V}{\partial z^2} = -4\pi\rho,$$

where ρ is the density of the attracting body and is also a function of x, y, and z. Though (31) is still called Poisson's equation, his proof that it holds was not rigorous, as he himself recognized, even by the standards of that time.

In this same paper Poisson called attention to the utility of this function V in electrical investigations, remarking that its value over the surface of any conductor must be constant when electrical charge is allowed to distribute itself over the surface. In other papers he solved a number of problems calling for the distribution of charge on the surfaces of conducting bodies when the bodies are near each other. His basic principle was that the resultant electrostatic force in the interior of any one of the conductors must be zero.

Despite the work of Laplace, Poisson, Gauss, and others, almost nothing was known in the 1820s about the general properties of solutions of the potential equation. It was believed that the general integral must contain two arbitrary functions, of which one gives the value of the solution on the boundary and the other, the derivative of the solution on the boundary. Yet it was known in the case of steady-state heat conduction, in which the temperature satisfies the potential equation, that the temperature or heat distribution throughout the three-dimensional body is determined when the temperature alone is specified on the surface. Hence one of the

7. *Nouv. Bull. de la Soc. Philo.*, 3, 1813, 388–92.

arbitrary functions in the supposed general solution of the potential equation must somehow be fixed by some other condition.

At this point George Green (1793–1841), a self-taught English mathematician, undertook to treat static electricity and magnetism in a thoroughly mathematical fashion. In 1828 Green published a privately printed booklet, *An Essay on the Application of Mathematical Analysis to the Theories of Electricity and Magnetism*. This was neglected until Sir William Thomson (Lord Kelvin, 1824–1907) discovered it, recognized its great value, and had it published in the *Journal für Mathematik*.[8] Green, who learned much from Poisson's papers, also carried over the notion of the potential function to electricity and magnetism.

He started with (30) and proved the following theorems. Let U and V be any two continuous functions of x, y, and z whose derivatives are not infinite at any point of an arbitrary body. The major theorem asserts that (we shall used ΔV for the left hand side of [30], though it was not used by Green)

$$(32) \qquad \iiint U \, \Delta V \, dv + \iint U \frac{\partial V}{\partial n} \, d\sigma$$

$$= \iiint V \, \Delta U \, dv + \iint V \frac{\partial U}{\partial n} \, d\sigma,$$

where n is the surface normal of the body directed inward and $d\sigma$ is a surface element. Theorem (32), incidentally, was also proved by Michel Ostrogradsky (1801–61), a Russian mathematician, who presented it to the St. Petersburg Academy of Sciences in 1828.[9]

Green then showed that the requirement that V and each of its first derivatives be continuous in the interior of the body can be imposed instead of a boundary condition on the derivatives of V. In light of this fact, Green represented V in the interior of the body in terms of its value $\overline{V}$ on the boundary (which function would be given) and in terms of another function U which has the properties: (a) U must be 0 on the surface; (b) at a fixed but undetermined point P in the interior, U becomes infinite as $1/r$ where r is the distance of any other point from P; (c) U must satisfy the potential equation (30) in the interior. If U is known, and it might be found more readily because it satisfies simpler conditions than V, then V can be represented at every interior point by

$$4\pi V = -\iint \overline{V} \frac{\partial U}{\partial n} \, d\sigma,$$

8. *Jour. für Math.*, 39, 1850, 73–89; 44, 1852, 356–74; and 47, 1854, 161–221 = *Green's Mathematical Papers*, 1871, 3–115.
9. *Mém. Acad. Sci. St. Peters.*, (6), 1, 1831, 39–53.

where the integral extends over the surface, and $\partial U/\partial n$ is the derivative of U in the direction perpendicular to the surface and into the body. It is understood that the coordinates of P are contained in $\partial U/\partial n$ and are the arguments at P. This function U, introduced by Green, which Riemann later called the Green's function, became a fundamental concept of partial differential equations. Green himself used the term "potential function" for this special function U as well as for V. His method of obtaining solutions of the potential equation, as opposed to the method of using series of special functions, is called the method of singularities. There is unfortunately no general expression for the function U, nor is there a general method for finding it. Green was content in this matter to give the physical meaning of U for the case of the potential created by electric charges.

Green applied his theorem and concepts to electrical and magnetic problems. He also took up in 1833[10] the problem of the gravitational potential of ellipsoids of variable densities. In this work Green showed that when V is given on the boundary of a body, there is just one function that satisfies $\Delta V = 0$ throughout the body, has no singularities, and has the given boundary values. To make his proof, Green assumed the existence of a function that minimizes

$$(33) \qquad \iiint \left[\left(\frac{\partial V}{\partial x}\right)^2 + \left(\frac{\partial V}{\partial y}\right)^2 + \left(\frac{\partial V}{\partial z}\right)^2 \right] dv.$$

This is the first use of the Dirichlet principle (cf. Chap. 27, sec. 8).

In this 1835 paper Green did much of the work in n dimensions instead of three and also gave important results on what are now called ultra-spherical functions, which are a generalization to n variables of Laplace's spherical surface harmonics. Because Green's work did not become well known for some time, other men did some of this work independently.

Green is the first great English mathematician to take up the threads of the work done on the Continent after the introduction of analysis to England. His work inspired the great Cambridge school of mathematical physicists which included Sir William Thomson, Sir Gabriel Stokes, Lord Rayleigh, and Clerk Maxwell.

Green's achievements were followed by Gauss's masterful work of 1839,[11] "Allgemeine Lehrsätze in Beziehung auf die im verkehrten Verhältnisse des Quadrats der Entfernung wirkenden Anziehungs-und Abstossungs-kräfte" (General Theorems on Attractive and Repulsive Forces Which Act According to the Inverse Square of the Distance). Gauss proved rigorously Poisson's result, namely, that $\Delta V = -4\pi\rho$ at a point inside the acting mass, under the condition that ρ is continuous at that point and in a

10. *Trans. Camb. Phil. Soc.*, 5_3, 1835, 395–430 = *Mathematical Papers*, 187–222.
11. *Resultate aus den Beobachtungen des magnetischen Vereins*, Vol. 4, 1840 = *Werke*, 5, 197–242.

small domain around it. This condition is not fulfilled on the surface of the acting mass. On the surface the quantities $\partial^2 V/\partial x^2$, $\partial^2 V/\partial y^2$, and $\partial^2 V/\partial z^2$ have jumps.

The work thus far on the potential equation and on Poisson's equation assumed the existence of a solution. Green's proof of the existence of a Green's function rested entirely on a physical argument. From the existence standpoint the fundamental problem of potential theory was to show the existence of a potential function V, which William Thomson about 1850 called a harmonic function, whose values are given on the boundary of a region and which satisfies $\Delta V = 0$ in the region. One might establish this directly, or establish the existence of a Green's function U and from that obtain V. The problem of establishing the existence of the Green's function or of V itself is known as the Dirichlet problem or the first boundary-value problem of potential theory, the most basic and oldest existence problem of the subject. The problem of finding a V satisfying $\Delta V = 0$ in a region when the normal derivative of V is specified on the boundary is called the Neumann problem, after Carl G. Neumann (1832–1925), a professor at Leipzig. This problem is called the second fundamental problem of potential theory.

One approach to the problem of establishing the existence of a solution of $\Delta V = 0$, which Green had already used (see [33]), was brought into prominence by William Thomson. In 1847[12] Thomson announced the theorem or principle which in England is named after him and on the Continent is called the Dirichlet *principle* because Riemann so named it. Though Thomson stated it in a somewhat more general form, the essence of the principle may be put thus: Consider the class of all functions U that have continuous derivatives of the second order in the interior and exterior domains T and T' respectively separated by a surface S. The U's are to be continuous everywhere and assume on S the values of a continuous function f. The function V that minimizes the Dirichlet integral

$$(34) \qquad I = \iiint_T \left[\left(\frac{\partial U}{\partial x} \right)^2 + \left(\frac{\partial U}{\partial y} \right)^2 + \left(\frac{\partial U}{\partial z} \right)^2 \right] dv$$

is the one that satisfies $\Delta V = 0$ and takes on the value f on the boundary S. The connection between (34) and ΔV is that the first variation of I in the sense of the calculus of variations is ΔV, and this must be 0 for a minimizing V. Since for real U, I cannot be negative, it seemed clear that a minimizing function V must exist, and it is then not difficult to prove it is unique. The Dirichlet principle is then *one* approach to the Dirichlet problem of potential theory.

12. *Jour. de Math.*, 12, 1847, 493–96 = *Cambridge and Dublin Math. Jour.*, 3, 1848, 84–87 = *Math. and Physical Papers*, 1, 93–96.

Riemann's work on complex functions gave a new importance to the Dirichlet problem and the principle itself. Riemann's "proof" of the existence of V in his doctoral thesis used the two-dimensional case of the Dirichlet principle, but it was not rigorous, as he himself realized.

When Weierstrass in his paper of 1870[13] presented a critique of the Dirichlet principle, he showed that the *a priori* existence of a minimizing U was not supported by proper arguments. It was correct that for all continuous differentiable functions U that go continuously from the interior onto the prescribed boundary values the integral has a *lower bound*. But whether there is a function U_0 in the class of continuous, differentiable functions that furnishes the lower bound was not established.

Another technique for the solution of the potential equation employs complex function theory. Though d'Alembert in his work of 1752 (Chap. 27, sec. 2) and Euler in special problems had used this technique to solve the potential equation, it was not until the middle of the nineteenth century that complex function theory was vitally employed in potential theory. The relevance of function theory to potential theory rests on the fact that if $u + iv$ is an analytic function of z, then both u and v satisfy Laplace's equation. Moreover, if u satisfies Laplace's equation, then the conjugate function v such that $u + iv$ is analytic necessarily exists (Chap. 27, sec. 8).

Where the equation $\Delta u = 0$ is used in fluid flow, the function $u(x, y)$ is what Helmholtz called the velocity potential, and then $\partial u/\partial x$ and $\partial u/\partial y$ represent the components of the velocity of the fluid at any point (x, y). In the case of static electricity, u is the electrostatic potential and $\partial u/\partial x$ and $\partial u/\partial y$ are the components of electric force. In both cases the curves $u =$ const. are equipotential lines and the curves $v =$ const., which are orthogonal to $u =$ const., are the flow or stream lines (lines of force for electricity). The function $v(x, y)$ is called the stream function. The introduction of this function is clearly helpful because of its physical significance.

One advantage of the use of complex function theory in solving the potential equation derives from the fact that if $F(z) = F(x + iy)$ is an analytic function, so that its real and imaginary parts satisfy $\Delta V = 0$, then the transformation of x and y to ξ and η by

$$\xi = f(x, y), \qquad \eta = g(x, y)$$

where

$$\zeta = \xi + i\eta$$

produces another analytic function $G(\zeta) = G(\xi + i\eta)$, and its real and imaginary parts also satisfy $\Delta V(\xi, \eta) = 0$. Now if the original potential

13. Chap. 27, sec. 9.

problem $\Delta V = 0$ has to be solved in some domain D, then by proper choice of the transformation the domain D' in which the transformed $\Delta V = 0$ has to be solved can be much simpler. Here the use of conformal transformations, such as the Schwarz-Christoffel transformation, is of great service.

We shall not pursue the uses of complex function theory in potential theory because the details of its use go far beyond any basic methodology in the solution of partial differential equations. It is, however, again worth noting that many mathematicians resisted the use of complex functions because they were still not reconciled to complex numbers. At Cambridge University, even in 1850, cumbrous devices were used to avoid involving complex functions. Horace Lamb's *Treatise on the Mathematical Theory of the Motion of Fluids*, published in 1879 and still a classic (now known as *Hydrodynamics*), was the first book to acknowledge the acceptance of function theory at Cambridge.

5. *Curvilinear Coordinates*

Green introduced a number of major ideas whose significance extended far beyond the potential equation. Gabriel Lamé (1795–1870), a mathematician and engineer concerned primarily with the heat equation, introduced another major technique, the use of curvilinear coordinate systems, which could also be used for many types of equations. Lamé pointed out in 1833[14] that the heat equation had been solved only for conducting bodies whose surfaces are normal to the coordinate planes $x =$ const., $y =$ const., and $z =$ const. Lamé's idea was to introduce new systems of coordinates and the corresponding coordinate surfaces. To a very limited extent this had been done by Euler and Laplace, both of whom used spherical coordinates ρ, θ, and ϕ, in which case the coordinate surfaces $\rho =$ const., $\theta =$ const., and $\phi =$ const. are spheres, planes, and cones respectively. Knowing the equations that transform from rectangular to spherical coordinates, one can, as Euler and Laplace did, transform the potential equation from rectangular to spherical coordinates.

The value of the new coordinate systems and surfaces is twofold. First, a partial differential equation in rectangular coordinates might not be separable into ordinary differential equations in this system but might be separable in some other system. Secondly, the physical problem might call for a boundary condition on, say, an ellipsoid. Such a boundary is represented simply in a coordinate system wherein one family of surfaces consists of ellipsoids, whereas in the rectangular system a relatively complicated equation must be used. Moreover, after separation of variables in the proper coordinate system is employed, this boundary condition becomes applicable to just one of the resulting ordinary differential equations.

14. *Jour. de l'Ecole Poly.*, 14, 1833, 194–251.

Lamé introduced several new coordinate systems for the express purpose of solving the heat equation in these systems.[15] His chief system was the three families of surfaces given by the equations

$$\frac{x^2}{\lambda^2} + \frac{y^2}{\lambda^2 - b^2} + \frac{z^2}{\lambda^2 - c^2} - 1 = 0$$

$$\frac{x^2}{\mu^2} + \frac{y^2}{\mu^2 - b^2} + \frac{z^2}{\mu^2 - c^2} - 1 = 0$$

$$\frac{x^2}{\nu^2} + \frac{y^2}{\nu^2 - b^2} + \frac{z^2}{\nu^2 - c^2} - 1 = 0,$$

where $\lambda^2 > c^2 > \mu^2 > b^2 > \nu^2$. These three families are ellipsoids, hyperboloids of one sheet, and hyperboloids of two sheets, all of which possess the same foci. Any surface of one family cuts all the surfaces of the other two orthogonally, and in fact cuts them in lines of curvature (Chap. 23, sec. 7). Any point in space accordingly has coordinates (λ, μ, ν), namely the λ, μ, and ν of the surfaces, one from each family, which go through that point. This new coordinate system is called ellipsoidal, though Lamé called it elliptical, a term now used for another system.

Lamé transformed the heat equation for the steady-state case (temperature independent of time), that is, the potential equation, to these coordinates, and showed that he could use separation of variables to reduce the partial differential equation to three ordinary differential equations. Of course these equations must be solved subject to appropriate boundary conditions. In a paper of 1839[16] Lamé studied further the steady-state temperature distribution in a three-axis ellipsoid and gave a complete solution of the problem treated in his 1833 paper. In this 1839 paper he also introduced another curvilinear coordinate system, now called the spheroconal system, wherein the coordinate surfaces are a family of spheres and two families of cones. This system too Lamé used to solve heat conduction problems. Lamé wrote many other papers on heat conduction using ellipsoidal coordinates, including a second one of 1839 in the same volume of the *Journal de Mathématiques*, in which he treats special cases of the ellipsoid.[17]

The subject of mutually orthogonal families of surfaces had such obvious importance in the solution of partial differential equations that it became a subject of investigation in and for itself. In a paper of 1834[18] Lamé considered the general properties of any three families of mutually orthogonal surfaces and gave a procedure for expressing a partial differen-

15. *Annales de Chimie et Physique*, (2), 53, 1833, 190–204.
16. *Jour. de Math.*, 4, 1839, 126–63.
17. *Jour. de Math.*, 4, 1839, 351–85.
18. *Jour de l'Ecole Poly.*, 14, 1834, 191–288.

tial equation in any orthogonal coordinate system, a technique used continually ever since.

(Heinrich) Eduard Heine (1821–81) followed in Lamé's tracks. Heine in his doctoral dissertation of 1842 [19] determined the potential (steady-state temperature) not merely for the interior of an ellipsoid of revolution when the value of the potential is given at the surface, but also for the exterior of such an ellipsoid and for the shell between confocal ellipsoids of revolution.

Lamé was so much impressed with what he and others accomplished by the use of triply orthogonal coordinate systems that he thought all partial differential equations could be solved by finding a suitable system. Later he realized that this was a mistake. In 1859 he published a book on the whole subject, *Leçons sur les coordonnées curvilignes*.

Though the use of mutually orthogonal families of surfaces as the coordinate surfaces did not solve all partial differential equations, it did open up a new technique that could be exploited to advantage in many problems. The use of curvilinear coordinates was carried over to other partial differential equations. Thus Emile-Léonard Mathieu (1835–1900), in a paper of 1868,[20] treated the vibrations of an elliptic membrane, which involves the wave equation, and here introduced elliptic cylinder coordinates and functions appropriate to these coordinates, now called Mathieu functions (Chap. 29, sec. 2). In the same year Heinrich Weber (1842–1913), working with the equation $\partial^2 u/\partial x^2 + \partial^2 u/\partial y^2 + k^2 u = 0$, solved it[21] for a domain bounded by a complete ellipse and also for the region bounded by two arcs of confocal ellipses and two arcs of hyperbolas confocal with the ellipses. The special case in which the ellipses and hyperbolas become confocal parabolas was also considered, and here Weber introduced functions appropriate to expansions in this coordinate system, now called Weber functions or parabolic cylinder functions. In his *Cours de physique mathématique* (1873), Mathieu took up new problems involving the ellipsoid and introduced still other new functions.

Our discussion of the idea initiated by Lamé, the use of curvilinear coordinates, describes just the beginning of this work. Many other coordinate systems have been introduced; corresponding special functions that result from solving the ordinary differential equations, which arise from separation of variables, have also been studied.[22] Most of this theory of special functions was created by physicists as they needed the functions and their properties in concrete problems (see also Chap. 29).

19. *Jour. für Math.*, 26, 1843, 185–216.
20. *Jour. de Math.*, (2), 13, 1868, 137–203.
21. *Math. Ann.*, 1, 1869, 1–36.
22. See William E. Byerly, *An Elementary Treatise on Fourier Series*, Dover (reprint), 1959, and E. W. Hobson, *The Theory of Spherical and Ellipsoidal Harmonics*, Chelsea (reprint), 1955.

6. The Wave Equation and the Reduced Wave Equation

Perhaps the most important type of partial differential equation is the wave equation. In three spatial dimensions the basic form is

$$(36) \qquad \frac{\partial^2 u}{\partial x^2} + \frac{\partial^2 u}{\partial y^2} + \frac{\partial^2 u}{\partial z^2} = a^2 \frac{\partial^2 u}{\partial t^2}.$$

As we know, this equation had already been introduced in the eighteenth century and had also been expressed in spherical coordinates. During the nineteenth century new uses of the wave equation were found, especially in the burgeoning field of elasticity. The vibrations of solid bodies of a variety of shapes with different initial and boundary conditions and the propagation of waves in elastic bodies produced a host of problems. Further work in the propagation of sound and light raised hundreds of additional problems.

Where separation of variables is possible, the technique of solving (37) is no different from what Fourier did with the heat equation or Lamé did after expressing the potential equation in some system of curvilinear coordinates. Mathieu's use of curvilinear coordinates to solve the wave equation by separation of variables is typical of hundreds of papers.

Quite another and important class of results dealing with the wave equation was obtained by treating the equation as an entity. The first of such major results deals with initial-value problems and goes back to Poisson, who worked on this equation during the years 1808 to 1819. His principal achievement[23] was a formula for the propagation of a wave $u(x, y, z, t)$ whose initial state is described by the initial conditions

$$(37) \qquad u(x, y, z, 0) = \phi_0(x, y, z), \qquad u_t(x, y, z, 0) = \phi_1(x, y, z)$$

and which satisfies the partial differential equation

$$(38) \qquad \frac{\partial^2 u}{\partial x^2} + \frac{\partial^2 u}{\partial y^2} + \frac{\partial^2 u}{\partial z^2} = \frac{1}{a^2} \frac{\partial^2 u}{\partial t^2}$$

wherein a is a constant. The solution u is given by

$$(39) \quad u(x, y, z, t)$$

$$= \frac{1}{4\pi a} \int_0^\pi \int_0^{2\pi} \phi_1(x + at \sin \phi \cos \theta, y + at \sin \phi \sin \theta,$$

$$z + at \cos \phi) at \sin \phi \, d\theta \, d\phi$$

$$+ \frac{1}{4\pi a} \frac{\partial}{\partial t} \int_0^\pi \int_0^{2\pi} \phi_0(x + at \sin \phi \cos \theta, y + at \sin \phi \sin \theta,$$

$$z + at \cos \phi) at \sin \phi \, d\theta \, d\phi$$

wherein θ and ϕ are the usual spherical coordinates. The domain of integration is the surface of a sphere S_{at} with radius at about the point P with coordinates x, y, and z.

23. *Mém. de l'Acad. des Sci., Paris*, (2), 3, 1818, 121–76.

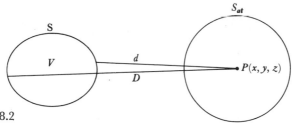

Figure 28.2

To obtain some indication of what Poisson's result means, let us consider a physical example. Suppose that the initial disturbance is set up by a body V (Fig. 28.2) with boundary S so that ϕ_0 and ϕ_1 are defined on V and are 0 outside of V. We say that the initial disturbance is localized in V. Physically a wave sets out from V and spreads out into space. Poisson's formula tells us what happens at any point $P(x, y, z)$ outside of V. Let d and D represent the minimum and maximum distances of P to the points of V. When $t < d/a$, the integrals in (39) are 0 because the domain of integration is the surface of the sphere S_{at} with radius at and center at P. Since ϕ_0 and ϕ_1 are 0 on S_{at}, then the function u is 0 at P. This means that the wave spreading out from S has not reached P. At $t = d/a$, the sphere S_{at} just touches S so the leading front of the wave emanating from S arrives at P. Between $t = d/a$ and $t = D/a$ the sphere S_{at} cuts V and so $u(P, t) \neq 0$. Finally for $t > D/a$, the sphere S_{at} will not cut S (the entire region V lies inside S_{at}); that is, the initial disturbance has passed through P. Hence again $u(P, t) = 0$. The instant $t = D/a$ corresponds to the passage of the trailing edge of the wave through P. At any given time t the leading edge of the wave takes the form of a surface which separates points not reached by the disturbance from those reached. This leading edge is the envelope of the family of spheres with centers on S and with radii at. The terminating edge of the wave at time t is a surface separating points at which the disturbance exists from those which the disturbance has passed. We see, then, that the disturbance which is localized in space gives rise at each point P to an effect that lasts only for a finite time. Moreover the wave (disturbance) has a leading and a terminating edge. This entire phenomenon is called Huygens's principle.

A quite different method of solving the initial-value problem for the wave equation was created by Riemann in the course of his work on the propagation of sound waves of finite amplitude.[24] Riemann considers a linear differential equation of second order that can be put in the form

$$(40) \qquad L(u) = \frac{\partial^2 u}{\partial x\, \partial y} + D\frac{\partial u}{\partial x} + E\frac{\partial u}{\partial y} + Fu = 0,$$

24. *Abh. der Ges. der Wiss. zu Gött.*, 8, 1858/59, 43–65 = *Werke*, 156–78.

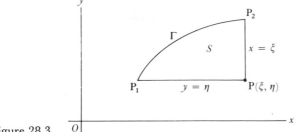

Figure 28.3

where D, E, and F are continuous and differentiable to second order functions of x and y. The problem calls for finding u at an arbitrary point P (Fig. 28.3) when one knows u and $\partial u/\partial n$ (which means knowing $\partial u/\partial x$ and $\partial u/\partial y$) along a curve Γ. His method depends on finding a function v (called a Riemann function or characteristic function)[25] that satisfies what is now called the adjoint equation

$$(41) \qquad M(v) = \frac{\partial^2 v}{\partial x\, \partial y} - \frac{\partial (Dv)}{\partial x} - \frac{\partial (Ev)}{\partial y} + Fv = 0$$

and other conditions we shall specify shortly.

Riemann introduced the segments PP_1 and PP_2 of the characteristics (he did not use the term) $x = \xi$ and $y = \eta$ through P. Now a generalized Green's theorem (in two dimensions) is applied to the differential expression $L(u)$. To express the theorem compactly, let us introduce

$$X = \frac{1}{2}\left(v\frac{\partial u}{\partial y} - u\frac{\partial v}{\partial y}\right) + Duv$$

$$Y = \frac{1}{2}\left(v\frac{\partial u}{\partial x} - u\frac{\partial v}{\partial x}\right) + Euv.$$

Then Green's theorem states

$$(42) \qquad \int_S [vL(u) - uM(v)]\, dS = \int_S \left(\frac{\partial X}{\partial x} + \frac{\partial Y}{\partial y}\right) dS$$

$$= \int_C \{X\cos(n, x) + Y\cos(n, y)\}\, ds,$$

where S is the area in the figure, C is the entire boundary of S, and $\cos(n, x)$ is the cosine of the angle between the normal to C and the x-axis.

25. v is not the same as a fundamental solution or a Green's function.

Beyond satisfying (41), Riemann requires of v that

$$\text{(a)} \quad v = 1 \quad \text{at} \quad P,$$

(43)

$$\text{(b)} \quad \frac{\partial v}{\partial y} - Dv = 0 \quad \text{on} \quad x = \xi,$$

$$\text{(c)} \quad \frac{\partial v}{\partial x} - Ev = 0 \quad \text{on} \quad y = \eta.^{26}$$

By using the condition that $M(v) = 0$ and the conditions (43), and by evaluation of the curvilinear integral over C, Riemann obtains

$$(44) \quad u(\xi, \eta) = \int_\Gamma \{X \cos (n, x) + Y \cos (n, y)\} \, ds + \frac{1}{2} \{(uv)_{P_1} + (uv)_{P_2}\}.$$

Thus the value of u at any arbitrary point P is given in terms of the values of u, $\partial u/\partial n$, v, and $\partial v/\partial n$ on Γ and the values of u and v at P_1 and P_2.

Now u is given at P_1 and P_2. The function v must itself be found by solving $M(v) = 0$ and meeting the conditions in (43). Hence what Riemann's method achieves is to change the original initial-value problem for u to another kind of initial-value problem, the one for v. The second problem is usually easier to solve. In Riemann's physical problem it was especially easy to find v. However the existence of such a v generally was not established by Riemann.

The Riemann method as just described is useful only for the type of equation exemplified by the wave equation (hyperbolic equations) in two independent variables and cannot be extended directly. The extension of the method to more than two independent variables meets with the difficulty that the Riemann function becomes singular on the boundary of the domain of integration and the integrals diverge. The method has been extended at the cost of increased complication.

Progress in the solution of the wave equation by other methods is intimately connected with what are called steady-state problems, which lead to the reduced wave equation. The wave equation, by its very form, involves the time variable. In many physical problems, where one is interested in simple harmonic waves, one assumes that $u = w(x, y, z)e^{ikt}$, and by substituting this into the wave equation one obtains

$$(45) \quad \Delta w + k^2 w = \frac{\partial^2 w}{\partial x^2} + \frac{\partial^2 w}{\partial y^2} + \frac{\partial^2 w}{\partial z^2} + k^2 w = 0.$$

This is the reduced wave equation or the Helmholtz equation. The equation $\Delta w + k^2 w = 0$ represents all harmonic, acoustic, elastic, and electromagnetic waves. While the older authors were satisfied to find particular integrals, Hermann von Helmholtz (1821–94), in his work on the oscillations

26. For two-dimensional problems v is a function of four variables, ξ, η, x, and y. It satisfies $M(v) = 0$ as a function of x and y.

of air in a tube (organ pipe) with an open end, gave the first general investigation of its solutions.[27] He was concerned with the acoustical problem in which w is the velocity potential of a harmonically moving air mass, k is a constant determined by the elasticity of the air and the oscillation frequency, and λ, which equals $2\pi/k$, is the wavelength. By applying Green's theorem he showed that any solution of $\Delta w + k^2 w = 0$ that is continuous in a given domain can be represented as the effect of single and double layers of excitation points on the surface of the domain. Using $e^{-ikr}/4\pi r$ as one of the functions in Green's theorem, he obtained

$$(46) \qquad w(P) = -\frac{1}{4\pi} \iint \frac{e^{-ikr}}{r} \frac{\partial w}{\partial n} \, dS + \frac{1}{4\pi} \iint w \frac{\partial}{\partial n} \left(\frac{e^{-ikr}}{r}\right) dS$$

wherein r denotes the distance from P to a variable point on the boundary. Thus w at any point P within the domain in which the solution is sought is given in terms of the values of w and $\partial w/\partial n$ on the boundary S.

The work of Helmholtz was used by Gustav R. Kirchhoff (1824–87), one of the great German nineteenth-century mathematical physicists, to obtain another solution of the initial-value problem for the wave equation. Let us suppose that $\Delta w + k^2 w = 0$ comes from

$$\frac{\partial^2 u}{\partial t^2} = c^2 \Delta u$$

wherein we have let $u = we^{i\sigma t}$ so that $k = \sigma/c$. Then (46) may be written as

$$(47) \quad u(P, t) = -\frac{1}{4\pi} \iint \frac{e^{i\sigma[t-(r/c)]}}{r} \frac{\partial u}{\partial n} \, dS + \frac{1}{4\pi} \iint u \frac{\partial}{\partial n} \left(\frac{e^{i\sigma[t-(r/c)]}}{r}\right) dS.$$

This formula was generalized by Kirchhoff. If we let $\phi(t)$ be the value of u at any point (x, y, z) of the boundary at the instant τ and let $f(\tau)$ be the corresponding value of $\partial u/\partial n$, then Kirchhoff showed[28] that

$$(48) \quad u(P, t) = -\frac{1}{4\pi} \iint \frac{f[t-(r/c)]}{r} \, dS + \frac{1}{4\pi} \iint \frac{\partial}{\partial n} \left(\frac{\phi[t-(r/c)]}{r}\right) dS,$$

provided that in the last term the differentiation with respect to n applies to r only insofar as it appears explicitly in both numerator and denominator. Thus u is obtained at P in terms of values of u and $\partial u/\partial n$ at earlier times at points of the closed surface surrounding P. This result is called Huygens's principle of acoustics and it is a generalization of Poisson's formula.

We have noted that Riemann used a somewhat generalized Green's theorem. The full generalization of Green's theorem that employs the adjoint differential equation and which is also called Green's theorem, comes from a paper by Paul Du Bois-Reymond (1831–89)[29] and from

27. *Jour. für Math.*, 57, 1860, 1–72 = *Wissenschaftliche Abhandlungen*, 1, 303–82.
28. *Sitzungsber. Akad. Wiss. zu Berlin*, 1882, 641–69 = *Ges. Abh.*, 2, 22 ff.
29. *Jour. für Math.*, 104, 1889, 241–301.

Darboux in his *Théorie générale des surfaces*;[30] both cite Riemann's paper of 1858/59. If the given equation is

$$L(u) = A\frac{\partial^2 u}{\partial x^2} + 2B\frac{\partial^2 u}{\partial x\,\partial y} + C\frac{\partial^2 u}{\partial y^2} + D\frac{\partial u}{\partial x} + E\frac{\partial u}{\partial y} + Fu = 0,$$

wherein the coefficients are functions of x and y, one integrates the product $vL(u)$ over an arbitrary domain R of the xy-plane under the assumption that u, v, and their first and second partial derivatives are continuous. Then integration by parts yields the generalized Green's theorem, which states that

$$\iint uM(v)\ dx\ dy = -\iint vL(u)\ dx\ dy - \int (Q\ dy - P\ dx),$$

where the double integrals are over the interior of R and the single integrals over the boundary of R,

$$M(v) = \frac{\partial^2(Av)}{\partial x^2} + 2\frac{\partial^2(Bv)}{\partial x\,\partial y} + \frac{\partial^2(Cv)}{\partial y^2} - \frac{\partial(Dv)}{\partial x} - \frac{\partial(Ev)}{\partial y} + Fu$$

$$P = B\left(v\frac{\partial u}{\partial x} - u\frac{\partial v}{\partial x}\right) + C\left(v\frac{\partial u}{\partial y} - u\frac{\partial v}{\partial y}\right) + \left(E - \frac{\partial B}{\partial x} - \frac{\partial C}{\partial y}\right)uv$$

$$Q = A\left(v\frac{\partial u}{\partial x} - u\frac{\partial v}{\partial x}\right) + B\left(v\frac{\partial u}{\partial y} - u\frac{\partial v}{\partial y}\right) + \left(D - \frac{\partial A}{\partial x} - \frac{\partial B}{\partial y}\right)uv.$$

$M(v)$ is the adjoint expression of $L(u)$ and $M(v) = 0$ is the adjoint differential equation. Conversely $L(u)$ is the adjoint of $M(v)$.

The significance of Green's theorem is that it can be used to obtain solutions of some partial differential equations. Thus, since the elliptic equation can always be put in the form

$$L(u) = \Delta u + a\frac{\partial u}{\partial x} + b\frac{\partial u}{\partial y} + cu = 0,$$

then

$$M(v) = \Delta v - \frac{\partial(av)}{\partial x} - \frac{\partial(bv)}{\partial y} + cv = 0.$$

Let v be a solution of the adjoint equation that becomes logarithmically infinite at an arbitrary point (ξ, η); that is, it behaves like

$$v = U\log r + V,$$

where r is the distance from (ξ, η) to (x, y); U and V are continuous in the domain R being considered; and U is normalized so that $U(\xi, \eta) = 1$. Now exclude (ξ, η) from the domain of integration by enclosing it in a

30. Vol. 2, Book IV, Chap. 4, 2nd ed., 1915.

circle. Then the generalized Green's theorem gives, when the circle is contracted to (ξ, η):

$$(49) \quad 2\pi u(\xi, \eta) = -\int\int vL(u) \; dx \; dy$$

$$+ \int \left[v \frac{\partial u}{\partial n} - u \frac{\partial v}{\partial n} + (a \cos (n, x) + b \cos (n, y)) \cdot uv \right] ds$$

where n is positive if directed to the outside of the domain; the single integral is taken counterclockwise over the boundary. Since u satisfies $L(u) = 0$, if we know v and if u and $\partial u/\partial n$ are given (both are not arbitrary) on the boundary, then we have u expressed as a single integral. The function v is called a Green's function, though often the condition that v vanish on the boundary of R is added in the definition of the Green's function. Various specializations and generalizations of this use of Green's theorem have been developed.

7. Systems of Partial Differential Equations

In the eighteenth century the differential equations of fluid motion presented the first important system of partial differential equations. In the nineteenth century three more fundamental systems, the fluid dynamical equations for viscous media, the equations of elastic media, and the equations of electromagnetic theory, were created.

The acquisition of the equations of fluid motion when viscosity is present (as it always is) took a tortuous path. Euler had given the equations of motion of a fluid that is nonviscous. Since the time of Lagrange the essential difference between the motion of a fluid when a velocity potential exists and when it does not had been recognized. Led by a formal analogy with the theory of elasticity and by the hypothesis of molecules animated by repulsive forces, Claude L. M. H. Navier (1785–1836), professor of mechanics at the Ecole Polytechnique and at the Ecole des Ponts et Chaussées, obtained the basic equations in 1821.[31] The Navier-Stokes equations, as they are now identified, are

$$\rho \frac{Du}{Dt} = \rho X - \frac{\partial p}{\partial x} + \frac{1}{3} \mu \frac{\partial \theta}{\partial x} + \mu \Delta u$$

$$(50) \qquad \rho \frac{Dv}{Dt} = \rho Y - \frac{\partial p}{\partial y} + \frac{1}{3} \mu \frac{\partial \theta}{\partial y} + \mu \Delta v$$

$$\rho \frac{Dw}{Dt} = \rho Z - \frac{\partial p}{\partial z} + \frac{1}{3} \mu \frac{\partial \theta}{\partial z} + \mu \Delta w$$

$$\theta = \frac{\partial u}{\partial x} + \frac{\partial v}{\partial y} + \frac{\partial w}{\partial z},$$

31. *Mém de l'Acad. des Sci., Paris*, (2), 1827, 375–94.

where Δ has the usual meaning; ρ is the density of the fluid; p, the pressure; u, v, and w are the components of velocity of the fluid at any x, y, z, and t; X, Y, and Z are the components of an external force; the constant μ, which depends on the nature of the fluid, is called the coefficient of viscosity; and the derivative D/Dt has the meaning explained in Chapter 22, section 8. For an incompressible fluid, $\theta = 0$.

The equations were also obtained in 1829 by Poisson.[32] They were then rederived in 1845 on the basis of the mechanics of continua by George Gabriel Stokes (1819–1903), professor of mathematics at Cambridge University, in his essay "On the Theories of the Internal Friction of Fluids in Motion."[33] Stokes endeavored to account for the frictional action in all known liquids, which causes the motion to subside by converting kinetic energy into heat. Fluids, by virtue of their viscosity, stick to the surfaces of solids and thus exert tangential forces on them.

The subject of elasticity was founded by Galileo, Hooke, and Mariotte and was cultivated by the Bernoullis and Euler. But these men dealt with specific problems. To solve them they concocted *ad hoc* hypotheses on how beams, rods, and plates behaved under stresses, pressures, or loads. The theory proper is the creation of the nineteenth century. From the beginning of the nineteenth century on a number of great men worked persistently to obtain the equations that govern the behavior of elastic media, which includes the air. These men were primarily engineers and physicists. Cauchy and Poisson are the great mathematicians among them, though Cauchy was an engineer by training.

The problems of elasticity include the behavior of bodies under stress wherein one considers what equilibrium position they will assume, the vibrations of bodies when set in motion by an initial disturbance or by a continuously applied force, and, in the case of air and solid bodies, the propagation of waves through them. The interest in elasticity in the nineteenth century was heightened by the appearance about 1820 of a wave theory of light, initiated by the physician Thomas Young (1773–1829) and by Augustin-Jean Fresnel (1788–1827), an engineer. Light was regarded as a wave motion in ether, and ether was believed to be an elastic medium. Hence the propagation of light through ether became a basic problem. Another stimulus to a strong interest in elasticity in the early nineteenth century was Ernst F. F. Chladni's (1756–1827) experiments (1787) on the vibrations of glass and metal, in which he showed the nodal lines. These should be related to the sounds given off by, for example, a vibrating drumhead.

The work to obtain basic equations of elasticity was long and full of

32. *Jour. de l'Ecole Poly.*, 13, 1831, 1–74.
33. *Trans. Camb. Phil. Soc.*, 8, 1849, 287–319 = *Math. and Phys. Papers*, 1, 75–129.

pitfalls because little was known of the internal or molecular structure of matter; hence it was difficult to grasp any physical principles. The assumptions made as to solid bodies, the air, and ether varied from one writer to another and were disputed. In the case of ether, which presumably penetrated solid bodies because light passed through some and was absorbed by others, the relationship of the ether molecules to the molecules of the solid body also posed great difficulties. We do not intend to follow the physical theories of elastic bodies, nor is our understanding complete even today.

Navier[34] was the first (1821) to investigate the general equations of equilibrium and vibrations of elastic solids. The material was assumed to be isotropic, and the equations contained a single constant representing the nature of the solid. By 1822, stimulated by Fresnel's work, Cauchy had created another approach to the theory of elasticity.[35] Cauchy's equations contain two constants to represent the material of the body, and for an isotropic body are

$$(\lambda + \mu) \frac{\partial \theta}{\partial x} + \mu \Delta u = \rho \frac{\partial^2 u}{\partial t^2}$$

(51) $$(\lambda + \mu) \frac{\partial \theta}{\partial y} + \mu \Delta v = \rho \frac{\partial^2 v}{\partial t^2}$$

$$(\lambda + \mu) \frac{\partial \theta}{\partial z} + \mu \Delta w = \rho \frac{\partial^2 w}{\partial t^2}$$

$$\theta = \frac{\partial u}{\partial x} + \frac{\partial v}{\partial y} + \frac{\partial w}{\partial z}.$$

Here u, v, and w are components of displacements, θ is called the dilatation, and λ and μ are constants of the body or medium. For general anisotropic media the equations are quite complicated, and it may be pointless to write them in all generality. The equations are given by Cauchy.[36]

The most spectacular triumph of the nineteenth century, with an enormous impact on science and technology, was Maxwell's derivation in 1864 of the laws of electromagnetism.[37] Maxwell, utilizing the electrical and magnetic researches of numerous predecessors, notably Faraday, introduced the notion of a displacement current—radio waves are one form of displacement current—and with this notion formulated the laws of electromagnetic wave propagation. His equations, which are most conveniently stated in the vector form adopted later by Oliver Heaviside, are four in number and involve the electric field intensity **E**, the magnetic field

34. *Mém. de l'Acad. des Sci.*, Paris, (2), 7, 1827, 375–94.
35. *Exercices de math.*, 1828 = *Œuvres*, (2), 8, 195–226.
36. *Exercices de math.*, 1828 = *Œuvres*, (2), 8, 253–77.
37. *Phil. Trans.*, 155, 1865, 459–512 = *Scientific Papers*, 1, 526–97.

intensity **H**, the dielectric constant ε of the medium, the magnetic permeability μ of the medium, and the charge density ρ. The equations are

(52) $$\text{curl } \mathbf{H} = \frac{1}{c}\frac{\partial(\varepsilon\mathbf{E})}{\partial t}, \quad \text{curl } \mathbf{E} = -\frac{1}{c}\frac{\partial(\mu\mathbf{H})}{\partial t}$$

(53) $$\text{div } \varepsilon\mathbf{E} = \rho \quad \text{and} \quad \text{div } \mu\mathbf{H} = 0.[38]$$

The first two equations are the primary ones and amount to six scalar (non-vectorial) partial differential equations. The displacement current is the term $\partial(\varepsilon\mathbf{E})/\partial t$.

By working with just these equations, Maxwell predicted that electromagnetic waves travel through space and at the speed of light. On the basis of the identity of the two speeds, he dared to assert that light is an electromagnetic phenomenon, a prediction that has been amply confirmed since his time.

No general methods for solving any of the above systems of equations are known. However, the nineteenth-century men gradually realized that in the case of partial differential equations, whether single equations or systems, general solutions are not nearly so useful as the solutions for specific problems where the initial and boundary conditions are given, and where experimental work might also aid one in making useful simplifying assumptions. The writings of Fourier, Cauchy, and Riemann furthered this realization. The work on the solution of the many initial- and boundary-value problems to which specializations of these systems gave rise is enormous, and almost all of the mathematicians of the century undertook such problems.

8. *Existence Theorems*

As the eighteenth- and nineteenth-century mathematicians created a vast number of types of differential equations, they found that methods of solving many of these equations were not available. Somewhat as in the case of polynomial equations, where efforts to solve equations of degree higher than four failed and Gauss turned to the proof of existence of a root (Chap. 25, sec. 2), so in the work on differential equations the failure to find explicit solutions, which of course *ipso facto* demonstrate existence, caused the mathematicians to turn to proof of the existence of solutions. Such proofs, even though they do not exhibit a solution or exhibit it in a useful form, serve several purposes. The differential equations were in nearly all cases the mathematical formulation of physical problems. No guarantees were available that the mathematical equations could be solved; hence a

38. For the meaning of curl and div see Chap. 32, sec. 5.

proof of the existence of a solution would at least insure that a search for a solution would not be attempting the impossible. The proof of existence would also answer the question: What must we know about a given physical situation, that is, what initial and boundary conditions insure a solution and preferably a unique one? Other objectives, perhaps not envisaged at the beginning of the work on existence theorems, were soon recognized. Does the solution change continuously with the initial conditions, or does some totally new phenomenon enter when the initial or boundary conditions are varied slightly? Thus a parabolic orbit that obtains for one value of the initial velocity of a planet may change to an elliptic orbit as a consequence of a slight change in the initial velocity. Such a difference in orbit is physically most significant. Further, some of the methodologies of solution, such as the use of the Dirichlet principle or Green's theorem, presupposed the existence of a particular solution. The existence of these particular solutions had not been established.

Before we give some brief indications of the work on existence theorems, it may be helpful to note a classification of partial differential equations that was actually made rather late in the century. Though some efforts toward classification by reducing these equations to normal or standard forms had been made by Laplace and Poisson, the classification introduced by Du Bois-Reymond has now become standard. In 1889[39] he classified the most general homogeneous linear equation of second order

$$(54) \qquad R\frac{\partial^2 u}{\partial x^2} + S\frac{\partial^2 u}{\partial x\, \partial y} + T\frac{\partial^2 u}{\partial y^2} + P\frac{\partial u}{\partial x} + Q\frac{\partial u}{\partial y} + Zu = 0,$$

where the coefficients are functions of x and y and they and their first and second derivatives are continuous, by means of the characteristics (Chap. 22, sec. 7). The projections of the characteristic curves onto the xy-plane (these projections are also called characteristics) satisfy

$$T\, dx^2 - S\, dx\, dy + R\, dy^2 = 0.$$

The characteristics are imaginary, real and distinct, or real and coincident according as

$$TR - S^2 > 0, \qquad TR - S^2 < 0, \qquad TR - S^2 = 0.$$

Du Bois-Reymond called these cases elliptic, hyperbolic, and parabolic, respectively. He then pointed out that by introducing new real independent variables

$$\xi = \phi(x, y), \qquad \eta = \psi(x, y),$$

39. *Jour. für Math.*, 104, 1889, 241–301.

the above equation can always be transformed into one of the three types of normal forms

(a) $\quad R'\left(\dfrac{\partial^2 u}{\partial \xi^2} + \dfrac{\partial^2 u}{\partial \eta^2}\right) + P'\dfrac{\partial u}{\partial \xi} + Q'\dfrac{\partial u}{\partial \eta} + Zu = 0$

(b) $\quad S'\dfrac{\partial^2 u}{\partial \xi\,\partial \eta} + P'\dfrac{\partial u}{\partial \xi} + Q'\dfrac{\partial u}{\partial \eta} + Zu = 0$

(c) $\quad R'\dfrac{\partial^2 u}{\partial \xi^2} + P'\dfrac{\partial u}{\partial \xi} + Q'\dfrac{\partial u}{\partial \eta} + Zu = 0$

respectively. The two families $\phi(x, y) = $ const. and $\psi(x, y) = $ const. are the equations of two families of characteristics.

The supplementary conditions that can be imposed differ for the three types of equations. In the elliptic case (a) one considers a bounded domain of the xy-plane and specifies the value of u on the boundary (or an equivalent condition) and asks for the value of u in the domain. For the initial-value problem of the hyperbolic differential equation (b) one must specify u and $\partial u/\partial n$ on some initial curve. There may also be boundary conditions. The proper initial conditions for the parabolic case (c) were not specified at this time, though it is now known that one initial condition and boundary conditions can be imposed. This classification of partial differential equations was extended to equations in more independent variables, higher-order equations and to systems. Though the classification and the supplementary conditions were not known early in the century, the mathematicians gradually became aware of the distinctions and these figured in the existence theorems they were able to prove.

The work on existence theorems became a major activity with Cauchy, who emphasized that existence can often be established where an explicit solution is not available. In a series of papers [40] Cauchy noted that any partial differential equation of order greater than one can be reduced to a system of partial differential equations, and he treated the existence of a solution for the system. He called his method the *calcul des limites* but it is known today as the method of majorant functions. The essence of the method is to show that a power series in the independent variables with a definite domain of convergence does satisfy the system of equations. We shall illustrate the method in connection with Cauchy's work on ordinary differential equations (Chap. 29, sec. 4). His theorem covers only the case of analytic coefficients in the equations and analytic initial conditions.

To obtain some concrete idea about Cauchy's work, we shall consider what it implies for the second order equation in two independent variables

(55) $\qquad\qquad\qquad r = f(z, x, y, p, q, s, t)$

40. *Comp. Rend.*, 14, 1842, 1020–25 = *Œuvres*, (1), 6, 461–67, and *Comp. Rend.*, 15, 1842, 44–59, 85–101, 131–38 = *Œuvres*, (1), 7, 17–33, 33–49, 52–58.

where, as usual, $r = \partial^2 z / \partial x^2$ and where f is analytic in its variables. In this case one must specify on the initial line $x = 0$ that

$$z(0, y) = z_0(y), \qquad \frac{\partial z}{\partial x}(0, y) = z_1(y),$$

where z_0 and z_1 are analytic. (The initial line may be replaced by a curve, in which case $\partial z / \partial x$ must be replaced by $\partial z / \partial n$.) If the above conditions are fulfilled, then the solution $z = z(x, y)$ exists and is unique and analytic in some domain starting at the initial line.

Cauchy's work on systems was done independently and in somewhat improved form by Sophie Kowalewsky (1850–91),[41] who was a pupil of Weierstrass and who pursued his ideas. Kowalewsky is one of the few women mathematicians of distinction. In 1816 Sophie Germain (1776–1831) had won a prize awarded by the French Academy for a paper on elasticity. Kowalewsky, too, won the Paris Academy's prize, for a work of 1888 on the integration of the equations of motion for a solid body rotating around a fixed point; in 1889 she became a professor of mathematics in Stockholm. The proofs of Cauchy and Kowalewsky were later improved by Goursat.[42]

If, instead of (55), the given second order equation is in the form

(56) $$G(z, x, y, p, q, r, s, t) = 0,$$

it is necessary to solve for r before it can be put in the form (55). To consider a simple but vital case, if the equation is

$$G = A \frac{\partial^2 z}{\partial x^2} + 2B \frac{\partial^2 z}{\partial x \, \partial y} + C \frac{\partial^2 z}{\partial y^2} + D \frac{\partial z}{\partial x} + E \frac{\partial z}{\partial y} + Fz = 0,$$

where $A, B, \ldots, F$ are functions of x and y, then $\partial G / \partial r$ must not be 0 to solve for r. In case $\partial G / \partial r = 0$, the solution of the Cauchy problem need not exist, and when it does is not unique. In the case of three or more independent variables (let us consider three), and if the equation is written as

(57) $$\sum_{i,k} A_{ik} \frac{\partial^2 u}{\partial x_i \, \partial x_k} + \sum_i B_i \frac{\partial u}{\partial x_i} + Cu = f,$$

where the coefficients are functions of the independent variables x_1, x_2, and x_3, then the exceptional case occurs when the initial surface S satisfies the first order partial differential equation

(58) $$\sum A_{ik} \frac{\partial S}{\partial x_i} \frac{\partial S}{\partial x_k} = 0.$$

41. *Jour. für Math.*, 80, 1875, 1–32.
42. *Bull. Soc. Math. de France*, 26, 1898, 129–34.

Along such surfaces two solutions of (57) can be tangent and even have higher-order contact. This property is the same as for the characteristic curves of the first order equation $f(x, y, u, p, q) = 0$ (Chap. 22, sec. 5), and so these surfaces S are also called characteristics. Physically the surfaces S are wave fronts.

This theory of characteristics for the case of two independent variables was known to Monge and to André-Marie Ampère (1775–1836). Its extension to the case of second order equations in more than two independent variables was first made by Albert Victor Bäcklund (1845–1922),[43] but was not widely known until it was redone by Jules Beudon.[44]

In his *Leçons sur la propagation des ondes* (1903), Jacques Hadamard (1865–1963), the leading French mathematician of this century, generalized the theory of characteristics to partial differential equations of any order. As an example, let us consider a system of three partial differential equations of the second order in the dependent variables ξ, η, and ζ and the independent variables $x_1, x_2, \ldots, x_n$. The Cauchy problem for this system is: Given the values of ξ, η, ζ, and $\partial\xi/\xi x_n$, $\partial\eta/\xi x_n$, and $\partial\zeta/\partial x_n$ on a "surface" M_{n-1} of $n - 1$ dimensions, to find the functions ξ, η, and ζ. The values of the second and higher derivatives of ξ, η, and ζ may then be computed unless M_{n-1} satisfies a first order partial differential equation of the sixth degree, say $H = 0$. All "surfaces" satisfying $H = 0$ are characteristic "surfaces." According to the theory of first order partial differential equations, the differential equation $H = 0$ has characteristic lines (curves) defined by

$$\frac{dx_1}{\partial H/\partial P_1} = \frac{dx_2}{\partial H/\partial P_2} = \cdots = \frac{dx_{n-1}}{\partial H/\partial P_{n-1}},$$

where $P_1, P_2, \ldots, P_{n-1}$ are the partial derivatives of x_n with respect to $x_1, x_2, \ldots, x_{n-1}$ taken along the "surface" M_{n-1}. These lines are called the bicharacteristics of the original second order system. In the theory of light they are the rays.

The characteristics now play a vital role in the theory of partial differential equations. For example, Darboux[45] has given a powerful method of integrating second order partial differential equations in two independent variables that rests on the theory of characteristics. It converts the problem to the integration of one or more ordinary differential equations and embraces the methods of Monge, Laplace, and others.

Another class of existence theorems dealt with the Dirichlet problem, that is, establishing the existence of a solution of $\Delta V = 0$ either directly or by means of the Dirichlet principle. The first existence proof of the Dirichlet

43. *Math. Ann.*, 13, 1878, 411–28.
44. *Bull. Soc. Math. de France*, 25, 1897, 108–20.
45. *Ann. de l'Ecole Norm. Sup.*, (1), 7, 1870, 175–80.

problem in two dimensions (but not of the Dirichlet principle of minimizing the Dirichlet integral) was given by Hermann Amandus Schwarz (1843–1921), a pupil of Weierstrass, whom he succeeded at Berlin in 1892 and who suggested the problem to him. Under general assumptions about the bounding curve, and by using a process called the alternating procedure,[46] he demonstrated the existence of a solution.[47]

In the same year, 1870, Carl G. Neumann gave another proof of the existence of a solution of the Dirichlet problem in three dimensions[48] by using the method of arithmetic means, though he too did not use the Dirichlet principle.[49] The principal exposition of his ideas is in his *Vorlesungen über Riemann's Theorie der Abel'schen Integrale.*[50]

Then Henri Poincaré[51] used the *methode de balayage*, the method of "sweeping out," which approaches the problem by building a succession of functions not harmonic in the domain R but taking on the correct boundary values, the functions becoming more and more harmonic.

Finally, David Hilbert reconstructed the calculus of variations method of Thomson and Dirichlet and established the Dirichlet *principle* as a method for proving the existence of a solution of the Dirichlet problem. In 1899[52] Hilbert showed that under proper conditions on the region, boundary values, and the admissible functions U, the Dirichlet principle does hold. He made the Dirichlet principle a powerful tool in function theory. In another publication, of work done in 1901,[53] Hilbert gave more general conditions.

The history of the Dirichlet principle is remarkable. Green, Dirichlet, Thomson, and others of their time regarded it as a completely sound method and used it freely. Then Riemann in his complex function theory showed it to be extraordinarily instrumental in leading to major results. All of these men were aware that the fundamental existence question was not settled, even before Weierstrass announced his critique in 1870, which discredited the method for several decades. The principle was then rescued by Hilbert and was used and extended in this century. Had the progress made with the use of the principle awaited Hilbert's work, a large segment

46. Schwarz's method is sketched in Felix Klein, *Vorlesungen über die Entwicklung der Mathematik im 19. Jahrhundert*, Chelsea (reprint), 1950, 1, p. 265, and given fully in A. R. Forsyth, *Theory of Functions*, Dover (reprint), 1965, 2, Chap. 17. Many references are given in the latter source.
47. *Monatsber. Berliner Akad.*, 1870, 767–95 = *Ges. Math. Abh.*, 2, 144–71.
48. *Königlich Sächsischen Ges. der Wiss. zu Leipzig*, 1870, 49–56, 264–321.
49. The method is described in O. D. Kellogg, *Foundations of Potential Theory*, Julius Springer, 1929, 281 ff.
50. Second ed., 1884, 238 ff.
51. *Amer. Jour. of Math.*, 12, 1890, 211–94 = *Œuvres*, 9, 28–113.
52. *Jahres. der Deut. Math.-Verein.*, 8, 1900, 184–88 = *Ges. Abh.*, 3, 10–14.
53. *Math. Ann.*, 59, 1904, 161–86 = *Ges. Abh.*, 3, 15–37.

of nineteenth-century work on potential theory and function theory would have been lost.

The Laplace equation $\Delta V = 0$ is the basic form of elliptic differential equations. Many more existence theorems were established for more general elliptic differential equations, such as

$$(59) \qquad \frac{\partial^2 u}{\partial x^2} + \frac{\partial^2 u}{\partial y^2} + a\,\frac{\partial u}{\partial x} + b\,\frac{\partial u}{\partial y} + cu = 0.$$

The variety of such theorems is vast. We shall mention just one key result. The existence and uniqueness of a solution of this equation (the value of the solution is prescribed on the boundary) was demonstrated by Picard[54] for domains of sufficiently small area. The result has been extended to more variables, large domains, and in other respects by Picard and others. Picard also established[55] that every equation of the above form (and even slightly more general) whose coefficients are analytic functions possesses only analytic solutions inside the domain in which the solution is sought, even though the solution assumes non-analytic boundary values.

The theorems discussed thus far have generally dealt with analytic differential equations and analytic initial or boundary data. However, such conditions are too restrictive for applications because the given physical data may not be analytic. Another major class of theorems deals with less stringent conditions. We shall give just one example. Riemann's method, which applies to the hyperbolic equation, relies upon the existence of his characteristic function v and, as we pointed out, the existence of v was not established by Riemann.

For this hyperbolic case (see [40]), Du Bois-Reymond in 1889 sought the proper conditions and obtained results[56] which, expressed for the case where $x = $ const. and $y = $ const. are the characteristics, read thus: Given the continuous functions u and $\partial u/\partial n$ along a curve AB that is cut not more than once by any characteristic line, then there exists one and only one solution u of the differential equation which takes on the given values of u and $\partial u/\partial n$ along AB. This solution is defined in the rectangle determined by the characteristics through A and through B. If instead the values of a continuous function u on two segments of characteristics which abut one another are given, then u is again uniquely determined in the rectangle determined by the characteristics. In terms of x, y, and u as spatial coordinates, the first result states that the surface $u(x, y)$ goes through a given space curve and with a given inclination. The second result means that the solution or surface is enclosed in the space determined by two intersecting space curves.

54. *Comp. Rend.*, 107, 1888, 939–41, *Jour. de Math.*, (4), 6, 1890, 145–210, *Jour. de Math.*, (5), 2, 1896, 295–304.
55. *Jour de l'Ecole Poly.*, 60, 1890, 89–105.
56. *Jour. für Math.*, 104, 1889, 241–301.

For the continuous initial conditions u and $\partial u/\partial n$ (and u in the second case), the solutions will be regular or satisfy the differential equation everywhere in the rectangles described above. Discontinuities in u and $\partial u/\partial n$ will propagate along the characteristics in the rectangle.

A great deal of work was done in the second half of the century on the existence of eigenvalues for $\Delta u + k^2 u = 0$ considered in a domain D. The main result is that for a given domain and under any one of the three boundary conditions $u = 0$, $\partial u/\partial n = 0$, $\partial u/\partial n + hu = 0$ ($h > 0$ when the positive normal is directed outside the domain), there are always an infinite number of discrete values of k^2, for each of which there is a solution. In two dimensions the vibrations of a membrane fixed along its boundary illustrate this theorem. The values of k are the frequencies of the infinitely many purely harmonic vibrations. The corresponding solutions give the deformation of the membrane in carrying out its characteristic oscillations.

The first major step was the proof by Schwarz[57] of the existence of the first eigenfunction of

$$\Delta u + \xi f(x, y) u = 0,$$

that is, the existence of a U_1 such that

$$\Delta U_1 + k_1^2 f(x, y) U_1 = 0$$

and $U_1 = 0$ on the boundary of the domain considered. His method gave a procedure for finding the solution and permitted the calculation of k_1^2. Picard[58] then established the existence of the second eigenvalue k_2^2.

Schwarz also showed in the 1885 paper that when the domain varies continuously, the value of k_1^2, the first characteristic number, also varies continuously; and as the domain becomes smaller, k_1^2 increases unboundedly. Thus a smaller membrane gives off a higher first harmonic.

In 1894 Poincaré[59] demonstrated the existence and the essential properties of all the eigenvalues of

(60) $$\Delta u + \lambda u = f,$$

λ complex, in a bounded, three-dimensional domain, with $u = 0$ on the boundary. The existence of u was demonstrated by a generalization of Schwarz's method. He proved next that $u(\lambda)$ is a meromorphic function of the complex variable λ and that the poles are real; these are just the eigenvalues λ_n. Then he obtained the characteristic solutions U_i, that is,

$$\Delta U_i + k_i^2 U_i = 0 \qquad \text{(in the interior)}$$
$$U_i = 0 \qquad \text{(on the boundary)}.$$

57. *Acta Soc. Fennicae*, 15, 1885, 315–62 = *Ges. Math. Abh.*, 1, 223–69.
58. *Comp. Rend.*, 117, 1893, 502–7.
59. *Rendiconti del Circolo Matematico di Palermo*, 8, 1894, 57–155 = *Œuvres*, 9, 123–96.

The k_i^2 are the charactcristic numbers (eigenvalues) and determine the frequencies of the respective characteristic solutions.

Physically Poincaré's result has the following significance. The function f in (60) can be thought of as an applied force. The free oscillations of a mechanical system are those at which the forced oscillations degenerate and become infinite. In fact, (60) is the equation of an oscillating system excited by a periodically varying force of amplitude f; and the characteristic solutions are the free oscillations of the system, which, once excited, continue indefinitely. The frequencies of the free oscillations, which are proportional to the k_i, are calculated by Poincaré's method as the values of $\sqrt{\lambda}$ for which the forced oscillation u becomes infinite.

At the end of the century the systematic theory of boundary- and initial-value problems for partial differential equations, which dates from Schwarz's fundamental 1885 paper, was still young. The work in this area expanded rapidly in the twentieth century.

Bibliography

Bacharach, Max: *Abriss der Geschichte der Potentialtheorie*, Vandenhoeck and Ruprecht, 1883.

Burkhardt, H.: "Entwicklungen nach oscillirenden Functionen und Integration der Differentialgleichungen der mathematischen Physik," *Jahres. der Deut. Math.-Verein.*, Vol. 10, 1908, 1–1804.

Burkhardt, H. and W. Franz Meyer: "Potentialtheorie," *Encyk. der Math. Wiss.*, B. G. Teubner, 1899–1916, II A7b, 464–503.

Cauchy, Augustin-Louis: *Œuvres complètes*, Gauthier-Villars, 1890, (2), 8.

Fourier, Joseph: *The Analytical Theory of Heat* (1822), Dover reprint of English translation, 1955.

———: *Œuvres*, 2 vols., Gauthier-Villars, 1888–90; Georg Olms (reprint), 1970.

Green, George: *Essay on the Application of Mathematical Analysis to the Theories of Electricity and Magnetism*, 1828, reprinted by Wczäta-Melins Aktiebolag, 1958; also in Ostwald's Klassiker #61 (in German), Wilhelm Engelmann, 1895.

———: *Mathematical Papers*, Macmillan, 1871; Chelsea (reprint), 1970.

Heine, Eduard: *Handbuch der Kugelfunktionen*, 2 vols., 1878–81, Physica Verlag (reprint), 1961.

Helmholtz, Hermann von: "Theorie der Luftschwingungen in Röhren mit offenen Enden," *Ostwald's Klassiker der exakten Wissenschaften*, Wilhelm Engelmann, 1896.

Klein, Felix: *Vorlesungen über die Entwicklung der Mathematik im 19. Jahrhundert*, 2 vols., Chelsea (reprint), 1950.

Langer, R. E.: "Fourier Series, the Genesis and Evolution of a Theory," *Amer. Math. Monthly*, 54, Part II, 1947, 1–86.

Pockels, Friedrich: *Über die partielle Differentialgleichung $\Delta u + k^2 u = 0$*, B. G. Teubner, 1891.

Poincaré, Henri: *Œuvres*, Gauthier-Villars, 1954, 9.

Rayleigh, Lord (Strutt, John William): *The Theory of Sound*, 2nd ed., 2 vols., Dover (reprint), 1945.

Riemann, Bernhard: *Gesammelte mathematische Werke*, 2nd ed., Dover (reprint), 1953, pp. 156–211.

Sommerfeld, Arnold: "Randwertaufgaben in der Theorie der partiellen Differentialgleichungen," *Encyk. der math. Wiss.*, B. G. Teubner, 1899–1916, II A7c, 504–70.

Todhunter, Isaac, and Karl Pearson: *A History of the Theory of Elasticity*, 2 vols., Dover (reprint), 1960.

Whittaker, Sir Edmund: *History of the Theories of Aether and Electricity*, rev. ed., Thomas Nelson and Sons, 1951, Vol. I.

29
Ordinary Differential Equations in the Nineteenth Century

La Physique ne nous donne pas seulement l'occasion de résoudre des problèmes . . . elle nous fait préssentir la solution.

HENRI POINCARÉ

1. *Introduction*

Ordinary differential equations arose in the eighteenth century in direct response to physical problems. By tackling more complicated physical phenomena, notably in the work on the vibrating string, the mathematicians arrived at partial differential equations. In the nineteenth century the roles of these two subjects were somewhat reversed. The efforts to solve partial differential equations by the method of separation of variables led to the problem of solving ordinary differential equations. Moreover, because the partial differential equations were expressed in various coordinate systems the ordinary differential equations that resulted were strange ones and not solvable in closed form. The mathematicians resorted to solutions in infinite series which are now known as special functions or higher transcendental functions as opposed to the elementary transcendental functions such as $\sin x$, e^x, and $\log x$.

After much work on the extended variety of ordinary differential equations, some deep theoretical studies were devoted to types of such equations. These theoretical investigations also differentiate the nineteenth-century work from that of the eighteenth century. The contributions of the new century were so vast that, as in the case of partial differential equations, we cannot hope to survey all of the major developments. Our topics are a sample of what was created during the century.

2. *Series Solutions and Special Functions*

As we have just observed, to solve the ordinary differential equations that resulted from the method of separation of variables applied to partial differential equations, the mathematicians, without worrying much about the existence and form the solutions should take, turned to the method of infinite series (Chap. 21, sec. 6). Of the ordinary differential equations that resulted from separation of variables the most important is the Bessel equation

$$(1) \qquad\qquad x^2 y'' + xy' + (x^2 - n^2)y = 0,$$

where n, a parameter, can be complex and x can also be complex. However, for Friedrich Wilhelm Bessel (1784–1846), a mathematician and director of the astronomical observatory in Königsberg, n and x were real. Special cases of this equation occurred as early as 1703, when James Bernoulli mentioned it as a particular solution in a letter to Leibniz, and thereafter in more extensive work by Daniel Bernoulli and Euler (Chap. 21, secs. 4 and 6, and Chap. 22, sec. 3). Special cases also occurred in the writings of Fourier and Poisson. The first systematic study of solutions of this equation was made by Bessel[1] while working on the motion of the planets. The equation has two independent solutions for each n, denoted today by $J_n(x)$ and $Y_n(x)$ and called Bessel functions of the first and second kind, respectively. Bessel, whose work on the equation dates from 1816, gave first the integral relation (for integral n)

$$J_n(x) = \frac{1}{2\pi} \int_0^{2\pi} \cos\,(nu - x \sin u)\,du$$

(he wrote I_k^h and his k is our x). Bessel also obtained the series

$$(2)$$

$$J_n(x) = \frac{x^n}{2^n \Gamma(n+1)} \left\{ 1 - \frac{x^2}{2^2 \cdot 1! \, (n+1)} + \frac{x^4}{2^4 \cdot 2! \, (n+1)(n+2)} - \cdots \right\}.$$

In 1818 Bessel proved that $J_0(x)$ has an infinite number of real zeros. In the 1824 paper Bessel also gave the recursion formula (for integral n)

$$xJ_{n+1}(x) - 2nJ_n(x) + xJ_{n-1}(x) = 0,$$

and many other relations involving the Bessel function of the first kind. The generalization of the Bessel function $J_n(x)$ to complex n and x was made by several men[2] with (2) remaining as the correct form.

Since there should be two independent solutions for a second order equation, many mathematicians sought it. When n is not an integer this second solution is $J_{-n}(x)$. For integral n the second solution was given by

1. *Abh. Konig. Akad. der Wiss. Berlin*, 1824, 1–52, pub. 1826 = *Werke*, 1, 84–109.
2. Chiefly by Eugen C. J. Lommel (1837–99) in his *Studien über die Bessel'schen Functionen* (1868).

Carl G. Neumann.[3] However, the form adopted most commonly today was given by Hermann Hankel (1839–73),[4] namely,

$$Y_n(z) = \sum_{r=0}^{\infty} \frac{(-1)^r[(1/2)z]^{n+2r}}{r!\,(n+r)!} \left\{ 2 \log\left(\frac{z}{2}\right) + 2\gamma - \sum_{m=1}^{n+r} \frac{1}{m} - \sum_{m=1}^{r} \frac{1}{m} \right\}$$
$$- \sum_{r=0}^{n-1} \frac{[(1/2)z]^{-n+2r}(n-r-1)!}{r!},$$

where γ is Euler's constant. Neumann[5] also gave the expansion of an analytic function $f(z)$, namely,

$$f(z) = \alpha_0 J_0(z) + \alpha_1 J_1(z) + \alpha_2 J_2(z) + \cdots,$$

where the α_i are constants and can be determined.

Many mathematicians, usually working in celestial mechanics, arrived independently at the Bessel functions and at hundreds of other relations and expressions for these functions. Some idea of the vast literature on these functions may be gained from perusing G. N. Watson's *A Treatise on the Theory of Bessel Functions*.[6]

The Legendre polynomials or spherical functions of one variable and the spherical surface functions, which are functions of two variables, had already been introduced by Legendre and Laplace (Chap. 22, sec. 4). The Legendre polynomials satisfy the Legendre differential equation

$$(1 - x^2)y'' - 2xy' + n(n+1)y = 0.$$

This equation, as we know, results from separation of variables applied to the potential equation expressed in spherical coordinates. In 1833, Robert Murphy (d. 1843), a fellow at Cambridge University, wrote a text, *Elementary Principles of the Theories of Electricity, Heat and Molecular Actions*. In it he put together some older results on Legendre polynomials and obtained some new ones. Since the major results were already known, we shall not present the details of Murphy's work except to point out that it was systematic and that he showed that "any" function $f(x)$ can be expanded in terms of the $P_n(x)$ by applying term-by-term integration and the orthogonality property (integral theorem).

Heine,[7] treating the problem of the potential for the exterior of an ellipsoid of revolution and for the shell between confocal ellipsoids of revolution (Chap. 28, sec. 5), introduced spherical harmonics of the second kind, usually denoted by $Q_n(x)$, which provide a second independent solution

3. *Theorie der Bessel'schen Funktionen*, 1867, 41.
4. *Math. Ann.*, 1, 1869, 467–501.
5. *Jour. für Math.*, 67, 1867, 310–14.
6. Cambridge University Press, 2nd ed., 1944.
7. *Jour. für Math.*, 26, 1843, 185–216.

of Legendre's equation. The Legendre functions, like the Bessel functions, have been extended to complex n and complex x, and a number of alternative representations and relationships among them have been obtained.[8]

The study of special functions that arise as series solutions of ordinary differential equations was furthered by Gauss in a famous paper of 1812 on the hypergeometric series.[9] In this paper Gauss made no use of the differential equation

$$x(1 - x)y'' + \{\gamma - (\alpha + \beta + 1)x\}y' - \alpha\beta y = 0,$$

but he did in unpublished material.[10] Of course the equation and the series solution

$$(3) \qquad F(\alpha, \beta, \gamma; x) = 1 + \frac{\alpha \cdot \beta}{1 \cdot \gamma} x + \frac{\alpha(\alpha + 1)\beta(\beta + 1)}{1 \cdot 2 \cdot \gamma(\gamma + 1)} x^2 + \cdots$$

were already known because they had been studied by Euler (Chap. 21, sec. 6). Gauss recognized that for special values of α, β, and γ the series included almost all the elementary functions then known and many higher transcendental functions such as the Bessel functions and spherical functions. In addition to proving any number of properties of the series, Gauss established the famous relation

$$F(\alpha, \beta, \gamma; 1) = \frac{\Gamma(\gamma)\Gamma(\gamma - \alpha - \beta)}{\Gamma(\gamma - \alpha)\Gamma(\gamma - \beta)}.$$

He also established the convergence of the series (cf. Chap. 40, sec. 5). The notation $F(\alpha, \beta, \gamma; x)$ is due to Gauss.

Another class of special functions was introduced by Lamé.[11] In Chapter 28, sec. 5, we pointed out that Lamé, working on a steady-state temperature distribution in an ellipsoid, separated Laplace's equation in ellipsoidal coordinates ρ, μ, and ν. This process gives the same ordinary differential equations in each of the three variables, namely,

$$(4) \quad (\rho^2 - h^2)(\rho^2 - k^2)\frac{d^2E(\rho)}{d\rho^2} + \rho(2\rho^2 - h^2 - k^2)\frac{dE(\rho)}{d\rho}$$

$$+ \{(h^2 + k^2)p - n(n + 1)\rho^2\}E(\rho) = 0,$$

with appropriate changes for μ and ν in place of ρ. Here h^2 and k^2 are the parameters in the equations of the families of the coordinate surfaces and p and n are constants. This equation is known as Lamé's differential equation.

8. See, for example, E. W. Hobson, *The Theory of Spherical and Ellipsoidal Harmonics*, 1931, Chelsea (reprint), 1955.
9. *Comm. Soc. Sci. Gott.*, 2, 1813 = *Werke*, 3, 123–62.
10. *Werke*, 3, 207–30.
11. *Jour. de Math.*, 2, 1837, 147–83; 4, 1839, 126–63.

The solutions $E(\rho)$ are called Lamé functions or ellipsoidal harmonics. For integral n these functions fall into four classes of the form

$$E_n^p(\rho) = a_0\rho^n + a_1\rho^{n-2} + \cdots$$

or such polynomials multiplied by $\sqrt{\rho^2 - h^2}$ or $\sqrt{\rho^2 - k^2}$ or both factors. For a given value of n, the number of such functions (to insure some properties of $E(\rho)$) is $2n + 1$.

The second solution of Lamé's equation (resulting from other conditions or properties of $E(\rho)$) is

$$F_n^p(\rho) = (2n + 1)E_n^p(\rho) \int_\rho^\infty \frac{d\rho}{\sqrt{(\rho^2 - h^2)(\rho^2 - k^2)}[E_n^p(\rho)]^2},$$

and such functions are called Lamé functions of the second kind. These were introduced by Liouville[12] and Heine.[13]

The differential equations

$$\frac{d^2\phi}{d\eta^2} + (a - 2k^2\cos 2\eta)\phi = 0 \qquad \frac{d^2\psi}{d\xi^2} - (a - 2k^2\cosh 2\xi)\psi = 0$$

arose in Mathieu's work on the vibrations of an elliptical membrane[14] and they also arise in problems on the potential of elliptical cylinders when separation of variables is applied to $\Delta u + k^2 u = 0$ expressed in elliptical cylindrical coordinates in two dimensions. These elliptic coordinates, incidentally, are related to rectangular coordinates by the equations

$$x = h\cosh\xi\cos\eta, \qquad y = h\sinh\xi\sin\eta,$$

where $x = \pm h, y = 0$ are the foci of the confocal ellipses and hyperbolas of the family of ellipses and the family of hyperbolas in the elliptic coordinate system. The variety of forms in which Mathieu's equation is written and the many notations for the solutions by different authors present a confusing picture. The functions defined by either differential equation were called by Heine functions of the elliptic cylinder and are now called Mathieu functions. Mathieu and Heine first got series expressions for the solutions. Then they sought to fix the parameter a so that one class of solutions is periodic and of period 2π. The problem of finding periodic solutions, which are the most important ones for physical applications, was pursued throughout the century. In 1883[15] Gaston Floquet (1847–1920) published a complete discussion of the existence and properties of the periodic solutions of a linear differential equation of the nth order having periodic coefficients which have the same period ω. The general properties of the solutions having been

12. *Jour. de Math.*, 10, 1845, 222–28.
13. *Jour. für Math.*, 29, 1845, 185–208.
14. *Jour. de Math.*, (2), 13, 1868, 137–203.
15. *Ann. de l'Ecole Norm. Sup.*, (2), 12, 1883, 47–88.

determined, later writers devoted considerable attention to the problem of discovering practical methods of finding them. No general methods were found (cf. sec. 7).

A widely studied class of special functions was introduced by Heinrich Weber (1842–1913) in 1868.[16] Weber was interested in integrating $\Delta u + k^2 u = 0$ in a domain bounded by two parabolas. He therefore changed from rectangular coordinates to parabolic coordinates (which are a limiting case of elliptic coordinates) through the transformation

$$x = \xi^2 - \eta^2, \qquad y = 2\xi\eta.$$

For $\xi = $ const. and for $\eta = $ const., the two families of curves are families of parabolas, with each member of one family cutting the members of the other orthogonally. The ordinary differential equations which Weber derived from the reduced wave equation by separation of variables are

$$\frac{d^2 E}{d\xi^2} + (k^2 \xi^2 + a)E = 0$$

$$\frac{d^2 H}{d\eta^2} + (k^2 \eta^2 - a)H = 0.$$

Weber gave four particular solutions of the second equation in the form of definite integrals. The solutions are called parabolic cylinder functions. Weber also showed that the only case in which separation of variables can be applied to $\Delta u + k^2 u = 0$, among all orthogonal coordinate systems, is that of confocal surfaces of the second degree or specializations thereof.

The class of special functions is far more extensive than we can indicate here. The above-mentioned types and many others that were introduced serve to solve differential equations in some bounded domain and to represent arbitrary functions (usually the initial functions of a partial differential equations problem) in that domain. The limitation to bounded domains is imposed by the orthogonality property. In the basic case of the trigonometric functions this domain is $(-\pi, \pi)$ because, for example,

$$\int_{-\pi}^{\pi} \sin mx \sin nx \, dx = 0 \quad \text{if } m \neq n.$$

The problem of solving ordinary differential equations over infinite intervals or semi-infinite intervals and of obtaining expansions of arbitrary functions over such intervals was also tackled by many men during the second half of the century and such special functions as Hermite functions first introduced by Hermite in 1864[17] and Nikolai J. Sonine (1849–1915) in 1880[18] serve to solve this problem.

16. *Math. Ann.*, 1, 1869, 1–36.
17. *Comp. Rend.*, 58, 1864, 93–100 and 266–73 = *Œuvres*, 2, 293–308.
18. *Math. Ann.*, 16, 1880, 1–80.

To work with all these types of special functions one must know their properties as intimately as one knows the properties of the elementary functions. Also because these special functions are more complicated the properties are likewise so. The literature both in original papers and texts is almost incredibly vast. Whole treatises have been devoted to Bessel functions, spherical functions, ellipsoidal functions, Mathieu functions, and other types.

3. *Sturm-Liouville Theory*

The problems involving partial differential equations of mathematical physics usually contain boundary conditions such as the condition that the vibrating string must be fixed at the endpoints. When the method of separation of variables is applied to a partial differential equation this equation is resolved into two or more ordinary differential equations and the boundary conditions on the desired solution become boundary conditions on one ordinary differential equation. The ordinary equation generally contains a parameter, which in fact results from the separation of variables procedure, and solutions can usually be obtained for particular values of the parameter. These values are called eigenvalues or characteristic values and the solution for any one eigenvalue is called an eigenfunction. Moreover, to meet the initial condition or conditions of the original problem it is necessary to express a given function $f(x)$ in terms of the eigenfunctions (see, for example, [11] of Chap. 28).

These problems of determining the eigenvalues and eigenfunctions of an ordinary differential equation with boundary conditions and of expanding a given function in terms of an infinite series of the eigenfunctions, which date from about 1750, became more prominent as new coordinate systems were introduced and new classes of functions such as Bessel functions, Legendre polynomials, Lamé functions, and Mathieu functions arose as the eigenfunctions of ordinary differential equations. Two men, Charles Sturm (1803–55), professor of mechanics at the Sorbonne, and Joseph Liouville (1809–82), a friend of Sturm and professor of mathematics at the Collège de France, decided to tackle the general problem for any second order ordinary differential equation. Sturm had been working since 1833 on problems of partial differential equations, primarily on the flow of heat in a bar of variable density, and hence was fully aware of the eigenvalue and eigenfunction problem.

The mathematical ideas he applied to this problem[19] are closely related to his investigations of the reality and distribution of the roots of algebraic equations. His ideas on differential equations, he says, came from the study of difference equations and a passage to the limit.

19. *Jour. de Math.*, 1, 1836, 106–86 and 373–444.

Liouville, informed by Sturm of the problems he was working on, took up the same subject.[20] The results in the several papers of these two men are quite detailed and are most conveniently summarized in modern notation as follows. They considered the general second order equation

(5) $$Ly'' + My' + \lambda Ny = 0,$$

where L, M, and N are continuous functions of x, L is not zero, and λ is a parameter. Such an equation can be transformed, by multiplying through by $L^{-1} e^{\int ML^{-1}\,dx}$, into

$$\frac{d}{dx}\left[p(x)\frac{dy}{dx}\right] + \lambda \rho(x)y = 0, \qquad p(x) > 0,\; \rho(x) > 0.$$

The boundary conditions to be satisfied by the original or transformed equation can have the general form

$$\begin{aligned} y'(a) - h_1 y(a) &= 0, \\ y'(b) + h_2 y(b) &= 0 \end{aligned} \qquad h_1 \geq 0,\; h_2 \geq 0,\; a < b.$$

Sturm and Liouville demonstrated the following fundamental results:

(a) The problem has a non-zero solution only when λ takes on any one of a sequence of values λ_n of positive numbers which increase to ∞.

(b) For each λ_n the solutions are multiples of one function v_n, which one can normalize by the condition $\int_a^b \rho v_n^2\,dx = 1$.

(c) The orthogonality property, $\int_a^b \rho v_m v_n\,dx = 0$ for $m \neq n$, holds.

(d) Each function f twice differentiable in (a, b) and satisfying the boundary conditions can be expanded in a uniformly convergent series

$$f(x) = \sum_{n=1}^{\infty} c_n v_n(x)$$

where

$$c_n = \int_a^b \rho f v_n(x)\,dx.$$

(e) The equality

$$\int_a^b \rho f^2\,dx = \sum_{n=1}^{\infty} c_n^2$$

obtains. This last equality, called the Parseval equality, had already been demonstrated purely formally by Marc-Antoine Parseval (?–1836) in 1799[21]

20. *Jour. de Math.*, 1, 1836, 253–65; 2, 1837, 16–35 and 418–36.
21. *Mém. des sav. étrangers*, (2), 1, 1805, 639–48.

for the set of trigonometric functions. From it there follows the inequality demonstrated by Bessel in 1828[22] also for trigonometric series, namely,

$$\sum_{n=1}^{\infty} c_n^2 \le \int_a^b |f(x)|^2 \, dx.$$

Actually the Sturm-Liouville results were not satisfactorily established in all respects. The proof that $f(x)$ can be represented as an infinite sum of the eigenfunctions was inadequate. One difficulty was the matter of the completeness of the set of eigenfunctions, which for a continuous $f(x)$ on (a, b) is the condition (e) above and which loosely means that the set of eigenfunctions is large enough to represent "any" $f(x)$. Also the question of the sense in which the series $\sum c_n v_n(x)$ converges to $f(x)$, whether pointwise, uniformly, or in some more general sense, was not covered though Liouville did give convergence proofs in some cases, using theory developed by Cauchy and Dirichlet.

4. Existence Theorems

We have already noted under this same topic in the chapter on nineteenth-century partial differential equations that as mathematicians found the problem of obtaining solutions for specific differential equations more and more difficult they turned to the question, Given a differential equation does it have a solution for given initial conditions and boundary conditions? The same movement, of course to be expected, occurred in ordinary differential equations. That the question of existence was neglected so long is partly due to the fact that differential equations arose in physical and geometrical problems, and it was intuitively clear that these equations had solutions.

Cauchy was the first to consider the question of the existence of solutions of differential equations and succeeded in giving two methods. The first, applicable to

(6) $$y' = f(x, y),$$

was created sometime between 1820 and 1830 and summarized in his *Exercices d'analyse.*[23]

This method, the essence of which may be found in Euler,[24] utilizes the same idea as is involved in the integral as a limit of a sum. Cauchy wished to show that there is one and only one $y = f(x)$ that satisfies (6) and which meets the given initial condition that $y_0 = f(x_0)$. He divided (x_0, x) into n parts $\Delta x_0, \Delta x_1, \ldots, \Delta x_{n-1}$ and formed

$$y_{i+1} = y_i + f(x_i, y_i) \, \Delta x_i,$$

22. *Astronom. Nach.*, 6, 1828, 333–48.
23. Vol. 1, 1840, 327 ff. = *Œuvres*, (2), 11, 399–465.
24. *Inst. Cal. Int.*, 1, 1768, 493.

where x_i is any value of x in Δx_i. Then by definition

$$y_n = y_0 + \sum_{i=0}^{n-1} f(x_i, y_i)\, \Delta x_i.$$

Now Cauchy shows that as n becomes infinite y_n converges to a unique function

$$y = y_0 + \int_{x_0}^{x} f(x, y)\, dx$$

and that this function satisfies (6) and the initial conditions.

Cauchy assumed that $f(x, y)$ and f_y are continuous for all real values of x and y in the rectangle determined by the intervals (x_0, x) and (y_0, y). In 1876 Rudolph Lipschitz (1832–1903) weakened the hypotheses of the theorem.[25] His essential condition was that for all (x, y_1) and (x, y_2) in the rectangle $|x - x_0| \le a$, $|y - y_0| \le b$, that is, for any two points with the same abscissa, there is a constant K such that

$$|f(x, y_1) - f(x, y_2)| < K(y_1 - y_2).$$

This condition is known as the Lipschitz condition, and the existence theorem is called the Cauchy-Lipschitz theorem.

Cauchy's second method of establishing the existence of solutions of differential equations, the method of dominant or majorant functions, is more broadly applicable than his first one and was applied by Cauchy in the complex domain. The method was presented in a series of papers in the *Comptes Rendus* during the years 1839–42.[26] The method was called by Cauchy the *calcul des limites* because it provides lower limits within which the solution whose existence is established is sure to converge. The method was simplified by Briot and Bouquet and their version[27] has become standard.

To illustrate the method let us note how it applies to

$$y' = f(x, y),$$

where f is analytic in x and y. The theorem to be established reads thus: If for

(7) $$\frac{dy}{dx} = f(x, y)$$

the function $f(x, y)$ is analytic in the neighborhood of $P_0 = (x_0, y_0)$, the differential equation then has a unique solution $y(x)$ which is analytic in a

25. *Bull. des Sci. Math.*, (1), 10, 1876, 149–59.
26. *Œuvres*, (1), Vols. 4 to 7 and 10. The most important papers are in the *Comptes Rendus* for Aug. 5 and Nov. 21, 1839, June 29, Oct. 26, Nov. 2, and Nov. 9 of 1840, and June 20 and July 4 of 1842.
27. *Comp. Rend.*, 39, 1854, 368–71.

neighborhood of x_0 and which reduces to y_0 when $x = x_0$. The solution can be represented by the series

(8) $\qquad y = y_0 + y_0'(x - x_0) + \dfrac{y_0''}{2!}(x - x_0)^2 + \dfrac{y_0'''}{3!}(x - x_0)^3 + \cdots$

wherein $y_0' = dy/dx$ at (x_0, y_0) and similarly for $y_0'', y_0''', \ldots$, and where the derivatives are determined by successive differentiation of the original differential equation in which y is treated as a function of x.

The method of proof, which we merely sketch, uses first the fact that because $f(x, y)$ is analytic in the neighborhood of (x_0, y_0), which for convenience we take to be $(0, 0)$, there is a circle of radius a about $x_0 = 0$ and a circle of radius b about $y_0 = 0$ in which $f(x, y)$ is analytic. Then $f(x, y)$ has an upper bound M for all values of x and y in the respective circles. Now the very method of obtaining the series (8) guarantees that it formally satisfies (7). The problem is to show that the series converges.

Toward this end one sets up the majorant function

$$F(x, y) = \sum \frac{M}{a^p b^q} x^p y^q,$$

which is the expansion of

(9) $\qquad\qquad F(x, y) = \dfrac{M}{(1 - x/a)(1 - y/b)}.$

One shows next that the series solution of

(10) $\qquad\qquad\qquad \dfrac{dY}{dx} = F(x, Y)$

namely,

(11) $\qquad\qquad Y = Y_0'x + Y_0'' \dfrac{x^2}{2!} + Y_0''' \dfrac{x^3}{3!} + \cdots,$

which is derived from (10) in the same way that (8) is derived from (7), dominates term for term the series (8). Hence if (11) converges then (8) does. To show that (11) converges one solves (10) explicitly using the value of F in (9) and shows that the series expansion of the solution, which must be (11), converges.

The method does not in itself determine the precise radius of convergence of the series for y. Numerous efforts were therefore devoted to showing that the radius can be extended. However, the papers do not give the full domain of convergence and are of little practical importance.

A third method of establishing the existence of solutions of ordinary differential equations, probably known to Cauchy, was first published by Liouville[28] for a second order equation. This is the method of successive

28. *Jour. de Math.*, (1), 3, 1838, 561–614.

approximation and is now credited to Emile Picard because he gave the
method a general form.[29] For the equation in real x and y,

$$y' = f(x, y),$$

wherein $f(x, y)$ is analytic in x and y and whose solution $y = f(x)$ is to pass
through (x_0, y_0), the method is to introduce the sequence of functions

$$y_1(x) = y_0 + \int_{x_0}^{x} f(t, y_0) \, dt$$

$$y_2(x) = y_0 + \int_{x_0}^{x} f(t, y_1(t)) \, dt$$

. . . .

$$y_n(x) = y_0 + \int_{x_0}^{x} f(t, y_{n-1}(t)) \, dt.$$

Then one proves that $y_n(x)$ tends to a limit $y(x)$ which is the one and only
continuous function of x satisfying the ordinary differential equation and
such that $y(x_0) = y_0$. The method as usually presented today presupposes
that $f(x, y)$ satisfies only the Lipschitz condition. The method was extended
to second order equations by Picard in the 1893 paper and has also been
extended to complex x and y.

The various methods described above were applied not only to higher
order ordinary differential equations but to systems of differential equations
for complex-valued variables. Thus Cauchy extended his second type of
existence theorem to systems of first order ordinary differential equations in
n dependent variables. He also extended this method of *calcul des limites* to
systems in the complex domain.[30] Cauchy's result reads as follows: Given the
system of equations

$$(12) \qquad \frac{dy_k}{dx} = f_k(x, y_0, \cdots, y_{n-1}), \qquad k = 0, 1, 2, \cdots, n - 1,$$

let $f_0, \ldots, f_{n-1}$ be monogenic (single-valued analytic) functions of their
arguments and let them be developable in the neighborhood of the initial
values

$$x = \xi, y_0 = \eta_0, \cdots, y_{n-1} = \eta_{n-1}$$

in positive integral powers of

$$x - \xi, y_0 - \eta_0, \cdots, y_{n-1} - \eta_{n-1}.$$

Then there are n power series in $x - \xi$ convergent in the neighborhood of
$x = \xi$ which when substituted for $y_0, \ldots, y_{n-1}$ in (12) satisfy the equations.

29. *Jour. de Math.*, (4), 6, 1890, 145–210; and (4), 9, 1893, 217–71.

30. For different existence proofs on systems of first order differential equations, see the
second reference to Painlevé in the bibliography at the end of this chapter.

These power series are unique. They give a regular solution of the system and take on the initial values. In this generality the result can be found in Cauchy's "Mémoire sur l'emploi du nouveau calcul, appelé calcul des limites, dans l'intégration d'un système d'équations différentielles."[31] Thus the idea was to content oneself with establishing the existence of and obtaining the solution in the neighborhood of one point in the complex plane. Weierstrass obtained the same result in the same year (1842) but did not publish it until his *Werke* came out in 1894.[32]

5. *The Theory of Singularities*

In the middle of the nineteenth century the study of ordinary differential equations took a new course. The existence theorems and Sturm-Liouville theory presuppose that the differential equations contain analytic functions or, at the very least, continuous functions in the domains in which solutions are considered. On the other hand some of the differential equations already considered, such as Bessel's, Legendre's, and the hypergeometric equation, when expressed so that the coefficient of the second derivative is unity, have coefficients that are singular, and the form of the series solutions in the neighborhood of the singular points, particularly that of the second solution, is peculiar. Hence mathematicians turned to the study of solutions in the neighborhood of singular points, that is, points at which one or more of the coefficients are singular. A point at which all the coefficients are at least continuous and usually analytic is called an ordinary point.

The solutions in the neighborhood of singular points are obtainable as series, and the knowledge of the proper form of the series must be at hand before calculating it. This knowledge can be obtained only from the differential equation. The new problem was described by Lazarus Fuchs (1833–1902) in a paper of 1866 (see below). "In the present condition of science the problem of the theory of differential equations is not so much to reduce a given differential equation to quadratures, as to deduce from the equation itself the behavior of its integrals at all points of the plane, that is, for all values of the complex variable." For this problem Gauss's work on the hypergeometric series pointed the way. The leaders were Riemann and Fuchs, the latter a student of Weierstrass and his successor at Berlin. The theory which resulted is called the Fuchsian theory of linear differential equations.

Attention in this new area concentrated on linear differential equations of the form

$$(13) \qquad y^{(n)} + p_1(z)y^{(n-1)} + \cdots + p_n(z)y = 0,$$

31. *Comp. Rend.*, 15, 1842, 14–25 = *Œuvres* (1), 7, 5–17.
32. *Math. Werke*, 1, 75–85.

where the $p_i(z)$ are single-valued analytic functions of complex z except at isolated singular points. This equation was emphasized because its solutions embrace all the elementary functions and even some higher functions, such as the modular and automorphic functions we shall encounter later.

Before considering solutions at and in the neighborhood of singular points, let us note a basic theorem that does follow from Cauchy's existence theorem on systems of ordinary differential equations but which was proven directly by Fuchs,[33] though he acknowledged his indebtedness to Weierstrass's lectures. If the coefficients $p_1, \ldots, p_n$ are analytic at a point a and in some neighborhood of that point, and if arbitrary initial conditions for y and its first $n - 1$ derivatives are given at $z = a$, then there is a unique power series solution for y in terms of z of the form

$$(14) \qquad y(z) = \sum_{r=0}^{\infty} \frac{1}{r!} y^{(r)}(a)(z - a)^r.$$

To Cauchy's result Fuchs added that the series is absolutely and uniformly convergent within any circle having a as a center and in which the $p_i(z)$ are analytic. It follows that the solutions can possess singularities only where the coefficients are singular.

The study of solutions in the neighborhood of singular points was initiated by Briot and Bouquet.[34] Since their results for first order linear equations were soon generalized we shall consider the more general treatments.

To get at the behavior of solutions in the neighborhood of singular points Riemann proposed an unusual approach. Though the $p_i(z)$ in (13) are assumed to be single-valued functions analytic except at isolated singular points, the solutions $y_i(z)$, analytic except possibly at the singular points, are not in general single-valued over the entire domain of z-values. Let us suppose that we have a fundamental set of solutions, $y_i(z)$, $i = 1, 2, \ldots, n$, that is, n independent solutions of the kind specified in the above theorem. Then the general solution is

$$y = c_1 y_1 + c_2 y_2 + \cdots + c_n y_n$$

wherein the c_i are constants.

If we now trace the behavior of an analytic y_i along a closed path enclosing a singular point, y_i will change its value to another branch of the same function though it remains a solution of the differential equation. Since any solution is a linear combination of n particular solutions, the altered y_i,

33. *Jour. für Math.*, 66, 1866, 121–60 = *Math. Werke*, 1, 159 ff.
34. *Jour. d'Ecole Poly.*, (1), 21, 1856, 85–132, 133–198, 199–254.

say y'_i, is still a linear combination of the y_i. Thus we obtain

(15)
$$
\begin{aligned}
y'_1 &= c_{11}y_1 + \cdots + c_{1n}y_n \\
y'_2 &= c_{21}y_1 + \cdots + c_{2n}y_n \\
&\cdot \quad \cdot \quad \cdot \quad \cdot \quad \cdot \quad \cdot \quad \cdot \quad \cdot \\
y'_n &= c_{n1}y_1 + \cdots + c_{nn}y_n.
\end{aligned}
$$

That is, the $y_1, \ldots, y_n$ undergo a certain linear transformation when each is carried around a closed path enclosing a singular point. Such a transformation arises for any closed path around each of the singular points or combination of singular points. The set of transformations forms a group,[35] which is called the monodromy group of the differential equation, a term introduced by Hermite.[36]

Riemann's approach to obtaining the character of solutions in the neighborhood of singular points appeared in his paper of 1857 "Beiträge zur Theorie der durch die Gauss'sche Reihe $F(\alpha, \beta, \gamma, x)$ darstellbaren Functionen."[37] The hypergeometric differential equation, as Gauss knew, has three singular points, 0, 1, and ∞. Now Riemann showed that for complex x, to obtain conclusions about the behavior of the particular solutions around singular points of the second order equation, one does not need to know the differential equation itself but rather how two independent solutions behave as the independent variable traces closed paths around the three singular points. That is, we must know the transformations

$$
y'_1 = c_{11}y_1 + c_{12}y_2, \qquad y'_2 = c_{21}y + c_{22}y_2
$$

for each singular point.

Thus Riemann's idea in treating functions defined by differential equations was to derive the properties of the functions from a knowledge of the monodromy group. His 1857 paper dealt with the hypergeometric differential equation but his plan was to treat nth-order linear ordinary differential equations with algebraic coefficients. In a fragment written in 1857 but not published until his collected works appeared in 1876,[38] Riemann considered more general equations than the second order with three singular points. He accordingly assumes he has n functions uniform, finite, and continuous except at certain arbitrarily assigned points (the singular points) and undergoing an arbitrarily assigned linear substitution when z describes a closed circuit around such a point. He then shows that such a system of

35. By Riemann's time the algebraic notion of a group was known. It will be introduced in this book in Chap. 31. However, all that one needs to know here is that the application of two successive transformations is a transformation of the set and that the inverse of each transformation belongs to the set.
36. *Comp. Rend.*, 32, 1851, 458–61 = *Œuvres*, 1, 276–80.
37. *Werke*, 67–83.
38. *Werke*, 379–90.

functions will satisfy an nth-order linear differential equation. But he does not prove that the branch-points (singular points) and the substitutions may be chosen arbitrarily. His work here was incomplete and he left open a problem known as the Riemann problem: Given m points $a_1, \ldots, a_m$ in the complex plane with each of which is associated a linear transformation of the form (15), to prove on the basis of elementary assumptions about the behavior of the monodromy group associated with these singular points (so far as such behavior is not already determined) that a class of functions $y_1, \ldots, y_n$ is determined which satisfy a linear nth-order differential equation with the given a_i as singular (branch) points and such that when z traverses a closed path around a_i the y_i's undergo the linear transformation associated with a_i.

Guided by Riemann's 1857 paper on the hypergeometric equation the work on singularities was carried further by Fuchs. Beginning in 1865,[39] Fuchs and his students took up nth-order differential equations whereas Riemann had published only on Gauss's hypergeometric differential equation. Fuchs did not follow Riemann's approach but worked directly with the differential equation. Fuchs also brought not only linear differential equations but the entire theory of differential equations generally into the domain of complex function theory.

In the papers mentioned above Fuchs gave his major work on ordinary differential equations. He starts with the linear differential equation of the nth order whose coefficients are rational functions of x. By a careful examination of the convergence of the series which formally satisfy the equation he finds that the singular points of the equation are fixed, that is, independent of the constants of integration, and can be found before integrating because they are the poles of the coefficients of the differential equation.

He then shows that a fundamental system of solutions undergoes a linear transformation with constant coefficients when the independent variable z describes a circuit enclosing a singular point. From this behavior of the solutions he derives expressions for them valid in a circular region surrounding that point and extending to the next singular point. He thus establishes the existence of systems of n functions uniform, finite, and continuous except in the vicinity of certain points and undergoing linear substitutions with constant coefficients when the variable z describes closed circuits around these points.

Fuchs then considered what properties a differential equation of the form (13) must have in order that its solutions at a singular point $z = a$ have the form

$$(z - a)^s [\phi_0 + \phi_1 \log (z - a) + \cdots + \phi_\lambda \log^\lambda (z - a)],$$

where s is some number (which can be further specified) and the ϕ_i are single-valued functions in the neighborhood of $z = a$ which may have poles

39. *Jour. für Math.*, 66, 1866, 121–60; 68, 1868, 354–85.

of finite order. His answer was that a necessary and sufficient condition is that $p_r(z) = (z - a)^{-r}P(z)$, where $P(z)$ is analytic at and in the neighborhood of $z = a$. Thus $p_1(z)$ has a pole of order one and so on. Such a point a is called a regular singular point (Fuchs called it a point of determinateness).

Fuchs also studied a more specialized class of equations of the form (13). A homogeneous linear equation of this type is said to be of *Fuchsian type* when it has at worst regular singular points in the extended complex plane (including the point at ∞). In this case the $p_i(z)$ must be rational functions of z. For example, the hypergeometric equation has regular singular points at $z = 0$, 1, and ∞.

But the study of integrals of differential equations in the neighborhood of a given point does not necessarily furnish the integrals themselves. The study was taken as the point of departure for the investigation of the full integrals. Since the great researches of Fuchs, mathematicians have succeeded in extending the variety of linear ordinary differential equations that can be integrated explicitly. Previously only the nth-order linear equations with constant coefficients and Legendre's equation

$$(ax + b)^n \frac{d^n y}{dx^n} + A(ax + b)^{n-1} \frac{d^{n-1} y}{dx^{n-1}} + \cdots + L(ax + b) \frac{dy}{dx} + My = 0$$

could be integrated, the latter by the transformation $ax + b = e^t$. The new ones that can be integrated are those with integrals that are uniform (single-valued) functions of z. One recognizes that the integrals have this property by studying the singular points of the differential equation. The general integrals so obtained are usually new functions.

Beyond general results on the kinds of integrals which special classes of differential equations can have, there is the series approach to the solutions at a point $z = a$ where the equation has a regular singular point. If the origin is such a point then the equation must have the form

$$z^n \frac{d^n w}{dz^n} + z^{n-1}P_1(z) \frac{d^{n-1} w}{dz^{n-1}} + \cdots + zP_{n-1}(z) \frac{dw}{dz} + P_n(z)w = 0,$$

in which the $P_i(z)$ are analytic at and around $z = 0$. In this case one can obtain the n fundamental solutions in the form of series about $z = 0$ and show that the series converge for some range of z-values. The series are of the form

$$w = \sum_{v=0}^{\infty} c_v z^{\rho + v}$$

and the ρ and the c_v are determinable for each solution. The result is due to Georg Frobenius (1849–1917).[40]

40. *Jour. für Math.*, 76, 1874, 214–35 = *Ges. Abh.*, 1, 84–105.

The Riemann problem was also taken up during the latter part of the nineteenth century, but unsuccessfully until Hilbert in 1905[41] and Oliver D. Kellogg (1878-1932),[42] with the help of the theory of integral equations, which was developed in the meantime, gave the first complete solution. They showed that the generating transformation of the monodromy group can be prescribed arbitrarily.

6. Automorphic Functions

The theory of solutions of linear differential equations was tackled next by Poincaré and Felix Klein. The subject they introduced is called automorphic functions, which, though important for various other applications, play a major role in the theory of differential equations.

Henri Poincaré (1854-1912) was a professor at the Sorbonne. His publications, almost as numerous as Euler's and Cauchy's, cover a wide range of mathematics and mathematical physics. His physical researches, which we shall not have occasion to discuss, included capillary attraction, elasticity, potential theory, hydrodynamics, the propagation of heat, electricity, optics, electromagnetic theory, relativity, and above all, celestial mechanics. Poincaré had penetrating insight, and in every problem he studied he brought out its essential character. He focused sharply on a problem and examined it minutely. He also believed in a qualitative study of all aspects of a problem.

Automorphic functions are generalizations of the circular, hyperbolic, elliptic, and other functions of elementary analysis. The function $\sin z$ is unchanged in value if z is replaced by $z + 2m\pi$ where m is any integer. One can also say that the function is unaltered in value if z is subjected to any transformation of the group $z' = z + 2m\pi$. The hyperbolic function $\sinh z$ is unchanged in value if z be subjected to any transformation of the group $z' = z + 2\pi mi$. An elliptic function remains invariant in value under transformations of the group $z' = z + m\omega + m'\omega'$ where ω and ω' are the periods of the function. All of these groups are discontinuous (a term introduced by Poincaré); that is, all the transforms of any point under the transformations of the group are finite in number in any closed bounded domain.

The term automorphic function is now used to cover functions that are invariant under the group of transformations

$$(16) \qquad\qquad z' = \frac{az + b}{cz + d},$$

41. *Proc. Third Internat. Math. Cong.*, 1905, 233-40; and *Nachrichten König. Ges. der Wiss. zu Gött.*, 1905, 307-88. Also in D. Hilbert, *Grundzüge einer allgemeinen Theorie der linearen Integralgleichungen*, 1912, Chelsea (reprint), 1953, 81-108.
42. *Math. A..n.*, 60, 1905, 424-33.

where a, b, c, and d may be real or complex numbers, and $ad - bc = 1$, or under some subgroup of this group. Moreover the group must be discontinuous in any finite part of the complex plane.

The earliest automorphic functions to be studied were the elliptic modular functions. These functions are invariant under the modular group, which is that subgroup of (16) wherein a, b, c, and d are real integers and $ad - bc = 1$, or under some subgroup of this group. These elliptic modular functions derive from the elliptic functions. We shall not pursue them here because they do not bear on the basic theory of differential equations.

More general automorphic functions were introduced to study linear differential equations of the second order

$$(17) \qquad \frac{d^2\eta}{dz^2} + p_1 \frac{d\eta}{dz} + p_2\eta = 0,$$

where p_1 and p_2 were at first rational functions of z. A special case is the hypergeometric equation

$$(18) \qquad \frac{d^2\eta}{dz^2} + \frac{\gamma - (\alpha + \beta + 1)}{z(1 - z)} \frac{d\eta}{dz} + \frac{\alpha\beta}{z(z - 1)} \eta = 0$$

with the three singular points 0, 1, and ∞.

Riemann, in his lectures of 1858–59 on the hypergeometric series and in a posthumous paper of 1867 on minimal surfaces, and Schwarz[43] independently established the following. Let η_1 and η_2 be any two particular solutions of the equation (17). All solutions can then be expressed as

$$\eta = m\eta_1 + n\eta_2.$$

When z traverses a closed path around a singular point, η_1 and η_2 go over into

$$\eta_1^1 = a\eta_1 + b\eta_2, \qquad \eta_2^2 = c\eta_1 + d\eta_2$$

and by letting z traverse closed paths around all the singular points one obtains a whole group of such linear transformations, which is the monodromy group of the differential equation.

Now let $\zeta(z) = \eta_1/\eta_2$. The quotient ζ as z traverses a closed path is transformed to

$$(19) \qquad \zeta^1 = \frac{a\zeta + b}{c\zeta + d}.$$

From (17) we find that ζ satisfies the differential equation

$$(20) \qquad \frac{\zeta'''}{\zeta'} - \frac{3}{2} \cdot \left(\frac{\zeta''}{\zeta'}\right) = 2p_2 - \frac{1}{2}p_1^2 - p_1'.$$

43. *Jour. für Math.*, 75, 1873, 292–335 = *Ges. Abh.*, 2, 211–59.

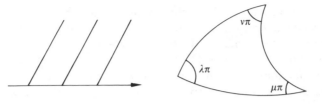

Figure 29.1

If we take for the p_1 and p_2 in (17) the particular functions in (18) we obtain

$$(21) \quad \frac{\zeta'''}{\zeta'} - \frac{3}{2}\left(\frac{\zeta''}{\zeta'}\right)^2 = \frac{1 - \lambda^2}{2z^2} + \frac{1 - \mu^2}{2(1 - z)^2} - \frac{\lambda^2 + \mu^2 - \nu^2 - 1}{2z(1 - z)},$$

where $\lambda^2 = 1 - \gamma^2$, $\mu^2 = (\gamma - \alpha - \beta)^2$, $\gamma^2 = (\alpha - \beta)^2$, and λ, μ, ν are taken positive (α, β, γ are real). The class of transformations (19) is the monodromy group of the differential equation (21).

Then Riemann and Schwarz showed that every particular solution $\zeta(z)$ of the equation (21), when λ, μ, and ν are real, is a conformal mapping of the upper half of the z-plane (Fig. 29.1) into a curvilinear triangle with circular arcs in the ζ-plane whose angles are $\lambda\pi$, $\mu\pi$, and $\nu\pi$.

In the case of a domain bounded by three arcs of circles, if the angles of the triangle satisfy certain conditions, the inverse function to $\zeta = \zeta(z)$ is an automorphic function $z = \phi(\zeta)$ whose entire domain of existence is a half-plane or a circle. This function remains invariant under transformation of ζ by elements of the group of linear transformations (19), which carry any curvilinear triangle of the form shown into another. The given "circular" triangle is the fundamental domain of the group. Under the group of transformations this domain is carried into analogous triangles whose union covers the half-plane or circle. The circular triangle is the analogue of the parallelogram in the case of elliptic functions.

The work of Poincaré and Klein carries on from this point. Klein did some basic work on automorphic functions before 1880. Then during the years 1881–82 he worked with Poincaré, who had also done previous work on the subject after his attention had been drawn to it by the above-described work of Fuchs. By 1884 Poincaré published five major papers on automorphic functions in the first five volumes of the *Acta Mathematica*. When the first of these was published in the first volume of the new *Acta Mathematica*, Kronecker warned the editor, Mittag-Leffler, that this immature and obscure article would kill the journal.

Guided by the theory of elliptic functions, Poincaré invented a new class of automorphic functions.[44] This class was obtained by considering the

44. *Acta Math.*, 1, 1882, 1–62 and 193–294 = *Œuvres*, 2, 108–68, 169–257.

inverse function of the ratio of two linearly independent solutions of the equation

$$\frac{d^2\eta}{dz^2} + P(w, z)\frac{d\eta}{dz} + Q(w, z)\eta = 0,$$

where w and z are connected by a polynomial equation $\phi(w, z) = 0$ and P and Q are rational functions. This is the class of Fuchsian automorphic functions and consists of uniform (single-valued) meromorphic functions within a circle (called the fundamental circle) which are invariant under the class of linear transformations of the form

$$(22) \qquad\qquad z' = \frac{az + b}{cz + d},$$

where a, b, c, and d are real and $ad - bc = 1$. These transformations, which leave the circle and its interior invariant, form a group called the Fuchsian group. Schwarz's function $\phi(\zeta)$ constitutes the simplest example of a Fuchsian function. Thus Poincaré demonstrated the existence of a class of automorphic functions more general than the elliptic modular functions.[45]

Poincaré's construction of automorphic functions (in the second paper of 1882) was based on his theta series. Let the transformations of the group (22) be

$$(23) \qquad z' = \frac{a_i z + b_i}{c_i z + d_i}, \qquad a_i d_i - b_i c_i = 1, \qquad i = 1, 2, \cdots.$$

Let $z_1, z_2, \ldots$ be the transforms of z under the various transformations of the group. Let $H(z)$ be a rational function (aside from other minor conditions). Then Poincaré's theta series is the function

$$(24) \qquad \theta(z) = \sum_{i=0}^{\infty} (c_i z + d_i)^{-2m} H(z_i), \qquad m > 1.$$

One can show that $\theta(z_j) = (c_j z + d_j)^{2m}\theta(z)$. Now let $\theta_1(z)$ and $\theta_2(z)$ be two theta series with the same m. These series are not only uniform functions but entire. Then

$$(25) \qquad\qquad F(z) = \frac{\theta_1(z)}{\theta_2(z)}$$

is an automorphic function of the group (23). Poincaré called the series (24) a theta-fuchsian series or a theta-kleinian series according as the group to which it belongs is Fuchsian or Kleinian (the latter will be described in a moment).

45. In this work on Fuchsian groups Poincaré used non-Euclidean geometry (Chap. 36) and showed that the study of Fuchsian groups reduces to that of the translation group of Lobatchevskian geometry.

Fuchsian functions are of two kinds, one existing in the entire plane, the other existing only in the interior of the fundamental circle. The inverse function of a Fuchsian function is, as we saw above, the ratio of two integrals of a second order linear differential equation with algebraic coefficients. Such an equation, which Poincaré called a Fuchsian equation, can be integrated by means of Fuchsian functions.

Then Poincaré[46] extended the group of transformations (22) to complex coefficients and considered several types of such groups, which he named Kleinian. We must be content here to note that a group is Kleinian if, essentially, it is not finite and not Fuchsian but, of course, is of the form (22) and discontinuous in any part of the complex plane. For these Kleinian groups Poincaré obtained new automorphic functions, that is, functions invariant under the Kleinian groups, which he called Kleinian functions. These functions have properties analogous to the Fuchsian ones; however, the fundamental region for the new functions is more complicated than a circle. Incidentally, Klein had considered Fuchsian functions while Lazarus Fuchs had not. Klein therefore protested to Poincaré. Poincaré responded by naming the next class of automorphic functions that he discovered Kleinian because, as someone wryly observed, they had never been considered by Klein.

Then Poincaré showed how to express the integrals of nth-order linear equations with *algebraic* coefficients having only regular singular points with the aid of the Kleinian functions. Thus this entire class of linear differential equations is solved by the use of these new transcendental functions of Poincaré.

7. *Hill's Work on Periodic Solutions of Linear Equations*

While the theory of automorphic functions was being created, the work in astronomy stimulated interest in a second order ordinary differential equation somewhat more general than Mathieu's equation. Since the n-body problem was not solvable explicitly and only complicated series solutions were at all available, the mathematicians turned to culling periodic solutions.

The importance of periodic solutions stems from the problem of the stability of a planetary or satellite orbit. If a planet is displaced slightly from its orbit and given a small velocity will it then oscillate about its orbit and perhaps return to the orbit after a time, or will it depart from the orbit? In the former case the orbit is stable, and in the latter unstable. Thus the question of whether the primary motion of the planets or any irregularities in the motions are periodic is vital.

As we know (Chap. 21, sec. 7), Lagrange had found special periodic solutions in the problem of three bodies. No new periodic solutions of the

46. *Acta Math.*, 3, 1883, 49–92; 4, 1884, 201–312 = *Œuvres*, 2, 258–99, 300–401.

three-body problem were found until George William Hill (1838–1914), the first great American mathematician, did his work on lunar theory. In 1877 Hill published privately a remarkably original paper on the motion of the moon's perigee.[47] He also published a very important paper on the motion of the moon in the *American Journal of Mathematics*.[48] His work founded the mathematical theory of homogeneous linear differential equations with periodic coefficients.

Hill's first fundamental idea (in his 1877 paper) was to determine a periodic solution of the differential equations for the motion of the moon that approximated the actual observed motion. He then wrote equations for variations from this periodic solution that led him to a fourth order system of linear differential equations with periodic coefficients. Knowing some integrals, he was able to reduce his fourth order system to a single linear differential equation of the second order

$$(26) \qquad \frac{d^2x}{dt^2} + \theta(t)x = 0,$$

with $\theta(t)$ periodic of period π and even. The form of Hill's equation can be put, by expanding $\theta(t)$ in a Fourier series, as

$$(27) \qquad \frac{d^2x}{dt^2} + x(q_a + 2q_1 \cos 2t + 2q_2 \cos 4t + \cdots) = 0.$$

Hill put $\zeta = e^{it}$, $q_{-\alpha} = q_\alpha$ and wrote (27) as

$$(28) \qquad \frac{d^2x}{dt^2} + x \sum_{-\infty}^{\infty} q_a \zeta^{2\alpha} = 0.$$

He then let

$$x = \sum_{j=-\infty}^{\infty} b_j \zeta^{\mu + 2j},$$

where μ and b_j were to be determined. By substituting this value of x in (28) and setting the coefficients of each power of ζ equal to 0 he obtained the doubly infinite system of linear equations

$$
\begin{aligned}
\cdots\ \cdot\ [-2]b_{-2} - q_1 b_{-1} - q_2 b_0 - q_3 b_1 - q_4 b_2 - \cdots\ = 0 \\
\cdots - q_1 b_{-2} + [-1]b_{-1} - q_1 b_0 - q_2 b_1 - q_3 b_2 - \cdots\ = 0 \\
\cdots - q_2 b_{-2} - q_1 b_{-1} + [0]b_0 - q_1 b_1 - q_2 b_2 - \cdots\ = 0 \\
\cdots - q_3 b_{-2} - q_2 b_{-1} - q_1 b_0 + [1]b_1 - q_1 b_2 - \cdots\ = 0 \\
\cdots - q_4 b_{-2} - q_3 b_{-1} - q_2 b_0 - q_1 b_1 + [2]b_2 - \cdots\ = 0
\end{aligned}
$$

47. This was reprinted in *Acta Math.*, 8, 1886, 1–36 = *Coll. Math. Works*, 1, 243–70.
48. *Amer. Jour. of Math.*, 1, 1878, 5–26, 129–47, 245–60 = *Coll. Math. Works*, 1, 284–335.

where

$$[j] = (\mu + 2j)^2 - q_0.$$

Hill set the determinant of the coefficients of the unknown b_j equal to 0. He first determined the properties of the infinitely many solutions for μ and gave explicit formulas for determining the μ. With these values of μ he then solved the system of an infinite number of linear homogeneous equations in the infinite number of b_j for the ratio of the b_j to b_0. Hill did show that the second order differential equation has a periodic solution and that the motion of the moon's perigee is periodic.

Hill's work was ridiculed until Poincaré[49] proved the convergence of the procedure and thereby put the theory of infinite determinants and infinite systems of linear equations on its feet. Poincaré's attention to and completion of Hill's efforts gave prominence to Hill and the subjects involved.

8. *Nonlinear Differential Equations: The Qualitative Theory*

A new approach to the search for periodic solutions of the differential equations governing planetary motion and the stability of planetary and satellite orbits was initiated by Poincaré under the stimulus of Hill's work. Because the relevant equations are nonlinear, Poincaré took up this class. Nonlinear ordinary differential equations had appeared practically from the beginnings of the subject as, for example, in the Riccati equation (Chap. 21, sec. 4), the pendulum equation, and the Euler equations of the calculus of variations (Chap. 24, sec. 2). No general methods of solving nonlinear equations had been developed.

In view of the fact that the equations for the motion of even three bodies cannot be solved explicitly in terms of known functions, the problem of stability cannot be solved by examining the solution. Poincaré therefore sought methods by which the problem could be answered by examining the differential equations themselves. The theory he initiated he called the qualitative theory of differential equations. It was presented in four papers all under substantially the same title, "Mémoire sur les courbes définies par une équation différentielle."[50] The questions he sought to answer were stated by him in these words: "Does the moving point describe a closed curve? Does it always remain in the interior of a certain portion of the plane? In other words, and speaking in the language of astronomy, we have inquired whether the orbit is stable or unstable."

49. *Bull. Soc. Math. de France*, 13, 1885, 19–27; 14, 1886, 77–90 = *Œuvres*, 5, 85–94, 95–107.
50. *Jour. de Math.*, (3), 7, 1881, 375–422; 8, 1882, 251–96; (4), 1, 1885, 167–244; 2, 1886, 151–217 = *Œuvres*, 1, 3–84, 90–161, 167–221.

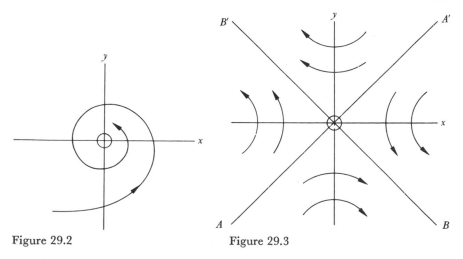

Figure 29.2   Figure 29.3

Poincaré started with nonlinear equations of the form

(29) $$\frac{dy}{dx} = \frac{P(x,y)}{Q(x,y)},$$

where P and Q are analytic in x and y. This form was chosen partly because some problems of planetary motion led to it and partly because it was the simplest mathematical type with which to commence the kind of investigation Poincaré had in mind. The solution of (29) is of the form $f(x,y) = 0$, and this equation is said to define a system of trajectories. In place of $f(x,y) = 0$ one can consider the parametric form $x = x(t), y = y(t)$.

In the analysis of the kinds of solutions equation (29) can have, Poincaré found that the singular points of the differential equation (the points at which P and Q both vanish) play a key role. These singular points are undetermined or irregular in Fuchs's sense. Here Poincaré took up earlier work by Briot and Bouquet (sec. 5) but limited himself to real values and to studying the behavior of the entire solution rather than just in the neighborhood of the singular points. He distinguished four types of singular points and described the behavior of solutions around such points.

The first type of singular point is the focus (*foyer*), the origin in Figure 29.2, and the solution spirals around and approaches the origin as t runs from $-\infty$ to ∞. This type of solution is considered stable. The second kind of singular point is the saddle point (*col*). It is the origin of Figure 29.3 and the trajectories approach this point and then depart from it. The lines AA' and BB' are the asymptotes of the trajectories. The motion is unstable. The third type of singular point, called a node (*nœud*), is a point where an infinity of solutions cross, and the fourth, called a center, is one around which closed trajectories exist, one enclosing the other and all enclosing the center.

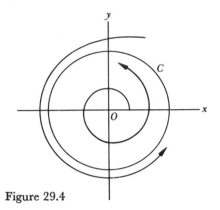

Figure 29.4

Among many results, Poincaré found that there can be closed curves which do not *touch* any of the curves satisfying the differential equation. He called these closed curves cycles without contact. A curve satisfying the differential equation cannot meet such a cycle in more than one point, and so if it crosses the cycle it cannot re-cross it. Such a curve, if it is an orbit of a planet, represents unstable motion.

Beyond cycles without contact there are closed curves that Poincaré called limit cycles. These are closed curves which satisfy the differential equation and which other solutions approach asymptotically, that is, without ever reaching the limit cycle. The approach can be from without or within the limit cycle C (Fig. 29.4). For some differential equations of the type (29) he determined the limit cycles and the regions in which they exist. In the case of limit cycles the trajectories approach a periodic curve, and the motion is again stable. If, however, the direction of the motions were away from the limit cycle the motion outside would be unstable and the motion inside would be a contracting spiral.

In the third of his papers on this subject, Poincaré studied first order equations of higher degree and of the form $F(x, y, y') = 0$, where F is a polynomial in x, y, and y'. To study these equations Poincaré regarded x, y, and y' as three Cartesian coordinates and considered the surface defined by the differential equation. If this surface has genus 0 (form of a sphere), then the integral curves have the same properties as in the case of first degree differential equations. For other genuses the results on integral curves can be quite different. Thus for a torus many new circumstances arise. Poincaré did not complete this study. In the fourth paper (1886) he studied second order equations and obtained some results analogous to those for first order equations.

While continuing his work on the types of solutions of the differential equation (29), Poincaré considered a more general theory directed to the three-body problem of astronomy. In a prize paper, "Sur le problème de trois corps et les équations de la dynamique,"[51] he considered the system of differential equations

(30) $$\frac{dx_i}{dt} = X_i(x_1, \cdots, x_n, \mu), \qquad i = 1, 2, \cdots, n.$$

He developed the X_i in powers of the small parameter μ and, supposing that the system had for $\mu = 0$ a known periodic solution

$$x_i = \phi_i(t), \qquad i = 1, 2, \cdots, n$$

of period T, he proposed to find the periodic solution of the system which for $\mu = 0$ reduces to the $\phi_i(t)$. The existence of periodic solutions for the three-body problem had been discovered by Hill, and Poincaré made use of this fact.

The details of Poincaré's work are too specialized to consider here. He first generalized earlier work of Cauchy on solutions of systems of ordinary differential equations wherein the latter had used his *calcul des limites*. Poincaré then demonstrated the existence of the periodic solutions he sought and applied what he learned to the study of periodic solutions of the three-body problem for the case where the masses of two of the bodies (but not that of the sun) are small. Thus one obtains such solutions by supposing that the two small masses move in concentric circles about the sun and are in the same plane. One can obtain others by supposing that for $\mu = 0$ the orbits are ellipses and that their periods are commensurable. With these solutions, and by using the theory he had developed for the system, he obtained other periodic solutions. In sum he showed that there is an infinity of initial positions and initial velocities such that the *mutual distances* of the three bodies are periodic functions of the time. (Such solutions are also called periodic.)

Poincaré in this paper of 1890 drew many other conclusions about periodic and almost periodic solutions of the system (30). Among them is the very notable discovery for such a system of a new class of solutions previously unknown. These he called asymptotic solutions. There are two kinds. In the first the solution approaches the periodic solution asymptotically as t approaches $-\infty$ or as t approaches $+\infty$. The second kind consists of doubly asymptotic solutions, that is, solutions that approach a periodic solution as t approaches $-\infty$ and $+\infty$. There is an infinity of such doubly asymptotic solutions. All of the results in this paper of 1890 and many others can be found also in Poincaré's *Les Méthodes nouvelles de la mécanique céleste*.[52]

Poincaré's work on the problem of stability of the solar system was only partially successful. The stability is still an open question. As a matter of

51. *Acta Math.*, 13, 1890, 1–270 = *Œuvres*, 7, 262–479.
52. Three volumes, 1892–99.

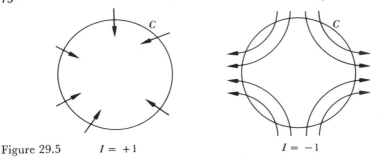

Figure 29.5 $I = +1$ $I = -1$

fact so is the question of whether the orbit of the moon is stable; most scientists today believe it is not.

The stability of the solutions of (29) can be analyzed by means of what is called the characteristic equation, namely,

$$(31) \qquad \begin{vmatrix} Q_x(x_0, y_0) - \lambda & P_x(x_0, y_0) \\ Q_y(x_0, y_0) & P_y(x_0, y_0) - \lambda \end{vmatrix} = 0,$$

where (x_0, y_0) is a singular point of (29). The stability in the neighborhood of (x_0, y_0), according to a theorem of the distinguished Russian mathematician Alexander Liapounoff (1857–1918) depends upon the roots of this characteristic equation.[53] The analysis of possible cases is detailed and includes many more types than those given in the discussion above of Poincaré's work. The basic result, according to Liapounoff, whose work on stability problems continued into the early years of this century, is that the solutions are stable in the neighborhood of a singular point when and only when the roots of the equation (31) in λ have negative real parts.

The qualitative study of nonlinear equations was advanced by Poincaré's introduction of topological arguments (in the first of the four papers in the *Journal de Mathématiques*). To describe the nature of a singular point he introduced the notion of an index. Consider a singular point P_0 and a simple closed curve C surrounding it. At each intersection of C with the solutions of

$$(32) \qquad \frac{dy}{dx} = \frac{P(x, y)}{Q(x, y)}$$

there is a direction angle of the trajectory, which we shall denote by ϕ and which can have any value from 0 to 2π radians. If a point now moves in a counterclockwise direction around C (Fig. 29.5), the angle ϕ will vary; and after completion of the circuit around C, ϕ will have the value $2\pi I$ where I is an integer or zero (since the direction angle of the trajectories has

53. *Ann. Fac. Sci. de Toulouse*, (2), 9, 1907, 203–474; originally published in Russian in 1892.

returned to its original value). The quantity I is the index of the curve. It can be proved that the index of a closed curve that contains several singularities is the algebraic sum of their indices. The index of a closed trajectory is $+1$ and conversely.

The nature of the trajectories can be determined by the characteristic equation, and so the index I of a curve should be determinable by knowing just the differential equation. One can prove that

$$I = \frac{1}{2\pi} \int_C d\left(\text{arc tan}\, \frac{P}{Q}\right) = \frac{1}{2\pi} \int_C \frac{P\, dQ - Q\, dP}{P^2 + Q^2},$$

where the path of integration is the closed curve C.

After Poincaré, the most significant work on solutions of equations of the form (32) is due to Ivar Bendixson (1861–1935). One of his major results[54] provides a criterion by means of which, in certain regions, one can show that no closed trajectory exists. Let D be a region in which $\partial Q/\partial x + \partial P/\partial y$ has the same sign. Then the equation (32) has no periodic solution in D.

The theorem now named after Poincaré and Bendixson, which is in the latter's 1901 paper, provides a positive criterion for the existence of a periodic solution of (32). If P and Q are defined and regular in $-\infty < x, y < \infty$ and if as t approaches ∞ a solution $x(t), y(t)$ remains within a bounded region of the (x, y)-plane without approaching singular points, then there exists at least one closed solution curve of the differential equation.

The study of nonlinear equations that Poincaré launched was broadened in various directions. One more topic begun in the nineteenth century will be mentioned. The linear differential equations studied by Fuchs have the property that the singular points are fixed and are, in fact, determined by the coefficients of the differential equation. In the case of nonlinear equations the singular points may vary with the initial conditions and are called movable singular points. Thus the equation $y' + y^2 = 0$ has the general solution $y = 1/(x - c)$ where c is arbitrary. The location of the singularity in the solution depends on the value of c. This phenomenon of movable singular points was discovered by Fuchs.[55] The study of movable singular points and of nonlinear second order equations with and without such singular points was taken up by many men, notably by Paul Painlevé (1863–1933). One interesting feature is that many of the types of second order equations of the form $y'' = f(x, y, y')$ require for their solution new types of transcendental functions now called Painlevé transcendents.[56]

The interest in nonlinear equations has become strong in the twentieth century. The applications have moved from astronomy to problems of

54. *Acta Math.*, 24, 1901, 1–88.
55. *Sitzungsber. Akad. Wiss. zu Berlin*, 1884, 699–710 = *Werke*, 2, 364 ff.
56. *Comp. Rend.*, 143, 1906, 1111–17.

communications, servomechanisms, automatic control systems, and electronics. The study has also moved from the qualitative stage to quantitative investigations.

Bibliography

Acta Mathematica, Vol. 38, 1921. This entire volume is devoted to articles on Poincaré's work by various leading mathematicians.

Bôcher, M.: "Randwertaufgaben bei gewöhnlichen Differentialgleichungen," *Encyk. der Math. Wiss.*, B. G. Teubner, 1899–1916, II, A7a, 437–63.

————: "Boundary Problems in One Dimension," *Internat. Cong. of Math.*, Proc., Cambridge, 1912, 1, 163–95.

Burkhardt, H.: "Entwicklungen nach oscillirenden Functionen und Integration der Differentialgleichungen der mathematischen Physik," *Jahres. der Deut. Math.-Verein.*, 10, 1908, 1–1804.

Cauchy, A. L.: *Œuvres complètes*, (1), Vols. 4, 7, and 10, Gauthier-Villars, 1884, 1892, and 1897.

Craig, T.: "Some of the Developments in the Theory of Ordinary Differential Equations Between 1878 and 1893," *N.Y. Math. Soc. Bull.*, 2, 1893, 119–34.

Fuchs, Lazarus: *Gesammelte mathematische Werke*, 3 vols., 1904–09, Georg Olms (reprint), 1970.

Heine, Eduard: *Handbuch der Kugelfunktionen*, 2 vols., 1878–81, Physica Verlag (reprint), 1961.

Hilb, E.: "Lineare Differentialgleichungen im komplexen Gebiet," *Encyk. der Math. Wiss.*, B. G. Teubner, 1899–1916, II, B5.

————: "Nichtlineare Differentialgleichungen," *Encyk. der Math. Wiss.*, B. G. Teubner, 1899–1916, II, B6.

Hill, George W.: *Collected Mathematical Works*, 4 vols., 1905, Johnson Reprint Corp., 1965.

Klein, Felix: *Vorlesungen über die Entwicklung der Mathematik im 19. Jahrhundert*, Vol. I, Chelsea (reprint), 1950.

————: *Gesammelte mathematische Abhandlungen*, Julius Springer, 1923, Vol. 3.

Painlevé, P.: "Le Problème moderne de l'intégration des équations différentielles," *Third Internat. Math. Cong. in Heidelberg*, 1904, 86–99, B. G. Teubner, 1905.

————: "Gewöhnliche Differentialgleichungen, Existenz der Lösungen," *Encyk. der Math. Wiss.*, B. G. Teubner, 1899–1916, II, A4a.

Poincaré, Henri: *Œuvres*, 1, 2, and 5, Gauthier-Villars, 1928, 1916, and 1960.

Riemann, Bernhard: *Gesammelte mathematische Werke*, 2nd ed., 1892, Dover (reprint), 1953.

Schlesinger, L.: "Bericht über die Entwickelung der Theorie der linearen Differentialgleichungen seit 1865," *Jahres. der Deut. Math.-Verein.*, 18, 1909, 133–266.

Wangerin, A.: "Theorie der Kugelfunktionen und der verwandten Funktionen," *Encyk. der Math. Wiss.*, B. G. Teubner, 1899–1916, II, A10.

Wirtinger, W.: "Riemanns Vorlesungen über die hypergeometrische Reihe und ihre Bedeutung," *Third Internat. Math. Cong. in Heidelberg*, 1904, B. G. Teubner, 1905, 121–39.

30

The Calculus of Variations in the Nineteenth Century

> Although to penetrate into the intimate mysteries of nature and thence to learn the true causes of phenomena is not allowed to us, nevertheless it can happen that a certain fictive hypothesis may suffice for explaining many phenomena.
>
> LEONHARD EULER

1. *Introduction*

As we have seen, the calculus of variations was founded in the eighteenth century chiefly by Euler and Lagrange. Beyond mathematical and physical problems of various sorts there was one leading motivation for the study, namely, the Principle of Least Action, which in the hands of Maupertuis, Euler, and Lagrange became a leading principle of mathematical physics. The nineteenth-century men continued the work on least action and the greatest stimulus to the calculus of variations in the first half of the century came from this direction. Physically, the interest was in the science of mechanics and particularly in problems of astronomy.

2. *Mathematical Physics and the Calculus of Variations*

Lagrange's successful formulation of the laws of dynamics in terms of his Principle of Least Action suggested that the idea should be applicable to other branches of physics. Lagrange[1] gave a minimum principle for fluid dynamics (applicable to compressible and incompressible fluids) from which he derived Euler's equations for fluid dynamics (Chap. 22, sec. 8) and indeed he boasted that a minimum principle governed this field as it did the motion of particles and rigid bodies. Many problems of elasticity also were solved by the calculus of variations in the early part of the nineteenth century by Poisson, Sophie Germain, Cauchy, and others, and this work too helped to

1. *Misc. Taur.*, 2_2, 1760/61, 196–298, pub. 1762 = *Œuvres*, 1, 365–468.

739

keep the subject active, but no major new mathematical ideas of the calculus of variations are to be noted in this area or in Gauss's famous contribution to mechanics, The Principle of Least Constraint.[2]

The first new point worth noting is due to Poisson. Using Lagrange's generalized coordinates, he followed up immediately two papers by Lagrange and starts with[3] Lagrange's equations (Chap. 24, sec. 5)

$$(1) \qquad \frac{d}{dt}\left(\frac{\partial T}{\partial \dot{q}_i}\right) - \frac{\partial T}{\partial q_i} + \frac{\partial V}{\partial q_i} = 0, \qquad i = 1, 2, \cdots, n.$$

Here the kinetic energy T expressed in generalized coordinates is $2T = \sum_{i,j=1}^{n} a_{ij}\,\dot{q}_i\,\dot{q}_j$, V is the potential energy, and T and V are independent of t. He sets $L = T - V$. Where V depends only on the q_i and not on the $\dot{q}_i$ he can write

$$(2) \qquad \frac{\partial L}{\partial \dot{q}_i} = \frac{\partial T}{\partial \dot{q}_i},$$

so that the equations of motion read

$$(3) \qquad \frac{d}{dt}\left(\frac{\partial L}{\partial \dot{q}_i}\right) - \frac{\partial L}{\partial q_i} = 0, \qquad i = 1, 2, \cdots, n.$$

He also introduces

$$(4) \qquad p_i = \frac{\partial L}{\partial \dot{q}_i} = \frac{\partial T}{\partial \dot{q}_i},$$

and so from (3) he has

$$(5) \qquad \dot{p}_i = \frac{\partial L}{\partial q_i}, \qquad i = 1, 2, \cdots, n.$$

The p_i are momentum components when the q_i are rectangular coordinates of position. Equation (5) is a step in the direction we shall now look into.

The big change in the formulation of least action principles which is important for the calculus of variations and for ordinary and partial differential equations was made by William R. Hamilton. Hamilton came to dynamics through optics. His goal in optics was to fashion a deductive mathematical structure in the manner of Lagrange's treatment of mechanics.

Hamilton too started with a least action principle and was to deduce new ones. However, his attitude toward such principles was profoundly different from that of Maupertuis, Euler, and Lagrange. In a paper published in the *Dublin University Review*[4] he says, "But although the law of least action has thus attained a rank among the highest theorems of physics, yet its pretensions to a cosmological necessity, on the ground of economy in the

2. *Jour. für Math.*, 4, 1829, 232–35 = *Werke*, 5, 23–28.
3. *Jour. de l'Ecole Poly.*, 8, 1809, 266–344.
4. 1833, 795–826 = *Math. Papers*, 1, 311–32.

universe, are now generally rejected. And the rejection appears just, for this, among other reasons, that the quantity pretended to be economized is in fact often lavishly expended." Because in some phenomena of nature, even simple ones, the action is maximized, Hamilton preferred to speak of a principle of stationary action.

In a series of papers of the period 1824 to 1832 Hamilton developed his mathematical theory of optics and then carried over ideas he had introduced there to mechanics. He wrote two basic papers.[5] It is the second of these which is more pertinent. Here he introduces the action integral, namely, the time integral of the difference between kinetic and potential energies

$$(6) \qquad S = \int_{P_1, t_1}^{P_2, t_2} (T - V) \, dt.$$

The quantity $T - V$ is called the Lagrangian function though it was introduced by Poisson; P_1 stands for q_1^1, q_2^1, ..., q_n^1, and P_2 for $q_1^{(2)}$, $q_2^{(2)}$, ..., $q_n^{(2)}$. Now Hamilton generalizes the principle of Euler and Lagrange by allowing comparison paths that are not restricted except that the motion along them must start at P_1 at time t_1 and end at P_2 at time t_2. Also, the law of conservation of energy need not hold whereas in the Euler-Lagrange principle conservation of energy is presupposed, and as a consequence the time required by an object to traverse any one of the comparison paths differs from the time taken to traverse the actual path.

The Hamiltonian principle of least action asserts that the actual motion is the one that makes the action stationary. For conservative systems, that is, where the components of force are derivable from a potential that is a function of position only, $T + V = $ const. Hence $T - V = 2T - $ const. and so Hamilton's principle reduces to Lagrange's, but, as noted, Hamilton's principle also holds for nonconservative systems. Also the potential energy V can be a function of time and even of the velocities; that is, in generalized coordinates $V = V(q_1, \ldots, q_n, \dot{q}_1, \ldots, \dot{q}_n, t)$.

If, setting $T - V$ equal to L, we write the action integral (6) as

$$(7) \qquad S = \int_{t_1}^{t_2} L(q_1, \cdots, q_n, \dot{q}_1, \cdots, \dot{q}_n, t) \, dt,$$

with the condition that all comparison $q_i(t)$ must have the same given values at t_1 and t_2, then the problem is to determine the q_i as functions of t from the condition that the true q_i make the integral stationary. The Euler equations, which express the condition that the first variation of S is 0, become a system of simultaneous second order ordinary differential equations, namely,

$$(8) \qquad \frac{\partial L}{\partial q_k} - \frac{d}{dt}\left(\frac{\partial L}{\partial \dot{q}_k}\right) = 0, \qquad k = 1, 2, \cdots, n,$$

5. *Phil. Trans.*, 1834, Part II, 247–308; 1835, Part I, 95–144 = *Math. Papers*, 2, 103–211.

and the equations are to be solved in $t_1 \leq t \leq t_2$. These equations are still called the Lagrangian equations of motion even though L is now a different function. The choice of coordinate system is arbitrary and usually utilizes generalized coordinates. This is an essential advantage of variational principles.

Now introduce (see (4))

$$p_i = \frac{\partial L}{\partial \dot{q}_i}.$$

Then the equations (8) become

$$\dot{p}_i = \frac{\partial L}{\partial q_i}.$$

The introduction of the p_i as a new set of independent variables is credited to Hamilton though it was first done by Poisson. We now have the symmetrical system of differential equations

(9) $$p_i = \frac{\partial L}{\partial \dot{q}_i}, \qquad \dot{p}_i = \frac{\partial L}{\partial q_i}, \qquad i = 1, 2, \cdots, n.$$

This is a system of $2n$ first order ordinary differential equations in p_i and $\dot{p}_i$. However, the $\dot{p}_i$ are dp_i/dt.

In his second (1835) paper Hamilton simplifies this system of equations. He introduces a new function H which is defined by

(10) $$H(p_i, q_i, t) = -L + \sum_{i=1}^{n} p_i \dot{q}_i.$$

This function is physically the total energy, for the summation can be shown to be equal to $2T$. The transformation from L to H is called a Legendre transformation because it was used by Legendre in his work on ordinary differential equations. That H is a function of the p_i, q_i, and t, whereas $L = T - V$ is a function of the q_i, $\dot{q}_i$, and t, results from the fact that since $p_i = \partial L/\partial \dot{q}_i$ we can solve for the $\dot{q}_i$ and substitute in L.

With (10) the differential equations of motion (9) can be shown to take the form

(11) $$\dot{q}_i = \frac{\partial H}{\partial p_i}, \qquad \dot{p}_i = -\frac{\partial H}{\partial q_i}, \qquad i = 1, 2, \cdots, n.$$

The function H is assumed to be known in the application to physical problems. These equations are a system of $2n$ first order ordinary differential equations in the $2n$ dependent variables p_i and q_i as functions of t, whereas Lagrange's equations (1) are a system of n second order equations in the $q_i(t)$. Later Jacobi called Hamilton's equations the canonical differential equations. They are the variational equations (Euler equations) for the integral

$$S = \int_{P_1, t_1}^{P_2, t_2} (T - V) \, dt = \int L \, dt = \int \left\{ \sum_{i=1}^{n} p_i \dot{q}_i - H(p_i, q_i, t) \right\} dt.$$

This set of equations appears in one of Lagrange's papers of 1809 which deals with the perturbation theory of mechanical systems. However, whereas Lagrange did not recognize the basic connection of these equations with the equations of motion, Cauchy in an unpublished paper of 1831 did. Hamilton in 1835 made these equations the basis of his mechanical investigations.

To use Hamilton's equations of motion it is often possible to express H in an appropriate p and q coordinate system so that the system of equations (11) is solvable for the p_i and q_i as functions of the time. In particular, if we can choose coordinates so that H depends only on the p_i, the system is solvable.

In a paper of 1837[6] and in lectures on dynamics of the years 1842 and 1843, which were published in 1866 in the classic *Vorlesungen über Dynamik*, Jacobi showed that one can reverse Hamilton's process. In Hamilton's theory if one knows the action S or the Hamiltonian H one can form the $2n$ canonical differential equations and attempt to solve the system. Jacobi's thought was to try to find coordinates P_j and Q_j so that H is as simple as possible, and then the differential equations (11) would be easily integrated. Specifically, he sought a transformation

(12)
$$Q_j = Q_j(p_i, q_i, t)$$
$$P_j = P_j(p_i, q_i, t)$$

such that

$$\delta \int_{t_1}^{t_2} \left(\sum_{i=1}^{n} p_i \dot{q}_i - H(p_i, q_i, t) \right) dt = 0$$

goes over by the transformation (12) into

$$\delta \int_{t_1}^{t_2} \left(\sum_{i=1}^{n} P_i \dot{Q}_i - K(P_i, Q_i, t) \right) dt = 0,$$

and so that the Hamiltonian differential equations become

(13)
$$\dot{Q}_i = \frac{\partial K}{\partial P_i}, \qquad \dot{P}_i = -\frac{\partial K}{\partial Q_i},$$

where $K(P_i, Q_i, t)$ is the new Hamiltonian. This path leads to

$$K = H(p_i, q_i, t) + \frac{\partial \Omega}{\partial t}(Q_i, q_i, t),$$

6. *Jour. für Math.*, 17, 1837, 97–162 = *Ges. Werke*, 4, 57–127.

where Ω is a new function, called the generating function of the transformation. Jacobi chose $K = 0$ so that by (13)

$$\dot{Q}_i = 0, \qquad \dot{P}_i = 0,$$

or Q_i and P_i are constants. Moreover,

(14)
$$H + \frac{\partial \Omega}{\partial t} = 0,$$

and it can be shown that

$$p_i = \frac{\partial \Omega}{\partial q_i}.$$

Hence by (14), in view of the variables in H,

(15)
$$H\left(\frac{\partial \Omega}{\partial q_i}, q_i, t\right) + \frac{\partial \Omega}{\partial t}(Q_i, q_i, t) = 0.$$

Since $\dot{Q}_i = 0$, $Q_i = \alpha_i$, and so the equation is of first order in Ω with the independent variables q_i and t. With this change equation (15) is the Hamilton-Jacobi partial differential equation for the function Ω. If this equation can be solved for a complete Ω, that is, one containing n arbitrary constants, the solution would have the form

$$\Omega(\alpha_1, \alpha_2, \cdots, \alpha_n, q_1, q_2, \cdots, q_n, t).$$

Now it is a fact of the Jacobi transformation theory that

$$P_i = -\frac{\partial \Omega}{\partial \alpha_i}$$

and that $P_i = \beta_i$, a constant, because $\dot{P}_i = 0$. Hence one solves the algebraic equations

$$\frac{\partial \Omega}{\partial \alpha_i} = -\beta_i, \qquad i = 1, 2, \cdots, n$$

for the q_i. These solutions

$$q_i = f_i(\alpha_1, \cdots, \alpha_n, \beta_1, \cdots, \beta_n, t), \qquad i = 1, 2, \cdots, n,$$

are the solutions of Hamilton's canonical equations. Thus Jacobi had shown that one can solve the system of ordinary equations (11) by solving the partial differential equation (15). Jacobi himself found the proper Ω for many problems of mechanics.

Hamilton's work was the culmination of a series of efforts to provide a broad principle from which the laws of motion of various problems of mechanics could be derived. It inspired efforts to obtain similar variational principles in other branches of mathematical physics such as elasticity, electromagnetic theory, relativity, and quantum theory. The principles that have been derived, and even Hamilton's principle, are not necessarily more practical approaches to the solution of particular problems. Rather the attractiveness of such broad formulations lies in philosophic and aesthetic interests though scientists no longer infer that the existence of a maximum-minimum principle is evidence of God's wisdom and efficiency.

From the standpoint of the history of mathematics the work of Hamilton and Jacobi is significant because it prompted further research not only in the calculus of variations but also on systems of ordinary differential equations and first order partial differential equations.

3. Mathematical Extensions of the Calculus of Variations Proper

We may recall that even in the simplest case, maximizing or minimizing the integral

$$(16) \qquad\qquad J = \int_a^b f(x, y, y') \, dx,$$

the results of Euler and Legendre provided only necessary conditions (Chap. 24). For about fifty years after the work of Legendre, mathematicians explored further the first and second variations but no decisive results were obtained. In 1837[7] Jacobi found out how to sharpen Legendre's condition so that it might yield a sufficient condition. His chief discovery in this connection was the concept of conjugate point. Let us note first what this is.

Consider the curves which satisfy Euler's (characteristic) equation; such curves are called extremals. For the basic problem of the calculus of variations, there is a one-parameter family of extremals passing through a given point A. Suppose now that A is one of the two endpoints between which we seek a maximizing or minimizing curve. Given any one extremal, the limiting point of intersection of other extremals as they come closer to that extremal is the conjugate point to A on that extremal. Another way of putting it is that we have a family of curves, and this family may have an envelope. The point of contact of any one extremal and the envelope of the family is the point conjugate to A on that extremal. Then Jacobi's condition is that if $y(x)$ is an extremal between the endpoints A and B of the original problem, no conjugate point must lie on that extremal between A and B or even be B itself.

Just what this means in a concrete case may be seen from an example. It can be shown that the parabolic paths of all trajectories (Fig. 30.1)

7. *Jour. für. Math.*, 17, 1837, 68–82 = *Ges. Werke*, 4, 39–55.

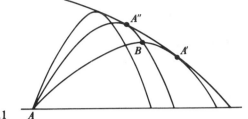

Figure 30.1 A

emanating from A with constant velocity v but with varying angles of fire are extremals of the problem of minimizing or maximizing the action integral

$$\frac{m}{2} \int_A^B v \, ds.$$

The problem of minimizing the action between two points A and B does in general have two solutions, the parabola $AA''B$ and the parabola ABA'. It is also the case that the family of parabolas through A has an envelope that touches the two parabolas at A'' and A'. The conjugate point on $AA''B$ is A'' and the one on ABA' is A'. According to Jacobi's condition, the extremal $AA''B$ could not furnish a maximum or a minimum but the extremal ABA' could.

Jacobi reconsidered the second variation $\delta^2 J$ (Chap. 24, sec. 4). If we write $y + \epsilon t(x)$ in place of Lagrange's $y + \delta y$ and if a and b are the abscissas of A and B then

(17) $$\delta^2 J = \frac{\epsilon^2}{2} \int_a^b (t^2 f_{yy} + 2tt' f_{yy'} + t'^2 f_{y'y'}) \, dx.$$

Jacobi showed that

$$\delta^2 J = \frac{\epsilon^2}{2} \int_a^b f_{y'y'} \left(t' - t \frac{u'}{u} \right)^2 dx,$$

where u is a solution of Jacobi's accessory equation

(18) $$\left\{ f_{yy} - \frac{d}{dx} f_{yy'} \right\} u - \frac{d}{dx} (f_{y'y'} u') = 0$$

and where the partial derivatives are evaluated along an extremal joining the two endpoints A and B. Now $u(x)$ is required to pass through A. Then all other points on the extremal $y(x)$ through A and B at which $u(x)$ vanishes are the conjugate points of A on that extremal. If $u = \beta_1 u_1 + \beta_2 u_2$ is the general solution of the accessory equation (18), then one can show that

$$\frac{u_1(x)}{u_2(x)} = \frac{u_1(a)}{u_2(a)},$$

where a is the abscissa of the point A, is the equation for the abscissas of all points conjugate to A.

Jacobi also showed that one need not solve the accessory equation. Since one must solve the Euler equation in any case, let $y = y(x, c_1, c_2)$ be the general solution of that equation, that is, the family of extremals. Then u_1 can be taken to be $\partial y/\partial c_1$, and u_2 to be $\partial y/\partial c_2$.

From his work on conjugate points Jacobi drew two conclusions. The first was that if along the extremal from A to B a conjugate point to A occurs then a maximum or a minimum is impossible. In this conclusion Jacobi was essentially correct.

On the basis of his considerations of conjugate points Jacobi also concluded that an extremal (a solution of Euler's equation) taken between A and B for which $f_{y'y'} > 0$ along the curve and for which no conjugate point exists between A and B (or at B) furnishes a minimum for the original integral. The corresponding statement with $f_{y'y'} < 0$, he asserted, holds for a maximum. Actually, these sufficient conditions were not correct, as we shall see in a few moments. In this 1837 paper Jacobi stated results and gave brief indications of proofs. The full proofs of the correct statements were supplied by later workers.

Aside from the specific value of Jacobi's results for the existence of a maximizing or minimizing function, his work made clear that progress in the calculus of variations could not be guided by the theory of maxima and minima of the ordinary calculus.

For thirty-five years both of Jacobi's conclusions were accepted as correct. During this period the papers on the subject were imprecise in statement and dubious in proof; problems were not sharply formulated and all sorts of errors were made. Then Weierstrass undertook work on the calculus of variations. He presented his material in his lectures at Berlin in 1872 but did not publish it himself. His ideas aroused a new interest, stimulated further activity in the subject, and sharpened the thinking, as did Weierstrass's work in other domains.

Weierstrass's first point was that the criteria for a minimum or a maximum hitherto established—Euler's, Legendre's, and Jacobi's—were limited because the supposed minimizing or maximizing curve $y(x)$ was compared with other curves $y(x) + \varepsilon t(x)$, wherein it was actually supposed that $\varepsilon t(x)$ and $\varepsilon t'(x)$, or what Lagrange called δy and $\delta y'$, were both small along the x-range from A to B. That is, $y(x)$ was being compared with a limited class of other curves, and by satisfying the three criteria it did better than any other one of *these* comparison curves. Such variations $\varepsilon t(x)$ were called by Adolf Kneser (1862–1930) weak variations. However, to find the curve that really maximizes or minimizes the integral J one must compare it with *all* other curves joining A and B, including those whose derivatives may not approach the derivatives of the maximizing (or minimizing) curve as

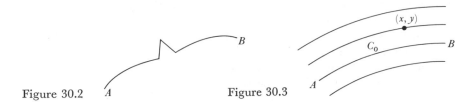

Figure 30.2   A   Figure 30.3

these comparison curves come closer in position to the maximizing curve. Thus a comparison curve may have a sharp corner (Fig. 30.2) at one or several places along the x-range from A to B. The comparison curves envisioned by Weierstrass are what Kneser called strong variations.

Weierstrass did prove in 1879 that for weak variations the three conditions, that a curve be an extremal (a solution of Euler's equation), that $f_{y'y'} > 0$ along the extremal, and that any conjugate point to A must lie beyond B, are indeed sufficient conditions that the extremal furnish a minimum of the integral J (and $f_{y'y'} < 0$ for a maximum).

Then Weierstrass considered strong variations. For these variations Weierstrass first introduced a fourth necessary condition. He introduced a new function called the E-function, or Excess function, defined by

$$(19) \qquad E(x, y, y', \tilde{p}) = f(x, y, \tilde{p}) - f(x, y, y') - (\tilde{p} - y')f_{y'}(x, y, y'),$$

and his result was: The fourth necessary condition that $y(x)$ furnish a minimum is that $E(x, y, y', \tilde{p}) \geq 0$ along the extremal $y(x)$ for every finite value of $\tilde{p}$. For a maximum $E \leq 0$.

Weierstrass then (1879) turned his attention to sufficient conditions for a maximum (or a minimum) when strong variations are permitted. To formulate his sufficient conditions it is necessary to introduce Weierstrass's concept of a field. Consider any one-parameter family (Fig. 30.3) of extremals $y = \Phi(x, \gamma)$ in which the particular extremal joining A and B is included, say for $\gamma = \gamma_0$. Aside from some details on the continuity and differentiability of $\Phi(x, \gamma)$, the essential fact about this family of extremals is that in a region about the extremal through A and B there passes through any point (x, y) of the region one and only one extremal of the family. A family of extremals satisfying this essential condition is called a field.

Given a field surrounding the extremal C_0 joining A and B (Fig. 30.4) then if at every point (x, y) lying between $x = a$ and $x = b$ and in the region covered by the field, $E(x, y, p(x, y), \tilde{p}) \geq 0$, where $p(x, y)$ denotes the slope at (x, y) of the extremal passing through (x, y) and $\tilde{p}$ is any finite value, then C_0 minimizes the integral J with respect to any other C lying within the field and joining A and B. (For a maximum, $E \leq 0$.)

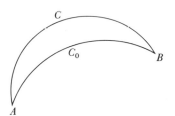

Figure 30.4

In 1900 Hilbert[8] introduced his invariant integral theory which greatly simplified the sufficiency proof. Hilbert asked the question: Is it possible to determine the function $p(x, y)$ so that the integral

$$(20) \qquad I = \int_{x_1}^{x_2} \left\{ f(x, y, p) - (y' - p)f_{y'}(x, y, p) \right\} dx$$

is independent of the path in a region of (x, y) values? He found that if $p(x, y)$ is so determined then the solutions of the differential equation

$$\frac{dy}{dx} = p(x, y)$$

are extremals of a field. Conversely, if $p(x, y)$ is the slope function of a field F then I is independent of the path in F. From this theorem Hilbert derived Weierstrass's sufficiency condition for strong variations.

4. Related Problems in the Calculus of Variations

Our exposition of the history of the calculus of variations has been concentrated largely on the integral

$$J = \int_{x_1}^{x_2} f(x, y, y') \, dx.$$

Some mention has been made of other problems, the isoperimetric problems, the problems of several functions of one variable such as arise in the Principle of Least Action, and the case of multiple integrals, which Lagrange first treated and which occurs in the minimal surface problem (Chap. 24, sec. 4). There are numerous types of related problems, such as those in which the minimizing or maximizing curve is treated in a parameteric representation $x = x(t)$ and $y = y(t)$—this problem was thoroughly discussed by Weierstrass—and problems largely in dynamics in which the variables that appear in the integrand are restricted by auxiliary or subsidiary equations,

8. *Nachrichten König. Ges. der Wiss. zu Gött.*, 1900, 291–96 = *Ges. Abh.*, 3, 323–29. There is an English translation by Mary Winston Newsom in the *Amer. Math. Soc. Bull.*, 8, 1902, 472–78. This material is part of Hilbert's famous paper of 1900, "Mathematical Problems."

called constraints. The last type of problem is somewhat related to the isoperimetric problem because there too a subsidiary condition, namely, the length of the curve bounding the maximum area, is specified—though in that problem the subsidiary condition is in the form of an integral that expresses the length of the curve, whereas in the case of dynamical constraints the subsidiary condition or conditions are in the form of equations involving the independent and dependent variables or even the differentials of the dependent variables. There is also the major problem called the Dirichlet problem, which has already been discussed (Chap. 28, secs. 4 and 8).

We shall not trace the detailed history of these problems because no major feature of the development of mathematics emerged from it, although the problems are significant and considerable work has been done on them down to the present time. It is perhaps worth noting that the subject of minimal surfaces, which calls for solving the equation

$$(1 + q^2)r - 2pqs + (1 + p^2)t = 0,$$

had been dormant after a paper by Ampère of 1817 until the Belgian physicist Joseph Plateau (1801–83) in a book of 1873, *Statique expérimentale et théorique des liquides soumis aux seules formes moléculaires*, showed that if one dips wires having the shapes of closed curves into a glycerine solution (or soapy water) and then withdraws them, a soap film which has the shape of the surface of least area will span the wire boundary. The mathematicians thus received a new stimulus to consider minimal surfaces bounded by a closed space curve. Since the boundary curve or curves can be quite complicated, the actual analytical explicit solution for the minimal surface may be impossible to obtain. This problem, now known as Plateau's problem, led to work on proving at least the existence of solutions, from which some properties of the solutions can be deduced.

Bibliography

Dresden, Arnold: "Some Recent Work in the Calculus of Variations," *Amer. Math. Soc. Bull.*, 32, 1926, 475–521.

Duren, W. L., Jr.: "The Development of Sufficient Conditions in the Calculus of Variations," *University of Chicago Contributions to the Calculus of Variations*, I, 1930, 245–349, University of Chicago Press, 1931.

Hamilton, W. R.: *The Mathematical Papers*, 3 vols., Cambridge University Press, 1931, 1940, and 1967.

Jacobi, C. G. J.: *Gesammelte Werke*, G. Reimer, 1886 and 1891, Chelsea (reprint), 1968, Vols. 4 and 7.

————: *Vorlesungen über Dynamik* (1866), Chelsea (reprint), 1968. Also in Vol. 8 of Jacobi's *Gesammelte Werke*.

McShane, E. J.: "Recent Developments in the Calculus of Variations," *Amer. Math. Soc. Semicentennial Publications*, II, 1938, 69–97.

Porter, Thomas Isaac: "A History of the Classical Isoperimetric Problem," *University of Chicago Contributions to the Calculus of Variations*, II, 475–517, University of Chicago Press, 1933.

Prange, Georg: "Die allgemeinen Integrationsmethoden der analytischen Mechanik," *Encyk. der Math. Wiss.*, B. G. Teubner, 1904–35, IV, 2, 509–804.

Todhunter, Isaac: *A History of the Calculus of Variations in the Nineteenth Century*, Chelsea (reprint), 1962.

Weierstrass, Karl: *Werke*, Akademische Verlagsgesellschaft, 1927, Vol. 7.

31
Galois Theory

1. Introduction

The basic problem of algebra, the solution of polynomial equations, continued to occupy the center of the stage in the algebra of the early nineteenth century. In this period, the broad question of which equations are solvable by algebraic operations was definitely and comprehensively answered by Galois. Moreover, he not only created the first significant coherent body of algebraic theory but he introduced new notions which were to be developed into still other broadly applicable theories of algebra. In particular the concepts of a group and a field emerged from his work and Abel's.

2. Binomial Equations

We have already discussed (Chap. 25, sec. 2) the fruitless efforts of Euler, Vandermonde, Lagrange, and Ruffini to solve algebraically equations of degree greater than 4 and the binomial equation $x^n - 1 = 0$. A major success was achieved by Gauss. In the last section of his *Disquisitiones Arithmeticae*[1] Gauss considered the equation

$$(1) \qquad\qquad x^p - 1 = 0,$$

where p is a prime.[2] This equation is often called the cyclotomic equation or the equation for the division of a circle. The latter term refers to the fact that the roots of this equation are, by de Moivre's theorem,

$$(2) \qquad x_j = \cos\frac{k2\pi\theta}{p} + i\sin\frac{k2\pi\theta}{p}, \qquad k = 1, 2, \ldots, p,$$

1. 1801, *Werke*, I.
2. The case of p prime takes care of $x^n - 1 = 0$ for if $n = pq$, let $y = x^q$. But $y^p - 1 = 0$ is solvable. Hence $x^q = $ const. can be solved if q is a prime and if not q can be decomposed in the same manner that n is.

and the complex numbers x_j when plotted geometrically are the vertices of a p-sided regular polygon that lie on the unit circle.

Gauss showed that the roots of this equation may be rationally expressed in terms of the roots of a sequence of equations

$$(3) \qquad\qquad Z_1 = 0, Z_2 = 0, \cdots,$$

whose coefficients are rational in the roots of the preceding equations of the sequence. The degrees of the equations (3) are precisely the prime factors of $p - 1$. There is a Z_i for each factor even if repeated. Moreover, each of the $Z_i = 0$ can be solved by radicals and so equation (1) can also be so solved.

This result is of course of great significance for the problem of solving the general nth-degree equation algebraically. It shows that some equations of high degree can be solved by radicals, for example a fifth degree equation if 5 is a factor of $p - 1$ or a seventh degree equation if 7 is a factor.

The result is of major importance also for the geometric problem of constructing regular polygons of p sides. If $p - 1$ contains no factors other than 2, then the polygon is constructible with straightedge and compass because the degrees of the equations (3) are each 2 and each of its roots is constructible in terms of its coefficients. Thus we are able to construct all polygons of a prime number of sides p if $p - 1$ is a power of 2. Such primes are 3, 5, 17, 257, 65537, ... Alternatively, a regular polygon can be constructed if p is a prime of the form $2^{2^h} + 1$.[3] Gauss remarks (Art. 365) that although the geometric construction of regular polygons of 3, 5, and 15 sides and those immediately derivable from them—for example, 2^n, $2^n \cdot 3$, $2^n \cdot 5$, $2^n \cdot 15$, wherein n is a positive integer—was known in Euclid's time, in an interval of 2000 years no new constructible polygons had been discovered and geometers had been unanimous in declaring that no others could be constructed.

Gauss thought that his result might lead to all sorts of attempts to find new constructible polygons of a prime number of sides. He warns then: "As often as $p - 1$ contains other prime factors besides 2, we arrive at higher equations, namely, to one or more cubic equations if 3 enters once or oftener as a factor of $p - 1$, to equations of the fifth degree if $p - 1$ is divisible by 5, etc. And we can prove with all rigor that these higher equations cannot be avoided or made to depend upon equations of lower degree; and although the limits of this work do not permit us to give the demonstration here, we still thought it necessary to note this fact in order that one should not seek to construct other polygons [of prime number of sides] than those given by our theory, as for example, polygons of 7, 11, 13, 19 sides, and so employ his time in vain."

Gauss then considers (Art. 366) polygons of any number of sides n and asserts that a regular polygon of n sides is constructible if and only if

3. In a prime of the form $2^\mu + 1$, μ is necessarily of the form 2^h, but $2^{2^h} + 1$ is not necessarily prime.

$n = 2^l p_1 \cdot p_2 \cdots p_n$, where $p_1, p_2, \ldots, p_n$ are distinct primes of the form $2^{2^h} + 1$ and where l is any positive integer or 0. The sufficiency of this condition does follow readily from Gauss's work on polygons of a prime number of sides but the necessity is not at all obvious and was not proven by Gauss.[4]

The construction of regular polygons had interested Gauss since 1796 when he conceived the first proof that the 17-sided polygon is constructible. There is a story about this discovery which may bear repeating. This construction problem was already famous. One day Gauss approached his professor A. G. Kästner at the University of Göttingen with the proof that this polygon is constructible. Kästner was incredulous and sought to dismiss Gauss, much as university teachers today dismiss angle-trisectors. Rather than take the time to examine Gauss's proof and find the supposed error in it, Kästner told Gauss the construction was unimportant because practical constructions were known. Of course Kästner knew that the existence of practical or approximate constructions was irrelevant for the theoretical problem. To interest Kästner in his proof Gauss pointed out that he had solved a seventeenth degree algebraic equation. Kästner replied that the solution was impossible. But Gauss rejoined that he had reduced the problem to solving an equation of lower degree. "Oh well," scoffed Kästner, "I have already done this." Later Gauss repaid Kästner, who also prided himself on his poetry, by lauding Kästner as the best poet among mathematicians and the best mathematician among poets.

3. Abel's Work on the Solution of Equations by Radicals

Abel read Lagrange's and Gauss's work on the theory of equations and while still a student in high school tackled the problem of the solvability of higher degree equations by following Gauss's treatment of the binomial equation. At first Abel thought he had solved the general fifth degree equation by radicals. But soon convinced of his error he tried to prove that such a solution was not possible (1824–26). First he succeeded in proving the theorem: The roots of an equation solvable by radicals can be given such a form that each of the radicals occurring in the expressions for the roots is expressible as a rational function of the roots of the equation and certain roots of unity. Abel then used this theorem to prove[5] the impossibility of solving by radicals the general equation of degree greater than four.

Abel's proof, done in ignorance of Ruffini's work (Chap. 25, sec. 2), is

4. See James Pierpont, "On an Undemonstrated Theorem of the *Disquisitiones Arithmeticae*," *Amer. Math. Soc. Bull.*, 2, 1895–96, 77–83. This article gives the proof. The fact that the Gauss condition is necessary was first proved by Pierre L. Wantzel (1814–48), *Jour. de Math.*, 2, 1837, 366–72.

5. *Jour. für Math.*, 1, 1826, 65–84 = *Œuvres*, 1, 66–94.

roundabout and unnecessarily complicated. His paper also contained an error in a classification of functions, which fortunately was not essential to the argument. He later published two more elaborate proofs. A simple, direct, and rigorous proof based on Abel's idea was given by Kronecker in 1879[6].

Thus the question of the solution of general equations of degree higher than four was settled by Abel. He also considered some special equations. He took up[7] the problem of the division of the lemniscate (solving $x^n - 1 = 0$ is the equivalent of the problem of the division of the circle into n equal arcs) and arrived at a class of algebraic equations, now called Abelian equations, that are solvable by radicals. The cyclotomic equation (1) is an example of an Abelian equation. More generally an equation is called Abelian if all its roots are rational functions of one of them, that is, if the roots are $x_1, \theta_1(x_1), \theta_2(x_1), \ldots, \theta_{n-1}(x_1)$ where the θ_i are rational functions. There is also the condition that $\theta_\alpha(\theta_\beta(x_1)) = \theta_\beta(\theta_\alpha(x_1))$ for all values of α and β from 1 to $n - 1$.

In this last work he introduced two notions (though not the terminology), field and polynomial irreducible in a given field. By a field of numbers he, like Galois later, meant a collection of numbers such that the sum, difference, product, and quotient of any two numbers in the collection (except division by 0) are also in the collection. Thus the rational numbers, real numbers, and complex numbers form a field. A polynomial is said to be reducible in a field (usually the field to which its coefficients belong) if it can be expressed as the product of two polynomials of lower degrees and with coefficients in the field. If the polynomial cannot be so expressed it is said to be irreducible.

Abel then tackled the problem of characterizing all equations which are solvable by radicals and had communicated some results to Crelle and to Legendre just before death overtook him in 1829.

4. Galois's Theory of Solvability

After Abel's work the situation was as follows: Although the general equation of degree higher than four was not solvable by radicals, there were many special equations, such as the binomial equations $x^p = a$, p a prime, and Abelian equations that were solvable by radicals. The task now became to determine which equations are solvable by radicals. This task, just begun by Abel, was taken up by Evariste Galois (1811–32). Born to well-to-do and educated parents he attended one of the celebrated lycées of Paris and started to study mathematics at the age of fifteen. This subject became his passion and he studied carefully the works of Lagrange, Gauss, Cauchy, and Abel.

6. *Monatsber. Berliner Akad.*, 1879, 205–29 = *Werke*, 4, 73–96. Kronecker's proof is explained by James Pierpoint: "On the Ruffini-Abelian Theorem, *Amer. Math. Soc. Bull.*, 2, 1895–96, 200–21.
7. *Jour. für Math.*, 4, 1829, 131–56 = *Œuvres*, 1, 478–507.

The other subjects he neglected. Galois sought to enter the Ecole Poly-technique but possibly because he failed to explain in sufficient detail the questions he had to answer orally at the entrance examination or because the examining professors did not understand him he was rejected in two tries. He therefore entered the Ecole Préparatoire (the name then for the Ecole Normale and a much inferior school at that time.) During the 1830 Revolu-tion, which drove Charles X from the throne and installed Louis Philippe, Galois publicly criticized the director of his school for failing to support the Revolution and was expelled. He was twice arrested for political offenses, spent most of the last year and a half of his life in prison, and was killed in a duel on May 31, 1832.

During his first year at the Ecole Galois published four papers. In 1829 he submitted two papers on the solution of equations to the Academy of Sciences. These were entrusted to Cauchy, who lost them. In January of 1830 he presented to the Academy of Sciences another carefully written paper on his research. This was sent to Fourier, who died soon after, and this paper too was lost. At the suggestion of Poisson Galois wrote (1831) a new paper on his research. This article, "Sur les conditions de résolubilité des équations par radicaux,"[8] the only finished article on his theory of the solution of equations, was returned by Poisson as unintelligible, with the recommendation that a fuller exposition be written. The night before his death Galois drew up a hastily written account of his researches which he entrusted to his friend, August Chevalier. This account has been preserved.

In 1846 Liouville edited and published in the *Journal de Mathématiques*[9] part of Galois's papers including a revision of the 1831 paper. Then Serret's *Cours d'algèbre supérieure* (3rd ed.) of 1866 gave an exposition of Galois's ideas. The first full and clear presentation of Galois theory was given in 1870 by Camille Jordan in his book *Traité des substitutions et des équations algébriques*.

Galois approached the problem of characterizing equations solvable by radicals by improving on Lagrange's ideas, though he also derived some suggestions from Legendre's, Gauss's, and Abel's work. He proposed to consider the general equation, which is, of course,

$$(4) \qquad x^n + a_1 x^{n-1} + \cdots + a_{n-1} x + a_n = 0$$

wherein, as in the work of Lagrange, the coefficients must be independent or completely arbitrary, and particular equations, such as

$$(5) \qquad x^4 + p x^2 + q = 0,$$

wherein only two coefficients are independent. Galois's main thought was to bypass the construction of the Lagrange resolvents (Chap. 25, sec. 2) of the

8. *Œuvres*, 1897, 33–50.
9. *Jour. de Math.*, 11, 1846, 381–444.

given polynomial equation, a construction requiring great skill and having no clear methodology.

Like Lagrange, Galois makes use of the notion of substitutions or permutations of the roots. Thus if x_1, x_2, x_3, and x_4 are the four roots of a fourth degree equation, the interchange of x_1 and x_2 in any expression involving the x_i is a substitution. This particular substitution is indicated by

$$\begin{pmatrix} x_1 & x_2 & x_3 & x_4 \\ x_2 & x_1 & x_3 & x_4 \end{pmatrix}.$$

A second substitution is indicated by

$$\begin{pmatrix} x_1 & x_2 & x_3 & x_4 \\ x_3 & x_4 & x_1 & x_2 \end{pmatrix}.$$

To perform the first substitution and then the second one is equivalent to performing the third substitution

$$\begin{pmatrix} x_1 & x_2 & x_3 & x_4 \\ x_4 & x_3 & x_1 & x_2 \end{pmatrix}$$

because, for example, by the first substitution x_1 is replaced by x_2; by the second substitution x_2 is replaced by x_4; and by the third substitution x_1 goes directly into x_4. One says that the *product* of the first two substitutions taken in the order just indicated is the third substitution. There are in all 4! possible substitutions. The set of substitutions is said to form a group because the product of any two substitutions is a member of the set. This notion, which is not of course a formal definition of an abstract group, is due to Galois.

To secure some grasp of Galois's ideas let us consider the equation[10]

$$x^4 + px^2 + q = 0,$$

where p and q are independent. Let R be the field formed by rational expressions in p and q and with coefficients in the field of rational numbers, a typical expression being $(3p^2 - 4q)/(q^2 - 7p)$. One says with Galois that R is the field obtained by adjoining the letters or indeterminates p and q to the rational numbers. This field R is the field or domain of rationality of the coefficients of the given equation and the equation is said to belong to the field R. Like Abel, Galois did not use the terms field or domain of rationality but he did use the concept.

10. Since Galois's own presentation of his ideas was not clear and he introduced so many new notions, we shall utilize an example due to Verriest (see the bibliography at the end of the chapter) to help make Galois's theory clear.

We happen to know that the roots of the fourth degree equation are

$$x_1 = \sqrt{\dfrac{-p + \sqrt{p^2 - 4q}}{2}}, \qquad x_2 = -\sqrt{\dfrac{-p + \sqrt{p^2 - 4q}}{2}},$$

$$x_3 = \sqrt{\dfrac{-p - \sqrt{p^2 - 4q}}{2}}, \qquad x_4 = -\sqrt{\dfrac{-p - \sqrt{p^2 - 4q}}{2}}.$$

Then it is true that the two relations with coefficients in R

$$x_1 + x_2 = 0 \quad \text{and} \quad x_3 + x_4 = 0$$

hold for the roots. Since our given equation is of the fourth degree there are twenty-four possible substitutions of the roots. The following eight substitutions

$$E = \begin{pmatrix} x_1 & x_2 & x_3 & x_4 \\ x_1 & x_2 & x_3 & x_4 \end{pmatrix}, \qquad E_1 = \begin{pmatrix} x_1 & x_2 & x_3 & x_4 \\ x_2 & x_1 & x_3 & x_4 \end{pmatrix},$$

$$E_2 = \begin{pmatrix} x_1 & x_2 & x_3 & x_4 \\ x_1 & x_2 & x_4 & x_3 \end{pmatrix}, \qquad E_3 = \begin{pmatrix} x_1 & x_2 & x_3 & x_4 \\ x_2 & x_1 & x_4 & x_3 \end{pmatrix},$$

$$E_4 = \begin{pmatrix} x_1 & x_2 & x_3 & x_4 \\ x_3 & x_4 & x_1 & x_2 \end{pmatrix}, \qquad E_5 = \begin{pmatrix} x_1 & x_2 & x_3 & x_4 \\ x_4 & x_3 & x_1 & x_2 \end{pmatrix},$$

$$E_6 = \begin{pmatrix} x_1 & x_2 & x_3 & x_4 \\ x_3 & x_4 & x_2 & x_1 \end{pmatrix}, \qquad E_7 = \begin{pmatrix} x_1 & x_2 & x_3 & x_4 \\ x_4 & x_3 & x_2 & x_1 \end{pmatrix}$$

leave the two relations true in R.[11] One could show that these eight are the only substitutions of the twenty-four which leave invariant *all* relations in R among the roots. These eight are the group of the equation in R. They are a subgroup of the full group. That is, the group of an equation with respect to a field R is the group or subgroup of substitutions on the roots which leave invariant *all* the relations with coefficients in R among the roots of the given equation (whether general or particular). One can say that the number of substitutions that leave all relations in R invariant is a measure of our

---

11. For the general equation of degree n, that is, with n independent quantities as co-efficients, a function of the roots is *invariant* under or unaltered by a substitution on the roots if and only if it remains identical with the original function. If the coefficients are all numerical then a function is unaltered if it remains numerically the same. Thus for $x^3 + x^2 + x + 1 = 0$ the roots are $x_1 = -1$, $x_2 = i$, and $x_3 = -i$. Consider x_2^2. The substitution x_3 for x_2 gives x_3^2. This has the same numerical value as x_2^2. Then x_2^2 is not altered by the substitution. If the coefficients contain some numerical values and some independent quantities then a function of the roots remains unaltered by a substitution on the roots if the function remains numerically the same for all values of the independent quantities (in the domain to which these may be limited) and the numerical values which the roots may assume.

ignorance of the roots because we can not distinguish them under these eight substitutions.

Now consider $x_1^2 - x_3^2$ which equals $\sqrt{p^2 - 4q}$. We adjoin this radical to R, and form the field R', that is, we form the smallest field containing R and $\sqrt{p^2 - 4q}$. Then

$$(6) \qquad x_1^2 - x_3^2 = \sqrt{p^2 - 4q}$$

is a relation in R'. Since $x_1 + x_2 = 0$ and $x_3 + x_4 = 0$ we also have

$$x_1^2 = x_2^2 \quad \text{and} \quad x_3^2 = x_4^2.$$

Then in view of these last two facts we can say that the first four of the above eight substitutions leave the relation (6) in R' true, but the last four do not. Then the four substitutions, if they leave *every* true relation in R' among the roots invariant, are the group of the equation in R'. The four are a subgroup of the eight.

Now suppose we adjoin to R' the quantity $\sqrt{(-p - D)/2}$ where $D = \sqrt{p^2 - 4q}$ and form the field R''. Then

$$x_3 - x_4 = 2\sqrt{\frac{-p - D}{2}}$$

is a relation in R''. This relation remains invariant only under the first two substitutions E and E_1 but not under the rest of the eight. Then the group of the equation in R'' consists of these two substitutions, provided every relation in R'' among the roots remains invariant under these two substitutions. The two are a subgroup of the four substitutions.

If we now adjoin to R'' the quantity $\sqrt{(-p + D)/2}$ we get R'''. In R''' we have

$$x_1 - x_2 = 2\sqrt{\frac{-p + D}{2}}.$$

Now the only substitution leaving all relations in R''' true is just E and this is the group of the equation in R'''.

We may see from the above discussion that the group of an equation is a key to its solvability because the group expresses the degree of indistinguishability of the roots. It tells us what we do not know about the roots.

There were many groups, or strictly a group of substitutions and successive subgroups, involved above. Now the *order* of a group (or subgroup) is the number of elements in it. Thus we had groups of the orders 24, 8, 4, 2, and 1. The order of a subgroup always divides the order of the group (sec. 6). The *index* of a subgroup is the order of the group in which it lies divided by the order of the subgroup. Thus the index of the subgroup of order 8 is 3.

The above sketch merely shows the ideas Galois dealt with. His work proceeded as follows: Given an equation general or particular, he showed first how one could find the group G of this equation in the field of the coefficients, that is, the group of substitutions on the roots that leaves invariant every relation among the roots with coefficients in that field. Of course one must find the group of the equation without knowing the roots. In our example above, the group of the quartic equation was of order 8 and the field of the coefficients was R. Having found the group G of the equation, one seeks next the largest subgroup H in G. In our example this was the subgroup of order 4. If there should be two or more largest subgroups we take any one. The determination of H is a matter of pure group theory and can be done. Having found H, one can find by a set of procedures involving only rational operations a function ϕ of the roots whose coefficients belong to R and which does not change value under the substitutions in H but does change under all other substitutions in G. In our example above the function was $x_1^2 - x_3^2$. Actually an infinity of such functions can be obtained. One must of course find such a function without knowing the roots. There is a method of constructing an equation in R one of whose roots is this function ϕ. The degree of this equation is the index of H in G. This equation is called a partial resolvent.[12] In our example the equation is $t^2 - (p^2 - 4q) = 0$ and its degree is 8/4 or 2.

One must now be able to solve the partial resolvent to find the root ϕ. In our example ϕ is $\sqrt{p^2 - 4q}$. One adjoins ϕ to R and obtains a new field R'. Then the group of the original equation with respect to the field R' can be shown to be H.

We now repeat the procedure. We have the group H, of order 4 in our example, and the field R'. We seek next the largest subgroup in H. In our example, it was the subgroup of order 2. Let us call this subgroup K. One can now obtain a function of the roots of the original equation whose coefficients belong to R' and whose value is unchanged by every substitution in K but is changed by other substitutions in H. In our example above this was $x_3 - x_4$. To find this function, ϕ_1 say, without knowing the roots, one constructs an equation in R' having the function ϕ_1 as a root. In our example this equation is $t^2 - 2(-p - \sqrt{p^2 - 4q}) = 0$. The degree of this equation is the index of K with respect to H, that is, 4/2 or 2. This equation is the second partial resolvent.

Now one must be able to solve this resolvent equation and get a root, the function ϕ_1, and adjoin this value to R' thereby forming the field R''. With respect to R'' the group of the equation is K.

We again repeat the process. We find the largest subgroup L in K. In our example this was just the identity substitution E. We seek a function of

---

12. This use of the word "resolvent" is different from Lagrange's use.

the roots (with coefficients in R'') which retains its value under E but not under the other substitutions of K. In our example such a function was $x_1 - x_2$. To obtain this ϕ_2 without knowing the roots we must construct an equation in R'' having the function ϕ_2 as a root. In our example the equation is $t^2 - 2(-p + \sqrt{p^2 - 4q}) = 0$. The degree of this equation is the index of L in K. In our example this index was $2/1$ or 2. This equation is the third partial resolvent. We must solve this equation to find the value ϕ_2.

By adjoining this root to R'' one gets a field R'''. Let us suppose that we have reached the final stage wherein the group of the original equation in R''' is the identity substitution, E.

Now Galois showed that when the group of an equation with respect to a given field is just E, then the roots of the equation are members of that field. Hence the roots lie in the field R''', and we know the field in which the roots lie because R''' was obtained from the known field R by successive adjunction of known quantities. There is next a straightforward process for finding the roots by rational operations in R'''.

Galois gave a method of finding the group of a given equation, the successive resolvents, and the groups of the equation with respect to the successively enlarged fields of coefficients that result from adjoining the roots of these successive resolvents to the original coefficient field, that is, the successive subgroups of the original group. These processes involve considerable theory but, as Galois pointed out, his work was not intended as a practical method of solving equations.

And now Galois applied the above theory to the problem of solving polynomial equations by rational operations and radicals. Here he introduced another notion of group theory. Suppose H is a subgroup of G. If one multiplies the substitutions of H by any element g of G then one obtains a new collection of substitutions which is denoted by gH, the notation indicating that the substitution g is performed first and then any element of H is applied. If $gH = Hg$ for every g in G then H is called a normal (self-conjugate or invariant) subgroup in G.

We recall that Galois's method of solving an equation called for finding and solving the successive resolvents. Galois showed that when the resolvent that serves to reduce the group of an equation, say from G to H, is a binomial equation $x^p = A$ of prime degree p, then H is a normal subgroup in G (and of index p), and conversely if H is a normal subgroup in G and of prime index p then the corresponding resolvent is a binomial equation of degree p or can be reduced to such. If all of the successive resolvents are binomial equations, then, in view of Gauss's result on binomial equations, we can solve the original equation by radicals. For, we know that we can pass from the initial field to the final field in which the roots lie by successive adjunctions of radicals. Conversely if an equation is solvable by radicals then the set of resolvent equations must exist, and these are binomial equations.

Thus the theory of solvability by radicals is in general outline the same as the theory of solution given earlier except that the series of subgroups

$$G, H, K, L, \cdots, E$$

must each be a maximum normal subgroup (not a subgroup of any larger normal subgroup) of the preceding group. Such a series is called a composition series. The indices of H in G, K in H, and so forth are called the indices of the composition series. If the indices are *prime* numbers the equation is solvable by radicals, and if the indices are not prime the equation is not solvable by radicals. As one proceeds to find the sequence of maximal normal subgroups there may be a choice; that is, there may be more than one maximal normal subgroup of highest order in a given group or subgroup. One may choose any one, though thereafter the subgroups may differ. But the very same set of indices will result though the order in which they appear may differ (see the Jordan-Hölder theorem below). The group G, which contains a composition series of prime indices, is said to be solvable.

How does the Galois theory show that the general nth-degree equation for $n > 4$ is not solvable by radicals whereas for $n \leq 4$ they are? For the general nth-degree equation the group is composed of all the $n!$ substitutions of the n roots. This group is called the symmetric group of degree n. Its order is of course $n!$. It is not difficult to find the composition series for each symmetric group. The maximal normal subgroup, which is called the alternating subgroup, is of order $n!/2$. The only normal subgroup of the alternating group is the identity element. Hence the indices are 2 and $n!/2$. But the number $n!/2$, for $n > 4$, is never prime. Hence the general equation of degree greater than 4 is not solvable by radicals. On the other hand, the quadratic equation can be solved with the aid of a single resolvent equation. The indices of the composition series consist of just the single number 2. The general third degree equation requires for its solution two resolvent equations of the form $y^2 = A$ and $z^3 = B$. These are of course binomial resolvents. The indices of the composition series are 2 and 3. The general fourth degree equation can be solved with four binomial resolvent equations, one of degree 3 and three of degree 2. The indices of the composition series are, then, 2, 3, 2, 2.

For equations with numerical coefficients as opposed to those with literal independent coefficients, Galois gave a theory similar to that described above. However, the process of determining the solvability by radicals is more complicated even though the basic principles are the same.

Galois also proved some special theorems. If one has an irreducible equation of prime degree whose coefficients lie in a field R and whose roots are all rational functions of two of the roots with coefficients in R then the equation is solvable by radicals. He also proved the converse: Every irreducible equation of prime degree which is solvable by radicals has the property that each of its roots is a rational function of two of them with coefficients in R. Such an equation is now called Galoisian. The simplest example of a Galoisian equation is $x^p - A = 0$. The notion is an extension of Abelian equations.

As a postscript Hermite[13] and Kronecker in a letter to Hermite[14] and in a subsequent paper[15] solved the general quintic equation by means of elliptic modular functions. This is analogous to the use of trigonometric functions to solve the irreducible case of the cubic equation.

5. The Geometric Construction Problems

The eighteenth-century mathematicians doubted that the famous construction problems could be solved. Galois's work supplied a criterion on constructibility that disposed of some of the famous problems.

Each step of a construction with straightedge and compass calls for finding a point of intersection either of two lines, a line and a circle, or two circles. With the introduction of coordinate geometry it was recognized that in algebraic terms such steps mean the simultaneous solution of either two linear equations, a linear and a quadratic, or two quadratic equations. In any case, the worst that is involved algebraically is a square root. Hence successive quantities found by successive steps or constructions are at worst the result of a chain of *square roots* applied to the given quantities. Accordingly, constructible quantities must lie in fields obtained by adjoining to the field containing the given quantities only square roots of the given or subsequently constructed quantities. We may call such extension fields quadratic extension fields.

In performing the successive constructions a few restrictions must be observed. For example, some of the processes permit the use of an arbitrary line or circle. Thus in bisecting a line segment we can use circles larger than half of the line segment. One must choose this circle in the field or the constructible extension field of the given elements. This can be done.

Also, the extension fields may involve complex elements because, for example, the square root of a negative coordinate may occur. These complex elements are constructible because the real and imaginary parts of the complex quantities that do occur are each roots of a real equation and these roots are constructible.

Given a construction problem, one first sets up an algebraic equation whose solution is the desired quantity. This quantity must belong to some quadratic extension field of the field of the given quantities. In the case of the regular polygon of 17 sides this equation is $x^{17} - 1 = 0$ and the given

13. *Comp. Rend.*, 46, 1858, 508–15 = *Œuvres*, 2, 5–12.
14. *Comp. Rend.*, 46, 1858, 1150–52 = *Werke*, 4, 43–48.
15. *Jour. für Math.*, 59, 1861, 306–10 = *Werke*, 4, 53–62.

quantity can be taken to be the radius of the unit circle. The relevant irreducible equation is $x^{16} + x^{15} + \cdots + 1 = 0$. In terms of Galois theory the necessary and sufficient condition for an equation to be solvable in *square roots* is that the order of the Galois group of the equation be a power of 2. This is the case for the equation $x^{16} + \cdots + 1 = 0$ and the composition series is 2, 2, 2, 2. This means that the resolvents are binomial equations of degree 2 so that only square roots are adjoined to the original rational field determined by the given radius of the unit circle. With this Galois criterion one can prove Gauss's statement that a regular polygon with a prime number p of sides can be constructed with straightedge and compass if and only if the prime number p has the form $2^{2^n} + 1$, that is, if $p = 3, 5, 17, 257, \ldots$ but not for $p = 7, 11, 13, 19, 23, 29, 31, \ldots$. Galois theory can be used to prove also that it is impossible to trisect an arbitrary angle or to double a given cube.

But Galois's criterion does not apply at all to the problem of squaring the circle. Here the given quantity is the radius of the circle. The equation to be solved is $x^2 = \pi r^2$. Though this equation is itself just a quadratic, it is not true that its solutions belong to a quadratic extension field of the field determined by the given quantity because π is not an algebraic irrational (Chap. 41, sec. 2). Galois's work, then, not only answered completely the question of which equations are solvable by algebraic operations but gave a general criterion to determine the constructibility with straightedge and compass of geometrical figures.

So far as the famous construction problems are concerned we should note that before Galois theory was applied to them, Gauss and Wantzel had determined which regular polygons are constructible (sec. 2), and Wantzel in the paper of 1837[16] showed that the general angle could not be trisected nor could a given cube be doubled. He proved that every constructible quantity must satisfy an equation of degree 2^n, and this is not true in the two problems just mentioned.

6. *The Theory of Substitution Groups*

In his work on the solvability of equations (Chap. 25, sec. 2) Lagrange introduced as the key to his analysis functions of the n roots which take the same value under some permutations of the roots. He therefore undertook in the very same papers to study functions for the purpose of determining the different values they can assume among the $n!$ possible values which the $n!$ permutations of the n variables (the roots) might give rise to. Subsequent work by Ruffini, Abel, and Galois lent increased importance to this topic. The fact that a rational function of n letters takes the same value under some

16. *Jour. de Math.*, 2, 1837, 366–72.

collection of permutations or substitutions of the roots means, as we have seen, that this collection is a subgroup of the entire symmetric group. This was observed explicitly by Ruffini in his *Teoria generale delle equazioni* (1799). Hence what Lagrange initiated was one way of studying subgroups of a group of substitutions. The more direct way is, of course, to study the group of substitutions itself and to determine its subgroups. Both methods of studying the structure or composition of substitution groups became an active subject which was pursued as an interest in and for itself though the connection with the solvability of equations was not ignored. The theory of substitution or permutation groups was the first major investigation that ultimately gave rise to the abstract theory of groups. Here we shall note some of the concrete theorems on substitution groups that were obtained during the nineteenth century.

Lagrange himself affirmed one major result which in modern language states that the order of a subgroup divides the order of the group. The proof of this theorem was communicated by Pietro Abbati (1768–1842) to Ruffini in a letter of September 30, 1802, which was published.[17]

Ruffini in his 1799 book introduced, though somewhat vaguely, the notions of transitivity and primitiveness. A permutation group is transitive if each letter of the group is replaced by every other letter under the various permutations of the group. If G is a transitive group and the n symbols or letters may be divided into r distinct subsets σ_i, $i = 1, 2, \ldots, r$, each subset containing s_i symbols such that any permutation of G either permutes the symbols of the σ_i among themselves or replaces these symbols by the symbols of σ_j, this holding for each $i = 1, 2, \ldots, r$, then G is called imprimitive. If no such separation of the n symbols is possible, then the transitive group is called a primitive group. Ruffini also proved that there does not exist a subgroup of order k for all k in a group of order n.

Cauchy, stimulated by the work of Lagrange and Ruffini, wrote a major paper on substitution groups.[18] With the theory of equations in mind, he proved that there is no group in n letters (degree n) whose index relative to the full symmetric group in n letters is less than the largest prime number which does not exceed n unless this index is 2 or 1. Cauchy stated the theorem in the language of function values: The number of different values of a nonsymmetric function of n letters cannot be less than the largest prime p smaller than n unless it is 2.

Galois made the largest step in introducing concepts and theorems about substitution groups. His most important concept was the notion of a normal (invariant or self-conjugate) subgroup. Another group concept due to Galois is that of an isomorphism between two groups. This is a one-to-one

17. *Memorie della Società Italiana delle Scienze*, 10, 1803, 385–409.
18. *Jour. de l'Ecole Poly.*, 10, 1815, 1–28 = *Œuvres*, (2), 1, 64–90.

correspondence between the elements of the two groups such that if $a \cdot b = c$ in the first one, then for the corresponding elements in the second, $a' \cdot b' = c'$. He also introduced the notions of simple and composite groups. A group having no invariant subgroup is simple; otherwise it is composite. Apropos of these notions, Galois expressed the conjecture[19] that the smallest simple group whose order is a composite number is the group of order 60.

Unfortunately Galois's work did not become known until Liouville published parts of it in 1846, and even that material was not readily understandable. On the other hand, Lagrange's and Ruffini's work on substitution groups, couched in the language of the values which a function of n letters can take on, became well known. Hence the subject of the solution of equations receded into the background, and when Cauchy returned to the theory of equations he concentrated on substitution groups. During the years 1844 to 1846 he wrote a host of papers. In the major paper[20] he systematized many earlier results and proved a number of special theorems on transitive primitive and imprimitive groups and on intransitive groups. In particular he proved the assertion by Galois that every finite (substitution) group whose order is divisible by a prime p contains at least one subgroup of order p. The major paper was followed by a great number of others published in the *Comptes Rendus* of the Paris Academy for the years 1844–46.[21] Most of this work was concerned with the formal (that is non-numerical) values that functions of n letters can take on under interchange of the letters and with finding functions that take on a given number of values.

After Liouville published some of Galois's work, Serret lectured on it at the Sorbonne and in the third edition of his *Cours* gave a better textual exposition of Galois theory. The work of clarifying Galois's ideas on the solvability of equations and the development of the theory of substitution groups thereafter proceeded hand in hand. In his text Serret gave an improved form of Cauchy's 1815 result. If a function of n letters has less than p values where p is the largest prime smaller than n, then the function cannot have more than two values.

One of the problems Serret stressed in the 1866 text asks for all the groups that one can form with n letters. This problem had already attracted the attention of Ruffini, and he, Cauchy, and Serret himself in a paper of 1850[22] gave a number of partial results, as did Thomas Penyngton Kirkman (1806–95). Despite many efforts and hundreds of limited results the problem is still unsolved.

After Galois, Camille Jordan (1838–1922) was the first to add signifi-

19. *Œuvres*, 1897 ed., 26.
20. *Exercices d'analyse et de physique mathématique*, 3, 1844, 151–252 = *Œuvres*, (2), 13, 171–282.
21. *Œuvres*, (1), Vols. 9 and 10.
22. *Jour. de Math.*, (1), 15, 1850, 45–70.

cantly to Galois theory. In 1869[23] he proved a basic result. Let G_1 be a maximal self-conjugate (normal) subgroup of G_0, G_2 a maximal self-conjugate subgroup of G_1, and so forth until the series terminates in the identity element. This series of subgroups is called a composition series of G_0. If G_{i+1} is any self-conjugate subgroup of order r in G_i whose order is p, then G_i may be decomposed into $\lambda = p/r$ classes. Two elements are in the same class if one is the product of the other and an element of G_{i+1}. If a is any element in one class and b any element in another, the product will be in the same third class. These classes form a group for which G_{i+1} is the identity element and the group is called the quotient group or factor group of G_i by G_{i+1}. It is denoted by G_i/G_{i+1}, a notation introduced by Jordan in 1872. The quotient groups G_0/G_1, G_1/G_2, ... are called the composition factor groups of G_0 and their orders are known as the composition factors or composition indices. There may be more than one composition series in G_0. Jordan proved that the set of composition factors is invariant except for the order in which they may occur and (Ludwig) Otto Hölder (1859–1937), a professor at the University of Leipzig, showed[24] that the quotient groups themselves were independent of the composition series; that is, the same set of quotient groups would be present for any composition series. The two results are called the Jordan-Hölder theorem.

The knowledge of (finite) substitution groups and their connection with the Galois theory of equations up to 1870 was organized in a masterful book by Jordan, his *Traité des substitutions et des équations algébriques* (1870). In this book Jordan, like almost all of the men preceding him, used as the definition of a substitution group that it is a collection of substitutions such that the product of any two members of the collection belongs to the collection. The other properties that we commonly postulate today in the definition of a group (Chap. 49, sec. 2) were utilized but were brought in as either obvious properties of such groups or as additional conditions but not specified in the definition. The *Traité* presented new results and made explicit for substitution groups the notions of isomorphism (*isomorphisme holoédrique*) and homomorphism (*isomorphisme mériédrique*), the latter being a many-to-one correspondence between two groups such that $a \cdot b = c$ implies $a' \cdot b' = c'$. Jordan added fundamental results on transitive and composite groups. The book also contains Jordan's solution of the problem posed by Abel, to determine the equations of a given degree that are solvable by radicals and to recognize whether a given equation does or does not belong to this class. The groups of the solvable equations are commutative. Jordan called them Abelian and the term Abelian was applied thereafter to commutative groups.

23. *Jour. de Math.*, (2), 14, 1869, 129–46 = *Œuvres*, 1, 241–48.
24. *Math. Ann.*, 34, 1889, 26–56.

Another major theorem on substitution groups was proved shortly after the *Traité* appeared by Ludwig Sylow (1832–1918), a Norwegian professor of mathematics. Cauchy had proved that every group whose order is divisible by a prime number p must contain one or more subgroups of order p. Sylow[25] extended Cauchy's theorem. If the order of a group is divisible by p^α, p being a prime, but not by $p^{\alpha+1}$ then the group contains one and only one system of conjugate subgroups of order p^α.[26] In this same paper Sylow also proved that every group of order p^α is solvable, that is, the indices of a sequence of maximal invariant subgroups are prime.

Quite another approach to substitution groups and ultimately to more general groups was suggested by a purely physical investigation. Auguste Bravais (1811–63), a physicist and mineralogist, studied groups of motions[27] to determine the possible structures of crystals. This study amounts mathematically to the investigation of the group of linear transformations in three variables

$$x'_i = a_{i1}x + a_{i2}y + a_{i3}z, \qquad i = 1, 2, 3,$$

of determinant $+1$ or -1 and it led Bravais to thirty-two classes of symmetric molecular structures that might occur in crystals.

Bravais's work impressed Jordan and he undertook to investigate what he called the analytic representation of groups and what is now called the representation for groups. Actually, Serret in the 1866 edition of his *Cours* had considered the representation of substitutions by transformations of the form

$$x' = \frac{ax + b}{cx + d}.$$

But the more useful representation of all types of groups was introduced by Jordan. He sought to represent substitutions by linear transformations of the form

$$(9) \qquad\qquad x'_i = \sum_{j=1}^{n} a_{ij}x_j, \qquad i = 1, 2, \cdots, n.$$

Since the substitution groups are finite some restriction had to be placed on the transformations so that the group of transformations would be finite. Galois had considered such transformations[28] and had limited them so that the coefficients and the variables take values in a finite field of prime order.

25. *Math. Ann.*, 5, 1872, 584–94.
26. If H is a subgroup of G and g is any element of G then $g^{-1}Hg$ is a subgroup conjugate to H. H and all its conjugates are said to form a system of conjugate subgroups of G or a complete conjugate set of subgroups.
27. *Jour. de Math.*, 14, 1849, 141–80.
28. *Œuvres*, 1897 ed., 21–23, 27–29.

Jordan in 1878[29] stated that a linear homogeneous substitution (9) of finite period p may be linearly transformed to the canonical form

$$y_i' = \varepsilon_i y_i, \qquad i = 1, 2, \cdots, n$$

where the ε_i's are pth roots of unity. The theorem was proved by a number of men.[30] This was the beginning of a vast amount of research on the determination of the possible linear substitution groups of a given order and of binary and ternary form (two and three variables). Also the determination of the subgroups of given linear substitution groups and the algebraic expressions left invariant by all the members of a group or subgroup stirred up much research.

Directly after noting Bravais's paper Jordan undertook the first major investigation of *infinite* groups. In his paper "Mémoire sur les groupes de mouvements,"[31] Jordan points out that the determination of all the groups of movements (he considered only translations and rotations) is equivalent to the determination of all the possible systems of molecules such that each movement of any one group transforms the corresponding system of molecules into itself. He therefore studied the various types of groups and classified them. The results are not as significant as the fact that his paper initiated the study of geometric transformations under the rubric of groups and the geometers were quick to pick up this line of thought (Chap. 38, sec. 5).

One other development of the middle nineteenth century is both noteworthy and instructive. Arthur Cayley, very much influenced by Cauchy's work, recognized that the notion of substitution group could be generalized. In three papers[32] Cayley introduced the notion of an *abstract* group. He used a general operator symbol θ applied to a system of elements $x, y, z, \ldots$ and spoke of θ so applied producing a function $x', y', z', \ldots$ of $x, y, z, \ldots$. He pointed out that in particular θ may be a substitution. The abstract group contains many operators $\theta, \phi, \ldots$. $\theta\phi$ is a compound (product) of two operations and the compound is associative but not necessarily commutative. His general definition of a group calls for a set of operators $1, \alpha, \beta, \ldots$, all of them different and such that the product of any two of them in either order or the product of any one and itself belongs to the set.[33] He mentions matrices under multiplication and quaternions (under addition) as constituting groups. Unfortunately, Cayley's introduction of the abstract group

29. *Jour. für. Math.*, 84, 1878, 89–215, p. 112 in particular = *Œuvres*, 2, 13–139, p. 36 in particular.
30. See E. H. Moore, *Math. Ann.*, 50, 1898, 215.
31. *Annali di Mat.*, (2), 2, 1868/69, 167–215 and 322–45 = *Œuvres*, 4, 231–302.
32. *Phil. Mag.*, (3), 34, 1849, 527–29 = *Coll. Math. Papers*, 1, 423–24 and (4), 7, 1854, 40–47 and 408–9 = *Papers*, 2, 123–30 and 131–32.
33. *Papers*, 2, 124.

concept attracted no attention at this time, partly because matrices and quaternions were new and not well known and the many other mathematical systems that could be subsumed under the notion of groups were either yet to be developed or were not recognized to be so subsumable. Premature abstraction falls on deaf ears whether they belong to mathematicians or to students.

Bibliography

Abel, N. H.: *Œuvres complètes* (1881), 2 vols., Johnson Reprint Corp., 1964.

Bachmann, P.: "Über Gauss' zahlentheoretische Arbeiten," *Nachrichten König. Ges. der Wiss. zu Gött.*, 1911, 455–518. Also in Gauss's *Werke*, 10, Part 2, 1–69.

Burkhardt, H.: "Endliche discrete Gruppen, *Encyk. der Math. Wiss.*, B. G. Teubner, 1903–15, I, Part 1, 208–26.

————: "Die Anfänge der Gruppentheorie und Paolo Ruffini," *Abhandlungen zur Geschichte der Mathematik*, Heft 6, 1892, 119–59.

Burns, Josephine E.: "The Foundation Period in the History of Group Theory," *Amer. Math. Monthly*, 20, 1913, 141–48.

Dupuy, P.: "La Vie d'Evariste Galois," *Ann. de l'Ecole Norm. Sup.*, (2), 13, 1896, 197–266.

Galois, Evariste: "Œuvres," *Jour. de Math.*, 11, 1846, 381–444.

————: *Œuvres mathématiques*, Gauthier-Villars, 1897.

————: *Ecrits et mémoires mathématiques* (ed. by R. Bourgne and J.-P. Azra), Gauthier-Villars, 1962.

Gauss, C. F.: *Disquisitiones Arithmeticae* (1801), *Werke*, Vol. 1, König. Ges. der Wiss., zu Göttingen, 1870, English translation by Arthur A. Clarke, S.J., Yale University Press, 1966.

Hobson, E. W.: *Squaring the Circle and Other Monographs*, Chelsea (reprint), 1953.

Hölder, Otto: "Galois'sche Theorie mit Anwendungen," *Encyk. der Math. Wiss.*, B. G. Teubner, 1898–1904, I, Part 1, 480–520.

Infeld, Leopold: *Whom the Gods Love: The Story of Evariste Galois*, McGraw-Hill, 1948.

Jordan, Camille: *Œuvres*, 4 vols., Gauthier-Villars, 1961–64.

————: *Traité des substitutions et des équations algébriques* (1870), Gauthier-Villars (reprint), 1957.

Kiernan, B. M.: "The Development of Galois Theory from Lagrange to Artin," *Archive for History of Exact Sciences*, 8, 1971, 40–154.

Lebesgue, Henri: *Notice sur la vie et les travaux de Camille Jordan*, Gauthier-Villars, 1923. Also in Lebesgue's *Notices d'histoire des mathématiques*, pp. 44–65, Institut de Mathématiques, Genève, 1958.

Miller, G. A.: "History of the Theory of Groups to 1900," *Collected Works*, Vol. 1, 427–67, University of Illinois Press, 1935.

Pierpont, James: "Lagrange's Place in the Theory of Substitutions," *Amer. Math. Soc. Bull.*, 1, 1894/95, 196–204.

————: "Early History of Galois's Theory of Equations," *Amer. Math. Soc. Bull.*, 4, 1898, 332–40.

Smith, David Eugene: *A Source Book in Mathematics*, Dover (reprint), 1959, Vol. 1, 232–52, 253–60, 261–66, 278–85.

Verriest, G.: *Œuvres mathématiques d'Evariste Galois* (1897 ed.), 2nd ed., Gauthier-Villars, 1951.

Wiman, A.: "Endliche Gruppen linearer Substitutionen," *Encyk. der Math. Wiss.*, B. G. Teubner, 1898–1904, I, Part 1, 522–54.

Wussing, H. L.: *Die Genesis des abstrakten Gruppenbegriffes*, VEB Deutscher Verlag der Wiss., 1969.

32
Quaternions, Vectors, and Linear Associative Algebras

Quaternions came from Hamilton after his really good work had been done; and though beautifully ingenious, have been an unmixed evil to those who have touched them in any way.... Vector is a useless survival, or offshoot from quaternions, and has never been of the slightest use to any creature.

LORD KELVIN

1. The Foundation of Algebra on Permanence of Form

Galois's work on equations solvable by algebraic processes closed a chapter of algebra and, though he introduced ideas such as group and domain of rationality (field) that would bear fruit, the fuller exploitation of these ideas had to await other developments. The next major algebraic creation, initiated by William R. Hamilton, opened up totally new domains while shattering age-old convictions as to how "numbers" must behave.

To appreciate the originality of Hamilton's work we must examine the logic of ordinary algebra as it was generally understood in the first half of the nineteenth century. By 1800 the mathematicians were using freely the various types of real numbers and even complex numbers, but the precise definitions of these various types of numbers were not available nor was there any logical justification of the operations with them. Expressions of dissatisfaction with this state of affairs were numerous but were submerged in the mass of new creations in algebra and analysis. The greatest uneasiness seemed to be caused by the fact that letters were manipulated as though they had the properties of the integers; yet the results of these operations were valid when any numbers were substituted for the letters. Since the logic of the various types of numbers was not developed, it was not possible to see that they possessed the same formal properties as the positive integers, and consequently that literal expressions which merely stood for any class of real or complex numbers must possess the same properties—that is, that ordinary algebra is just generalized arithmetic. It seemed as though the algebra of literal expressions possessed a logic of its own, which accounted for its effectiveness and correctness. Hence in the 1830s the mathematicians tackled the problem of justifying the operations with literal or symbolic expressions.

This problem was first considered by George Peacock (1791–1858), professor of mathematics at Cambridge University. To justify the operations with literal expressions that could stand for negative, irrational, and complex numbers he made the distinction between arithmetical algebra and symbolical algebra. The former dealt with symbols representing the positive integers and so was on solid ground. Here only operations leading to positive integers were permissible. Symbolical algebra adopts the rules of arithmetical algebra but removes the restrictions to positive integers. All the results deduced in arithmetical algebra, whose expressions are general in form but particular in value, are results likewise in symbolical algebra where they are general in value as well as in form. Thus $a^m a^n = a^{m+n}$ holds in arithmetical algebra when m and n are positive integers and it therefore holds in symbolical algebra for all m and n. Likewise the series for $(a + b)^n$ when n is a positive integer, if it be exhibited in a general form without reference to a final term, holds for all n. Peacock's argument is known as the principle of the permanence of form.

The explicit formulation of this principle was given in Peacock's "Report on the Recent Progress and Present State of Certain Branches of Analysis,"[1] in which he does not merely report but dogmatically affirms. In symbolical algebra, he says:

1. The symbols are unlimited both in value and in representation.

2. The operations on them, whatever they may be, are possible in all cases.

3. The laws of combination of the symbols are of such a kind as to coincide universally with those in arithmetical algebra when the symbols are arithmetical quantities, and when the operations to which they are subject are called by the same names as in arithmetical algebra.

From these principles he believed he could deduce the principle of permanence of form: "Whatever algebraical forms are equivalent when the symbols are general in form but specific in value [positive integers], will be equivalent likewise when the symbols are general in value as well as in form." Peacock used this principle to justify in particular the operations with complex numbers. He did try to protect his conclusion by the phrase "when the symbols are general in form." Thus one could not state special properties of particular whole numbers in symbolic form and insist that these symbolic statements are general. For example, the decomposition of a composite integer into a product of primes, though expressed symbolically, could not

1. *Brit. Assn. for Adv. of Sci.*, Rept. 3, 1833, 185–352.

be taken over as a statement of symbolical algebra. The principle sanctioned by fiat what was evidently empirically correct but not yet logically established.

This principle Peacock reaffirmed in the second edition of his *Treatise on Algebra*,[2] but here he also introduces a formal science of algebra. In this *Treatise* Peacock states that algebra like geometry is a deductive science. The processes of algebra have to be based on a complete statement of the body of laws that dictate the operations used in the processes. The symbols for the operations have, at least for the deductive science of algebra, no sense other than those given to them by the laws. Thus addition means no more than any process that obeys the laws of addition in algebra. His laws were, for example, the associative and commutative laws of addition and multiplication and the law that if $ac = bc$ and $c \neq 0$ then $a = b$.

Here the principle of permanence of form was derived from the adoption of axioms. This approach paved the way for more abstract thinking in algebra and in particular influenced Boole's thinking on the algebra of logic.

Throughout most of the nineteenth century the view of algebra affirmed by Peacock was accepted. It was supported, for example, by Duncan F. Gregory (1813–44), a great-great-grandson of the seventeenth-century James Gregory. Gregory wrote in a paper "On the Real Nature of Symbolical Algebra":[3]

> The light then in which I would consider symbolical algebra is that it is the science which treats of the combination of operations defined not by their nature, that is, by what they are or what they do, but by the laws of combination to which they are subject. . . . It is true that these laws have been in many cases suggested (as Mr. Peacock has aptly termed it) by the laws of the known operations of number, but the step which is taken from arithmetical to symbolical algebra is that, leaving out of view the nature of the operations which the symbols we use represent, we suppose the existence of classes of unknown operations subject to the same laws. We are thus able to prove certain relations between the different classes of operations, which, when expressed between the symbols, are called algebraical theorems.

In this paper Gregory emphasized the commutative and distributive laws, the terms having been introduced by François-Joseph Servois (1767–1847).[4]

The theory of algebra as the science of symbols and the laws of their combinations was carried further by Augustus De Morgan, who wrote several papers on the structure of algebra.[5] His *Trigonometry and Double Algebra* (1849) also contains his views. The words double algebra meant the algebra of complex numbers, whereas single algebra meant negative numbers. Prior

2. 1842–45; 1st ed., 1830.
3. *Transactions of the Royal Society of Edinburgh*, 14, 1840, 208–16.
4. *Ann. de Math.*, 5, 1814/15, 93–140.
5. *Trans. Camb. Phil. Soc.*, 1841, 1842, 1844, and 1847.

to single algebra is universal arithmetic, which covers the algebra of the positive real numbers. Algebra, De Morgan maintained, is a collection of meaningless symbols and operations with symbols. The symbols are 0, 1, +, −, ×, ÷, ( )$^{()}$, and letters. The laws of algebra are the laws which such symbols obey, for example, the commutative law, the distributive law, the laws of exponents, a negative times a positive is negative, $a - a = 0$, $a \div a = 1$, and derived laws. The basic laws are arbitrarily chosen.

By the middle of the nineteenth century the axioms of algebra generally accepted were:

1. Equal quantities added to a third yield equal quantities.
2. $(a + b) + c = a + (b + c)$.
3. $a + b = b + a$.
4. Equals added to equals give equals.
5. Equals added to unequals give unequals.
6. $a(bc) = (ab)c$.
7. $ab = ba$.
8. $a(b + c) = ab + ac$.

The principle of permanence of form rested on these axioms.

It is hard for us to see just what this principle means. It begs the question of why the various types of numbers possess the same properties as the whole numbers. But Peacock, Gregory, and De Morgan sought to make a science out of algebra independent of the properties of real and complex numbers and so regarded algebra as a science of uninterpreted symbols and their laws of combination. In effect it was the justification for assuming that the same fundamental properties hold for all types of numbers. This foundation was not only vague but inelastic. The men insisted on a parallelism between arithmetical and general algebra so rigid that, if maintained, it would destroy the generality of algebra. They do not seem to have realized that a formula that is true with one interpretation of the symbols might not be true with another.

The principle of permanence of form, an arbitrary dictum, could not serve as a solid foundation for algebra. In fact, the developments we shall deal with in this chapter undermined it. The first step, which merely obviated the need for this principle so far as complex numbers were concerned, was made by Hamilton when he founded the logic of complex numbers on the properties of real numbers.

Though by 1830 complex numbers were intuitively well grounded through their representation as points or as directed line segments in the plane, Hamilton, who was concerned with the logic of arithmetic, was not satisfied with just an intuitive foundation. In his paper "Conjugate Functions and on Algebra as the Science of Pure Time,"[6] Hamilton pointed out that a

6. *Trans. Royal Irish Academy*, 17, 1837, 293–422 = *Math. Papers*, 3, 3–96.

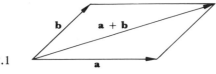

Figure 32.1

complex number $a + bi$ is not a genuine sum in the sense that $2 + 3$ is. The use of the plus sign is a historical accident and bi cannot be added to a. The complex number $a + bi$ is no more than an ordered couple (a, b) of real numbers. The peculiarity which i or $\sqrt{-1}$ introduces in the operations with complex numbers is incorporated by Hamilton in the definitions of the operations with ordered couples. Thus, if $a + bi$ and $c + di$ are two complex numbers then

$$(a, b) \pm (c, d) = (a \pm c, b \pm d)$$
$$(a, b) \cdot (c, d) = (ac - bd, ad + bc)$$

(1)
$$\frac{(a, b)}{(c, d)} = \left(\frac{ac + bd}{c^2 + d^2}, \frac{bc - ad}{c^2 + d^2} \right).$$

The usual associative, commutative, and distributive properties can now be deduced. Under this view of complex numbers, not only are these numbers logically founded on the basis of real numbers, but the hitherto somewhat mysterious $\sqrt{-1}$ is dispensed with entirely. Of course in practice it is still convenient to use the $a + bi$ form and to remember that $\sqrt{-1} \sqrt{-1} = -1$. Incidentally, Gauss did say in a letter of 1837 to Wolfgang Bolyai that he had had this notion of ordered couples in 1831. But it was Hamilton's publication that gave the ordered couple concept to the mathematical world.

2. The Search for a Three-Dimensional "Complex Number"

The notion of a vector, that is, a directed line segment that might represent the magnitude and direction of a force, a velocity, or an acceleration, entered mathematics quietly. Aristotle knew that forces can be represented as vectors and that the combined action of two forces can be obtained by what is commonly known as the parallelogram law; that is, the diagonal of the parallelogram formed by the two vectors **a** and **b** (Fig. 32.1), gives the magnitude and direction of the resultant force. Simon Stevin employed the parallelogram law in problems of statics, and Galileo stated the law explicitly.

After the geometric representation of complex numbers supplied by Wessel, Argand, and Gauss became somewhat familiar, the mathematicians realized that complex numbers could be used to represent and work with vectors in a plane. For example, if two vectors are represented respectively by say $3 + 2i$ and $2 + 4i$ then the sum of the complex numbers, namely,

$5 + 6i$, represents the sum of the vectors added by means of the parallelogram law. What the complex numbers do for vectors in a plane is to supply an algebra to represent the vectors and operations with vectors. One need not carry out the operations geometrically but can work with them algebraically much as the equation of a curve can be used to represent and work with curves.

This use of complex numbers to represent vectors in a plane became well known by 1830. However, the utility of complex numbers is limited. If several forces act on a body these forces need not lie in one plane. To treat these forces algebraically a three-dimensional analogue of complex numbers is needed. One could use the ordinary Cartesian coordinates (x, y, z) of a point to represent the vector from the origin to the point but there were no operations with the triples of numbers to represent the operations with vectors. These operations, as in the case of complex numbers, would seemingly have to include addition, subtraction, multiplication, and division and moreover obey the usual associative, commutative, and distributive laws so that algebraic operations could be applied freely and effectively. The mathematicians began a search for what was called a three-dimensional complex number and its algebra.

Wessel, Gauss, Servois, Möbius, and others worked on this problem. Gauss[7] wrote a short unpublished note dated 1819 on mutations of space. He thought of the complex numbers as displacements; $a + bi$ was a displacement of a units along one fixed direction followed by a displacement of b units in a perpendicular direction. Hence he tried to build an algebra of a three-component number in which the third component would represent a displacement in a direction perpendicular to the plane of $a + bi$. He arrived at a non-commutative algebra but it was not the effective algebra required by the physicists. Moreover, because he did not publish it, this work had little influence.

The creation of a useful spatial analogue of complex numbers is due to William R. Hamilton (1805–65). Next to Newton, Hamilton is the greatest of the English mathematicians and like Newton he was even greater as a physicist than as a mathematician. At the age of five Hamilton could read Latin, Greek, and Hebrew. At eight he added Italian and French; at ten he could read Arabic and Sanskrit and at fourteen, Persian. A contact with a lightning calculator inspired him to study mathematics. He entered Trinity College in Dublin in 1823 where he was a brilliant student. In 1822, at the age of seventeen he prepared a paper on caustics which was read before the Royal Irish Academy in 1824 but not published. Hamilton was advised to rework and expand it. In 1827 he submitted to the Academy a revision entitled "A Theory of Systems of Rays," which made a science of geometrical

7. *Werke*, 8, 357–62.

optics. Here he introduced what are called the characteristic functions of optics. The paper was published in 1828 in the *Transactions of the Royal Irish Academy*.[8]

In 1827, while still an undergraduate he was appointed Professor of Astronomy at Trinity College, an appointment that carried with it the title of Royal Astronomer of Ireland. His duties as professor were to lecture on science and to manage the astronomical observatory. He did not do much with the latter but he was a fine teacher.

In 1830 and 1832 he published three supplements to "A Theory of Systems of Rays." In the third paper[9] he predicted that a ray of light propagating in special directions in a biaxial crystal would give rise to a cone of refracted rays. This phenomenon was confirmed experimentally by Humphrey Lloyd, a friend and colleague. Hamilton then carried over to dynamics his ideas on optics and in the field of dynamics wrote two very famous papers (Chap. 30) in which he used the characteristic function concept which he had developed for optics. He also gave a system of complete and rigorous integrals for the differential equations of motion of a system of bodies. His major mathematical work was the subject of quaternions, which we shall discuss shortly. His final form of this work he presented in his *Lectures on Quaternions* (1853) and in a two-volume posthumously published *Elements of Quaternions* (1866)

Hamilton could use analogy skillfully to reason from the known to the unknown. Though he had a fine intuition, he did not have great flashes of ideas but worked long and hard on special problems to see what could be general. He was patient and systematic in working out many specific examples and was willing to undertake tremendous calculations to check or prove a point. However, in his publications one finds only the polished compressed general results.

He was deeply religious and this interest was most important to him. Next in order of importance were metaphysics, mathematics, poetry, physics, and general literature. He also wrote poetry. He thought that the geometrical ideas created in his time, the use of infinite elements and imaginary elements in the work of Poncelet and Chasles (Chap. 35), were akin to poetry. Though he was a modest man, he admitted and even emphasized that love of fame moves and cheers great mathematicians.

Hamilton's clarification of the notion of complex numbers enabled him to think more clearly about the problem of introducing a three-dimensional analogue to represent vectors in space. But the immediate effect was to frustrate his efforts. All of the numbers known to mathematicians at this time possessed the commutative property of multiplication and it was natural for

8. *Trans. Royal Irish Academy*, 15, 1828, 69–174 = *Math. Papers*, 1, 1–106.
9. *Trans. Royal Irish Academy*, 17, 1837, 1–144 = *Math. Papers*, 1, 164–293.

Hamilton to believe that the three-dimensional or three-component numbers he sought should possess this same property as well as the other properties that real and complex numbers possess. After some years of effort Hamilton found himself obliged to make two compromises. The first was that his new numbers contained four components and the second, that he had to sacrifice the commutative law of multiplication. Both features were revolutionary for algebra. He called the new numbers quaternions.

With hindsight one can see on geometric grounds that the new "numbers" had to contain four components. The new number regarded as an operator was expected to rotate a given vector about a given axis in space and to stretch or contract the vector. For these purposes, two parameters (angles) are needed to fix the axis of rotation, one parameter must specify the angle of rotation, and the fourth the stretch or contraction of the given vector.

Hamilton himself described his discovery of quaternions:[10]

> Tomorrow will be the fifteenth birthday of the Quaternions. They started into life, or light, full grown, on the 16th of October, 1843, as I was walking with Lady Hamilton to Dublin, and came up to Brougham Bridge. That is to say, I then and there felt the galvanic circuit of thought closed, and the sparks which fell from it were the fundamental equations between I, J, K; *exactly such* as I have used them ever since. I pulled out, on the spot, a pocketbook, which still exists, and made an entry, on which, *at the very moment*, I felt that it might be worth my while to expend the labour of at least ten (or it might be fifteen) years to come. But then it is fair to say that this was because I felt a *problem* to have been at that moment *solved*, an intellectual *want relieved*, which had *haunted* me for at least *fifteen years* before.

He announced the invention of quaternions in 1843 at a meeting of the Royal Irish Academy, spent the rest of his life developing the subject, and wrote many papers on it.

3. The Nature of Quaternions

A quaternion is a number of the form

(2) $$3 + 2\mathbf{i} + 6\mathbf{j} + 7\mathbf{k}$$

wherein the $\mathbf{i}$, $\mathbf{j}$, and $\mathbf{k}$ play somewhat the role that i does in complex numbers. The real part, 3 above, is called the scalar part of the quaternion and the remainder, the vector part. The three coefficients of the vector part are rectangular Cartesian coordinates of a point P while $\mathbf{i}$, $\mathbf{j}$, and $\mathbf{k}$ are called qualitative units that geometrically are directed along the three axes. The criterion of equality of two quaternions is that their scalar parts shall be

10. *North British Review*, 14, 1858, 57.

equal and that the coefficients of their $\mathbf{i}$, $\mathbf{j}$, and $\mathbf{k}$ units shall be respectively equal. Two quaternions are added by adding their scalar parts and adding the coefficients of each of the $\mathbf{i}$, $\mathbf{j}$, and $\mathbf{k}$ units to form new coefficients for those units. The sum of two quaternions is therefore itself a quaternion.

All the familiar algebraic rules of multiplication are supposed valid in operating with quaternions except that in forming products of the units $\mathbf{i}$, $\mathbf{j}$, and $\mathbf{k}$ the following rules, which abandon the commutative law, hold:

$$\mathbf{jk} = \mathbf{i}, \qquad \mathbf{kj} = -\mathbf{i}, \qquad \mathbf{ki} = \mathbf{j}, \qquad \mathbf{ik} = -\mathbf{j}, \qquad \mathbf{ij} = \mathbf{k}, \qquad \mathbf{ji} = -\mathbf{k}$$
$$(3) \qquad \mathbf{i}^2 = \mathbf{j}^2 = \mathbf{k}^2 = -1.$$

Thus if

$$\mathbf{p} = 3 + 2\mathbf{i} + 6\mathbf{j} + 7\mathbf{k} \quad \text{and} \quad \mathbf{q} = 4 + 6\mathbf{i} + 8\mathbf{j} + 9\mathbf{k}$$

then

$$\mathbf{pq} = (3 + 2\mathbf{i} + 6\mathbf{j} + 7\mathbf{k})(4 + 6\mathbf{i} + 8\mathbf{j} + 9\mathbf{k}) = -111 + 24\mathbf{i} + 72\mathbf{j} + 35\mathbf{k}$$

whereas

$$\mathbf{qp} = (4 + 6\mathbf{i} + 8\mathbf{j} + 9\mathbf{k})(3 + 2\mathbf{i} + 6\mathbf{j} + 7\mathbf{k}) = -111 + 28\mathbf{i} + 24\mathbf{j} + 75\mathbf{k}.$$

Hamilton proved that multiplication is associative. This is the first use of that term.[11]

Division of one quaternion by another can also be effected, but the fact that multiplication is not commutative implies that to divide the quaternion $\mathbf{p}$ by the quaternion $\mathbf{q}$ can mean to find $\mathbf{r}$ such that $\mathbf{p} = \mathbf{qr}$ or such that $\mathbf{p} = \mathbf{rq}$. The quotient $\mathbf{r}$ need not be the same in the two cases. The problem of division is best handled by introducing $\mathbf{q}^{-1}$ or $1/\mathbf{q}$. If $\mathbf{q} = a + b\mathbf{i} + c\mathbf{j} + d\mathbf{k}$ one defines $\mathbf{q}'$ to be $a - b\mathbf{i} - c\mathbf{j} - d\mathbf{k}$ and $N(\mathbf{q})$, called the norm of $\mathbf{q}$, to be $a^2 + b^2 + c^2 + d^2$. Then $N(\mathbf{q}) = \mathbf{qq}' = \mathbf{q}'\mathbf{q}$. By definition $\mathbf{q}^{-1} = \mathbf{q}'/N(\mathbf{q})$ and $\mathbf{q}^{-1}$ exists if $N(\mathbf{q}) \neq 0$. Also $\mathbf{qq}^{-1} = 1$ and $\mathbf{q}^{-1}\mathbf{q} = 1$. Now to find the $\mathbf{r}$ such that $\mathbf{p} = \mathbf{qr}$ we have $\mathbf{q}^{-1}\mathbf{p} = \mathbf{q}^{-1}\mathbf{qr}$ or $\mathbf{r} = \mathbf{q}^{-1}\mathbf{p}$. To find the $\mathbf{r}$ such that $\mathbf{p} = \mathbf{rq}$ we have $\mathbf{pq}^{-1} = \mathbf{rqq}^{-1}$ or $\mathbf{r} = \mathbf{pq}^{-1}$.

That quaternions can be used to rotate and stretch or contract a given vector into another given vector is readily shown. One must merely show that one can determine a, b, c, and d such that

$$(a + b\mathbf{i} + c\mathbf{j} + d\mathbf{k})(x\mathbf{i} + y\mathbf{j} + z\mathbf{k}) = (x'\mathbf{i} + y'\mathbf{j} + z'\mathbf{k}).$$

By multiplying out the left side as quaternions and equating corresponding coefficients of the left and right sides we get four equations in the unknowns a, b, c, and d. These four equations suffice to determine the unknowns.

Hamilton also introduced an important differential operator. The symbol ∇, which is an inverted Δ—which Hamilton termed "nabla"

11. *Proc. Royal Irish Academy*, 2, 1844, 424–34 = *Math. Papers*, 3, 111–16.

because it looks like an ancient Hebrew musical instrument of that name—stands for the operator

(4)
$$\nabla = \mathbf{i} \frac{\partial}{\partial x} + \mathbf{j} \frac{\partial}{\partial y} + \mathbf{k} \frac{\partial}{\partial z}.$$

When applied to a scalar point function $u(x, y, z)$ it produces the vector

(5)
$$\nabla u = \frac{\partial u}{\partial x} \mathbf{i} + \frac{\partial u}{\partial y} \mathbf{j} + \frac{\partial u}{\partial z} \mathbf{k}.$$

This vector, which varies from point to point of space, is now called the gradient of u. It represents in magnitude and direction the greatest space rate of increase of u.

Also letting $\mathbf{v} = v_1\mathbf{i} + v_2\mathbf{j} + v_3\mathbf{k}$ denote a continuous vector point function, wherein v_1, v_2, and v_3 are functions of x, y, and z, Hamilton introduced

(6)
$$\nabla \mathbf{v} = \left(\mathbf{i} \frac{\partial}{\partial x} + \mathbf{j} \frac{\partial}{\partial y} + \mathbf{k} \frac{\partial}{\partial z} \right) (v_1\mathbf{i} + v_2\mathbf{j} + v_3\mathbf{k})$$
$$= -\left(\frac{\partial v_1}{\partial x} + \frac{\partial v_2}{\partial y} + \frac{\partial v_3}{\partial k} \right) + \left(\frac{\partial v_3}{\partial y} - \frac{\partial v_2}{\partial z} \right)\mathbf{i} + \left(\frac{\partial v_1}{\partial z} - \frac{\partial v_3}{\partial x} \right)\mathbf{j}$$
$$+ \left(\frac{\partial v_2}{\partial x} - \frac{\partial v_1}{\partial y} \right)\mathbf{k}.$$

Thus the result of operating with ∇ on a vector point function $\mathbf{v}$ is to produce a quaternion; the scalar part of this quaternion (except for the minus sign) is what we now call the divergence of $\mathbf{v}$ and the vector part is now called curl $\mathbf{v}$.

Hamilton's enthusiasm for his quaternions was unbounded. He believed that this creation was as important as the calculus and that it would be the key instrument in mathematical physics. He himself made some applications to geometry, optics, and mechanics. His ideas were enthusiastically endorsed by his friend Peter Guthrie Tait (1831–1901), professor of mathematics at Queen's College and later professor of natural history at the University of Edinburgh. In many articles Tait urged physicists to adopt quaternions as the basic tool. He even became involved in long arguments with Cayley, who took a dim view of the usefulness of quaternions. But the physicists ignored quaternions and continued to work with conventional Cartesian coordinates. Nevertheless, as we shall see, Hamilton's work did lead indirectly to an algebra and analysis of vectors that physicists eagerly adopted.

Hamilton's quaternions proved to be of immeasurable importance for algebra. Once mathematicians realized that a meaningful, useful system of "numbers" could be built up which may fail to possess the commutative property of real and complex numbers, they felt freer to consider creations which departed even more from the usual properties of real and complex numbers. This realization was necessary before vector algebra and analysis could be developed because vectors violate more ordinary laws of algebra than do quaternions (sec. 5). More generally, Hamilton's work led to the theory of linear associative algebras (sec. 6). Hamilton himself began work on hypernumbers which contain n components or n-tuples,[12] but it was his work on quaternions which stimulated the new work on linear algebras.

4. Grassmann's Calculus of Extension

While Hamilton was developing his quaternions, another mathematician, Hermann Günther Grassmann (1809–77), who showed no talent for mathematics as a youth and who had no university education in mathematics but later became a teacher of mathematics in the *gymnasium* (high school) at Stettin, Germany, as well as an authority on Sanskrit, was developing an even more audacious generalization of complex numbers. Grassmann had his ideas before Hamilton but did not publish until 1844, one year after Hamilton announced his discovery of quaternions. In that year he published his *Die lineale Ausdehnungslehre* (The Calculus of Extension). Because he shrouded the ideas with mystic doctrines and the exposition was abstract, the more practical-minded mathematicians and physicists found the book vague and unreadable, and as a consequence the work, though highly original, remained little known for years. Grassmann issued a revised edition, entitled *Die Ausdehnungslehre*, in 1862. In it he simplified and amplified the original work but his style and lack of clarity still repelled readers.

Though Grassmann's exposition was almost inextricably bound up with geometrical ideas—he was in fact concerned with n-dimensional geometry—we shall abstract the algebraic notions that proved to be of lasting value. His basic notion, which he called an extensive quantity (*extensive Grösse*), is one type of hypernumber with n components. To study his ideas we shall discuss the case $n = 3$.

Consider two hypernumbers

$$\alpha = \alpha_1 e_1 + \alpha_2 e_2 + \alpha_3 e_3 \quad \text{and} \quad \beta = \beta_1 e_1 + \beta_2 e_2 + \beta_3 e_3,$$

where the α_i and β_i are real numbers and where e_1, e_2, and e_3 are primary or qualitative units represented geometrically by direct line segments of unit length drawn from a common origin so as to determine a right-handed orthogonal system of axes. The $\alpha_i e_i$ are multiples of the primary units and are represented geometrically by lengths α_i along the respective axes, while α is represented by a directed line segment in space whose projections on the axes are the lengths α_i. The same is true for the β_i and β. Grassmann called the directed line segments or line-vectors *Strecke*.

12. *Trans. Royal Irish Academy*, 21, 1848, 199–296 = *Math. Papers*, III, 159–226.

The addition and subtraction of these hypernumbers are defined by

(7) $$\alpha \pm \beta = (\alpha_1 \pm \beta_1)e_1 + (\alpha_2 \pm \beta_2)e_2 + (\alpha_3 \pm \beta_3)e_3.$$

Grassmann introduced two kinds of multiplications, the inner product and the outer product. For the inner product he postulated that

(8) $$e_i|e_i = 1, \; e_i|e_j = 0 \qquad \text{for } i \neq j.$$

For the outer product

(9) $$[e_ie_j] = -[e_je_i], \; [e_ie_i] = 0.$$

These brackets are called units of the second order and are not reduced by Grassmann (whereas Hamilton does) to units of the first order, that is, to the e_i, but are dealt with as though they were equivalent to first order units with $[e_1e_2] = e_3$, and so forth.

From these definitions it follows that the inner product $\alpha|\beta$ of α and β is given by

$$\alpha|\beta = \alpha_1\beta_1 + \alpha_2\beta_2 + \alpha_3\beta_3 \quad \text{and} \quad \alpha|\beta = \beta|\alpha.$$

The numerical value or magnitude a of a hypernumber α is defined as $\sqrt{\alpha|\alpha} = \sqrt{\alpha_1^2 + \alpha_2^2 + \alpha_3^2}$. Thus the magnitude of α is numerically equal to the length of the line-vector which represents it geometrically. If θ denotes the angle between the line-vectors α and β, then

$$\alpha|\beta = ab\left(\frac{\alpha_1\beta_1}{ab} + \frac{\alpha_2\beta_2}{ab} + \frac{\alpha_3\beta_3}{ab}\right) = ab \cos \theta.$$

With the aid of the outer product rule (9) the outer product P of the hypernumbers α and β can be expressed as follows:

(10) $$P = [\alpha\beta] = (\alpha_2\beta_3 - \alpha_3\beta_2)[e_2e_3] + (\alpha_3\beta_1 - \alpha_1\beta_3)[e_3e_1] \\ + (\alpha_1\beta_2 - \alpha_2\beta_1)[e_1e_2].$$

This product is a hypernumber of the second order and is expressed in terms of independent units of the second order. Its magnitude $|P|$ is obtained by means of a definition of the inner product of two hypernumbers of the second order and is

$$|P| = \sqrt{P|P} = \{(\alpha_2\beta_3 - \alpha_3\beta_2)^2 + (\alpha_3\beta_1 - \alpha_1\beta_3)^2 + (\alpha_1\beta_2 - \alpha_2\beta_1)^2\}^{1/2}$$

(11) $$= ab\left\{1 - \left(\frac{\alpha_1\beta_1}{ab} + \frac{\alpha_2\beta_2}{ab} + \frac{\alpha_3\beta_3}{ab}\right)^2\right\}^{1/2}$$

$$= ab \sin \theta.$$

Hence the magnitude $|P|$ of the outer product $[\alpha\beta]$ is represented geometrically by the area of the parallelogram constructed upon line-vectors which are the geometrical representations of α and β. This area together with a unit line-vector normal to it, whose direction is chosen so that if α is rotated into β about the normal, then the normal will point in the direction of a right-handed screw rotating from α to β, is now called a vectorial area. Grassmann's term was *Plangrösse*.

Grassmann's inner product of two primary hypernumbers for three dimensions is equivalent to the negative of the scalar part of Hamilton's quaternion product of two vectors; and again in the three-dimensional case, Grassmann's outer product, if we replace $[e_2e_3]$ by e_1 and so forth, is precisely Hamilton's quaternion product of two vectors. However, in the theory of quaternions the vector is a subsidiary part of the quaternion whereas in Grassman's algebra the vector appears as the basic quantity.

Another product was formed by Grassmann by taking the scalar (inner) product of a hypernumber γ with the vector (outer) product $[\alpha\beta]$ of two hypernumbers α and β. This product Q for the three-dimensional case is

$$Q = [\alpha\beta]\gamma = (\alpha_2\beta_3 - \alpha_3\beta_2)\gamma_1 + (\alpha_3\beta_1 - \alpha_1\beta_3)\gamma_2 + (\alpha_1\beta_2 - \alpha_2\beta_1)\gamma_3.$$

In determinant form

$$(12) \qquad\qquad Q = \begin{vmatrix} \alpha_1 & \beta_1 & \gamma_1 \\ \alpha_2 & \beta_2 & \gamma_2 \\ \alpha_3 & \beta_3 & \gamma_3 \end{vmatrix}.$$

Consequently Q can be interpreted geometrically as the volume of a parallelepiped constructed on the line-vectors that represent α, β, and γ. The volume may be positive or negative.

Grassmann considered (for n-component hypernumbers) not only the two types of products described above but also products of higher order. In a paper of 1855,[13] he gave sixteen different types of products for hypernumbers. He also gave the geometrical significance of the products and made applications to mechanics, magnetism, and crystallography.

It might seem as though Grassmann's treatment of hypernumbers of n parts was needlessly general since thus far at least the useful instances of hypernumbers contained at most four parts. Yet Grassmann's thinking helped to lead mathematicians into the theory of tensors (Chap. 48), for tensors, as we shall see, are hypernumbers. Other geometrical ideas and the idea of invariance had yet to make themselves known to mathematicians before the day of tensors was to come. Though thinking about hypernumbers

13. *Jour. für Math.*, 49, 1855, 10–20 and 123–41 = *Ges. Math. und Phys. Werke*, 2, Part I, 199–217.

did lead to various generalizations, no analysis (e.g. calculus) for Grassmann's *n*-dimensional hypernumbers was ever developed. The reason is simply that no applications for such an analysis were found. As we shall see, there is an extensive analysis for tensors but these have their origin in Riemannian geometry.

5. *From Quaternions to Vectors*

Grassmann's work remained neglected for a while whereas, as we have noted, quaternions did receive a great deal of attention almost at once. However, they were not quite what the physicists wanted. They sought a concept that was not divorced from but more closely associated with Cartesian coordinates than quaternions were. The first step in the direction of such a concept was made by James Clerk Maxwell (1831–79), the founder of electromagnetic theory, one of the greatest of mathematical physicists, and professor of physics at Cambridge University.

Maxwell knew Hamilton's work and though he had heard of Grassmann's work he had not seen it. He singled out the scalar and vector parts of Hamilton's quaternion and put the emphasis on these separate notions.[14] However, in his celebrated *A Treatise on Electricity and Magnetism* (1873) he made a greater concession to quaternions and speaks rather of the scalar and vector parts of a quaternion though he does treat these parts as separate entities. A vector, he says (p. 10), requires three quantities (components) for its specification and these can be interpreted as lengths along the three coordinate axes. This vector concept is the vector part of Hamilton's quaternion, and Maxwell states so. Hamilton had introduced a vector function $\mathbf{v}$ of x, y, and z with components v_1, v_2, and v_3, had applied to it the operator $\nabla = \mathbf{i}(\partial/\partial x) + \mathbf{j}(\partial/\partial y) + \mathbf{k}(\partial/\partial z)$, and obtained the result (6). Thus $\nabla \mathbf{v}$ is a quaternion. But Maxwell separated the scalar part from the vector part and indicated these by $S\nabla\mathbf{v}$ (the scalar part of $\nabla\mathbf{v}$) and $V\nabla\mathbf{v}$ (the vector part of $\nabla\mathbf{v}$). He called $S\nabla\mathbf{v}$ the convergence of $\mathbf{v}$ because this expression had already appeared many times in fluid dynamics and when $\mathbf{v}$ is velocity had the meaning of flux or the net quantity per unit volume per unit time which flows through a small area surrounding a point. And he called $V\nabla\mathbf{v}$ the rotation or curl of $\mathbf{v}$ because this expression too had appeared in fluid dynamics as twice the rate of rotation of the fluid at a point. Clifford later called $-S\nabla\mathbf{v}$ the divergence.

Maxwell then points out that the operator ∇ repeated, gives

$$\nabla^2 = -\frac{\partial^2}{\partial x^2} - \frac{\partial^2}{\partial y^2} - \frac{\partial^2}{\partial z^2}$$

14. *Proc. London Math. Soc.*, 3, 1871, 224–32 = *The Scientific Papers*, Vol. 2, 257–66.

which he calls Laplace's operator. He allows it to act on a scalar function to produce a scalar and on a vector function to produce a vector.[15]

Maxwell noted in his 1871 paper that the curl of a gradient of a scalar function and the divergence of the curl of a vector function are always zero. He also states that the curl of the curl of a vector function **v** is the gradient of the divergence of **v** minus the Laplacian of **v**. (This is true in rectangular coordinates only.)

Maxwell often used quaternions as the basic mathematical entity or he at least made frequent reference to quaternions, perhaps to help his readers. Nevertheless his work made clear that vectors were the real tool for physical thinking and not just an abbreviated scheme of writing as some maintained. Thus by Maxwell's time a great deal of vector analysis was created by treating the scalar and vector parts of quaternions separately.

The formal break with quaternions and the inauguration of a new independent subject, three-dimensional vector analysis, was made independently by Josiah Willard Gibbs and Oliver Heaviside in the early 1880s. Gibbs (1839–1903), professor of mathematical physics at Yale College but primarily a physical chemist, had printed (1881 and 1884) for private distribution among his students a small pamphlet on the *Elements of Vector Analysis*.[16] His viewpoint is set forth in an introductory note:

> The fundamental principles of the following analysis are such as are familiar under a slightly different form to students of quaternions. The manner in which the subject is developed is somewhat different from that followed in treatises on quaternions, being simply to give a suitable notation for those relations between vectors, or between vectors and scalars, which seem most important, and which lend themselves most readily to analytical transformations, and to explain some of these transformations. As a precedent for such a departure from quaternionic usage Clifford's *Kinematics* may be cited. In this connection the name Grassmann may be also mentioned, to whose system the following method attaches itself in some respects more closely than to that of Hamilton.

Although printed for private circulation, Gibbs's pamphlet on vector analysis became widely known. The material was finally incorporated in a book

15. Since in vector analysis $\nabla^2 = \nabla \cdot \nabla$, then $\nabla^2 q$ means physically the divergence of the gradient or the divergence of the maximum space rate of change of q. However, the physical meaning is clearer from the fact that the function q which satisfies $\nabla^2 q = 0$ minimizes the Dirichlet integral (see (34) of Chap. 28). The integral is the square of the magnitude of the gradient taken over some volume. Hence $\nabla^2 q = 0$ means that the minimum gradient holds at any point or that the departure from uniformity is a minimum. If $\nabla^2 q$ is not 0 there must be some departure from uniformity and there will be a restoring force. The various equations of mathematical physics, which contain $\nabla^2 q$ in one context or another, state in effect that nature always acts to restore uniformity.

The notation Δ for ∇^2 was introduced by Robert Murphy in 1833.

16. *The Scientific Papers*, 2, 17–90.

written by E. B. Wilson and based on Gibbs's lectures. The book, Gibbs and Wilson: *Vector Analysis*, appeared in 1901.

Oliver Heaviside (1850–1925) was in the early part of his scientific career a telegraph and telephone engineer. He retired to country life in 1874 and devoted himself to writing, principally on the subject of electricity and magnetism. Heaviside had studied quaternions in Hamilton's *Elements* but had been repelled by the many special theorems. He felt that quaternions were too hard for busy engineers to learn and so he built his own vector analysis, which to him was just a shorthand form of ordinary Cartesian coordinates. In papers written during the 1880s in the journal *Electrician* he used this vector analysis freely. Then in his three-volume work *Electromagnetic Theory* (1893, 1899, 1912) he gave a good deal of vector algebra in Volume I. The third chapter of about 175 pages is devoted to vector methods. His development of the subject was essentially in harmony with Gibb's although he did not like Gibbs's notation and adopted his own based on Tait's quaternionic notation.

As formulated by Gibbs and Heaviside, a vector is no more than the vector part of a quaternion but considered independently of any quaternions. Thus a vector $\mathbf{v}$ is

$$\mathbf{v} = a\mathbf{i} + b\mathbf{j} + c\mathbf{k},$$

where $\mathbf{i}$, $\mathbf{j}$, and $\mathbf{k}$ are unit vectors along the x-, y-, and z-axes respectively. The coefficients a, b, and c are real numbers and are called components. Two vectors are equal if the respective components are equal and the sum of two vectors is the vector which has as its components the sum of the respective components of the summands.

Two types of multiplication, both physically useful, were introduced. The first type known as scalar multiplication is defined thus: We multiply $\mathbf{v}$ and $\mathbf{v}' = a'\mathbf{i} + b'\mathbf{j} + c'\mathbf{k}$ as ordinary polynomials and, using the dot as the symbol for multiplication, sct

(13a)
$$\mathbf{i}\cdot\mathbf{i} = \mathbf{j}\cdot\mathbf{j} = \mathbf{k}\cdot\mathbf{k} = 1$$
$$\mathbf{i}\cdot\mathbf{j} = \mathbf{j}\cdot\mathbf{i} = \mathbf{i}\cdot\mathbf{k} = \mathbf{k}\cdot\mathbf{i} = \mathbf{j}\cdot\mathbf{k} = \mathbf{k}\cdot\mathbf{j} = 0.$$

Thus $\mathbf{v}\cdot\mathbf{v}' = aa' + bb' + cc'$. This product is no longer a vector but a real number or scalar and is called the scalar product. It possesses a new algebraic feature, for the product of two real numbers or complex numbers or quaternions is always a number of the same kind as we started with. Another surprising property of the scalar product is that it may be zero when neither factor is zero. For example, the product of the vectors $\mathbf{v} = 3\mathbf{i}$ and $\mathbf{v}' = 6\mathbf{j} + 7\mathbf{k}$ is zero.

The scalar product of two vectors is algebraically novel in still another respect—it does not permit an inverse process. That is, we cannot always find a vector or scalar q such that $\mathbf{v}/\mathbf{v}' = q$. Thus if $\mathbf{q}$ were a vector, $\mathbf{q}\cdot\mathbf{v}'$ would

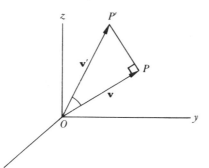

Figure 32.2

be a scalar and not equal to the vector **v**. On the other hand if q were a scalar then, though $q\mathbf{v}'$ is defined as $qa'\mathbf{i} + qb'\mathbf{j} + qc'\mathbf{k}$, it would be rare that $qa' = a$, $qb' = b$ and $qc' = c$ where a, b, and c are the coefficients in **v**. Despite the absence of a quotient the scalar product is useful.

The physical significance of the scalar product is readily shown to be the following: If $\mathbf{v}'$ is a force (Fig. 32.2) whose direction and magnitude are represented by the line segment from O to P' then the effectiveness of this force in pushing an object at O, say, in the direction of OP, where OP represents **v**, is the projection of OP' on OP or $OP' \cos \phi$ where ϕ is the angle between OP' and OP. If OP is of unit length, the projection of OP' is precisely the value of the product $\mathbf{v} \cdot \mathbf{v}'$.

The second kind of product of vectors, called the vector product, is defined as follows: We again multiply **v** and $\mathbf{v}'$ as polynomials but this time let

$$\mathbf{i} \times \mathbf{i} = \mathbf{j} \times \mathbf{j} = \mathbf{k} \times \mathbf{k} = 0$$

(13b) $\mathbf{i} \times \mathbf{j} = \mathbf{k}, \qquad \mathbf{j} \times \mathbf{i} = -\mathbf{k}, \qquad \mathbf{j} \times \mathbf{k} = \mathbf{i}, \qquad \mathbf{k} \times \mathbf{j} = -\mathbf{i},$
$\qquad\qquad \mathbf{k} \times \mathbf{i} = \mathbf{j}, \qquad \mathbf{i} \times \mathbf{k} = -\mathbf{j}.$

Thus the product, which is indicated by $\mathbf{v} \times \mathbf{v}'$, is

$$\mathbf{v} \times \mathbf{v}' = (bc' - b'c)\mathbf{i} + (ca' - ac')\mathbf{j} + (ab' - b'a)\mathbf{k}.$$

The vector product of two vectors is a vector. Its direction can readily be shown to be perpendicular to that of **v** and $\mathbf{v}'$ and pointed in the direction that a right-hand screw moves when it is turned from **v** to $\mathbf{v}'$ through the smaller angle. The vector product of two parallel vectors is zero, though neither factor is. Moreover, this product, like the quaternion product, is not commutative. Further, it is not even associative. For example, $\mathbf{i} \times \mathbf{j} \times \mathbf{j}$ can mean $(\mathbf{i} \times \mathbf{j}) \times \mathbf{j} = \mathbf{k} \times \mathbf{j} = -\mathbf{i}$ or $\mathbf{i} \times (\mathbf{j} \times \mathbf{j}) = \mathbf{i} \times 0 = 0$.

There is no inverse to vector multiplication. For the quotient of **v** by $\mathbf{v}'$ to be a vector **q** we would have to have

$$\mathbf{v} = \mathbf{v}' \times \mathbf{q}$$

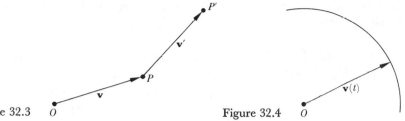

Figure 32.3 O Figure 32.4 O

and this would require, whatever $\mathbf{q}$ be, that $\mathbf{v}'$ be perpendicular to $\mathbf{v}$ which may not be the case to start with. If q were a scalar it would be accidental that $qa'\mathbf{i} + qb'\mathbf{j} + qc'\mathbf{k}$ would equal $\mathbf{v}$.

The vector product, like the scalar product, is suggested by physical situations. Let OP and PP' in Figure 32.3 be the lengths and directions of $\mathbf{v}$ and $\mathbf{v}'$. If $\mathbf{v}'$ is a force whose magnitude and direction are those of PP', the measure of the moment of force exerted by $\mathbf{v}'$ around O is the length of the vector $\mathbf{v} \times \mathbf{v}'$ and is usually taken to have the direction of $\mathbf{v} \times \mathbf{v}'$.

The algebra of vectors is extended to variable vectors and a calculus of vectors. For example, the variable vector $\mathbf{v}(t) = a(t)\mathbf{i} + b(t)\mathbf{j} + c(t)\mathbf{k}$, where $a(t)$, $b(t)$, and $c(t)$ are functions of t, is a vector function. If the various vectors one gets for various values of t are drawn from O as origin (Fig. 32.4) the endpoints of these vectors will trace out a curve. Hence the vector function of a scalar variable t plays a role analogous to the ordinary function

$$y = x^2 + 7,$$

say, and a calculus of vectors has been developed for these vector functions just as there is one for ordinary functions.

The concepts of gradient u,

$$(14) \qquad \nabla u = \frac{\partial u}{\partial x}\mathbf{i} + \frac{\partial u}{\partial y}\mathbf{j} + \frac{\partial u}{\partial z}\mathbf{k},$$

where u is a scalar function of x, y, and z, the divergence of a vector function $\mathbf{v}$,

$$(15) \qquad \nabla \cdot \mathbf{v} = \frac{\partial v_1}{\partial x} + \frac{\partial v_2}{\partial y} + \frac{\partial v_3}{\partial z},$$

where v_1, v_2, and v_3 are the components of $\mathbf{v}$, and the curl of $\mathbf{v}$,

$$(16) \qquad \nabla \times \mathbf{v} = \left(\frac{\partial v_3}{\partial y} - \frac{\partial v_2}{\partial z}\right)\mathbf{i} + \left(\frac{\partial v_1}{\partial z} - \frac{\partial v_3}{\partial x}\right)\mathbf{j} + \left(\frac{\partial v_2}{\partial x} - \frac{\partial v_1}{\partial y}\right)\mathbf{k},$$

were abstracted from quaternions.

Many basic theorems of analysis can be expressed in vector form. Thus in the course of solving the partial differential equation of heat,

Ostrogradsky[17] made use of the following conversion of volume integral to surface integral:

$$\iiint_V \left(\frac{\partial P}{\partial x} + \frac{\partial Q}{\partial y} + \frac{\partial R}{\partial z}\right) dx\, dy\, dz = \iint_S (P \cos \lambda + Q \cos \mu + R \cos \nu)\, dS.$$

Here P, Q, and R are functions of x, y, and z and are components of a vector, and λ, μ, and ν are the direction cosines of the normal to the surface S which bounds the volume V over which the left-hand integral is taken. This theorem, known as the divergence theorem (also Gauss's theorem and Ostrogradsky's theorem) can be expressed in vector form thus: If $\mathbf{F}$ is the vector whose components are P, Q, and R and $\mathbf{n}$ is the normal to S then

(17)
$$\iiint \nabla \cdot \mathbf{F}\, dv = \iint \mathbf{F} \cdot \mathbf{n}\, dS.$$

Likewise Stokes's theorem, which was first stated by him as a question in an examination for the Smith prize at Cambridge in 1854,[18] states in scalar form that

$$\iint_S \left\{\lambda\left(\frac{\partial R}{\partial y} - \frac{\partial Q}{\partial z}\right) + \mu\left(\frac{\partial P}{\partial z} - \frac{\partial R}{\partial x}\right) + \nu\left(\frac{\partial Q}{\partial x} - \frac{\partial P}{\partial y}\right)\right\} dS$$
$$= \int_c \left(P\frac{\partial x}{\partial s} + Q\frac{\partial y}{\partial s} + R\frac{\partial z}{\partial s}\right) ds,$$

where S is any portion of a surface, C is the curve bounding S, and $x(s), y(s)$, and $z(s)$ are the parametric representation of C. In vector form Stokes's theorem reads

(18)
$$\iint \operatorname{curl} \mathbf{F} \cdot \mathbf{n}\, dS = \int \mathbf{F} \cdot \frac{d\mathbf{r}}{ds}\, ds,$$

where $\mathbf{r}(s)$ is the vector whose components are $x(s), y(s)$, and $z(s)$.

When Maxwell wrote expressions and equations of electromagnetism, especially the equations which now bear his name, he usually wrote out the components of the vectors involved in grad u, div $\mathbf{v}$, and curl $\mathbf{v}$. However, Heaviside wrote Maxwell's equations in vector form (Chap. 28, (52)).

It is true that calculations with vectors and vector functions are often made by resort to Cartesian components but it is highly important to think also in terms of the single entity the vector and in terms of gradient, divergence, and curl. These have a direct physical significance, to say nothing of the fact that complicated technical steps can be performed directly with the

---

17. *Mém. Acad. Sci. St. Peters.*, (6), 1, 1831, 39–53.
18. The theorem was stated by Lord Kelvin in a letter to Stokes of July 1850.

vectors as when one replaces $\nabla \cdot (\mathbf{u}(x, y, z) \times \mathbf{v}(x, y, z))$ by its equivalent, $\mathbf{v} \cdot \nabla \times \mathbf{u} - \mathbf{u} \cdot \nabla \times \mathbf{v}$. Also integral definitions of gradient, divergence, and curl have been given which make these concepts independent of any coordinate definition. Thus in place of (14) we have, for example,

$$\operatorname{grad} u = \lim_{\Delta \tau \to 0} \frac{1}{\Delta \tau} \int_S u\mathbf{n} \, dS,$$

where S is the boundary of a volume element $\Delta \tau$ and $\mathbf{n}$ is the normal to the surface element dS of S.

While vector analysis was being created and afterward there was much controversy between the proponents of quaternions and the proponents of vectors as to which was more useful. The quaternionists were fanatical about the value of quaternions but the proponents of vector analysis were equally partisan. On one side were aligned the leading supporters of quaternions such as Tait and, on the other, Gibbs and Heaviside. Apropos of the controversy, Heaviside remarked sarcastically that for the treatment of quaternions, quaternions are the best instrument. On the other hand Tait described Heaviside's vector algebra as "a sort of hermaphrodite monster, compounded of the notations of Grassmann and Hamilton." Gibbs's book proved to be of inestimable value in promoting the vector cause.

The issue was finally decided in favor of vectors. Engineers welcomed Gibbs's and Heaviside's vector analysis, though the mathematicians did not. By the beginning of the present century the physicists too were quite convinced that vector analysis was what they wanted. Textbooks on the subject soon appeared in all countries and are now standard. The mathematicians finally followed suit and introduced vector methods in analytic and differential geometry.

The influence of physics in stimulating the creation of such mathematical entities as quaternions, Grassmann's hypernumbers, and vectors should be noted. These creations became part of mathematics. But their significance extends beyond the addition of new subjects. The introduction of these several quantities opened up a new vista—there is not just the one algebra of real and complex numbers but many and diverse algebras.

6. *Linear Associative Algebras*

From the purely algebraic standpoint quaternions were exciting because they furnished an example of an algebra that had the properties of real numbers and complex numbers except for commutativity of multiplication. During the second half of the nineteenth century many systems of hypernumbers were explored largely to see what varieties could be created and yet retain many properties of the real and complex numbers.

Cayley[19] gave an eight-unit generalization of real quaternions. His units were $1, e_1, e_2, \ldots, e_7$ with

$$e_i^2 = -1, \qquad e_i e_j = -e_j e_i \qquad \text{for } i, j = 1, 2, \cdots, 7 \quad \text{and} \quad i \neq j$$

$$e_1 e_2 = e_3, \qquad e_1 e_4 = e_5, \qquad e_1 e_6 = e_7$$

$$e_2 e_5 = e_7, \qquad e_2 e_4 = -e_6, \qquad e_3 e_4 = e_7, \qquad e_3 e_5 = e_6$$

and the fourteen equations obtained from these last seven by permuting each set of three subscripts cyclically; e.g. $e_2 e_3 = e_1$; $e_3 e_1 = e_2$.

A general number (octonion) x is defined by

$$x = x_0 + x_1 e_1 + \cdots + x_7 e_7,$$

wherein the x_i are real numbers. The norm of x, $N(x)$, is by definition

$$N(x) = x_0^2 + x_1^2 + \cdots + x_7^2.$$

The norm of a product equals the product of the norms. The associative law of multiplication fails in general (as does the commutative law of multiplication). Right- and left-hand division, except by zero, are always possible and unique, a fact overlooked by Cayley and proved by Leonard Eugene Dickson.[20] In later papers Cayley gave other algebras of hypernumbers differing somewhat from the one above.

Hamilton in his *Lectures on Quaternions*[21] also introduced biquaternions, that is, quaternions with complex coefficients. He noted that the product law does not hold for these biquaternions; that is, two non-zero biquaternions may have a product of zero.

William Kingdon Clifford (1845–79), professor of mathematics and mechanics at University College, London, created another type of hyper-number,[22] which he also called biquaternions. If q and Q are real quaternions and if ω is such that $\omega^2 = 1$, and ω commutes with every real quaternion, then $q + \omega Q$ is a biquaternion. Clifford's biquaternions satisfy the product law of multiplication but the multiplication is not associative. In the latter work Clifford introduced the algebras which bear his name. There are Clifford algebras with units, $1, e_1, \ldots, e_{n-1}$ such that the square of each $e_i = -1$ and $e_i e_j = -e_j e_i$ for $i \neq j$. Each product of two or more units is a new unit and so there are 2^n different units. All products are associative. A form is a scalar multiplied by a unit and an algebra is generated by the sum and product of forms.

The flood of new systems of hypernumbers continued to rise and the variety became enormous. In a paper "Linear Associative Algebra" read in

19. *Phil. Mag.*, (3), 26, 1845, 210–13 and 30, 1847, 257–58 = *Coll. Math. Papers*, 1, 127 and 301.

20. *Amer. Math. Soc. Trans.*, 13, 1912, 59–73.

21. 1853, p. 650.

22. *Proc. Lond. Math. Soc.*, 4, 1873, 381–95 = *Coll. Math. Papers*, 181–200 and *Amer. Jour. of Math.*, 1, 1878, 350–58 = *Coll. Math. Papers*, 266–76.

1870 and published in lithographed form in 1871,[23] Benjamin Peirce (1809–80), professor of mathematics at Harvard University, defined and gave a résumé of the linear associative algebras already known by that time. The word linear means that the product of any two primary units is reduced to one of the units just as i times j is replaced by k in quaternions and the word associative means that the multiplication is associative. Addition in these algebras has the usual properties of real and complex numbers. In this paper Peirce introduced the idea of a nilpotent element, that is, an element A such that $A^n = 0$ for some positive integral n, and an idempotent element, that is, $A^n = 1$ for some n. He also showed that an algebra in which at least one element is not nilpotent possesses an idempotent element.

The question of how much freedom there could be in the variety of such algebras had also occurred to mathematicians during the very period in which they were creating specific algebras. Gauss was convinced (*Werke*, 2, 178) that an extension of complex numbers that preserved the basic properties of complex numbers was impossible. It is significant that when Hamilton searched for a three-dimensional algebra to represent vectors in space and settled for quaternions with their lack of commutativity he could not prove that there was no three-dimensional commutative algebra. Nor did Grassman have such a proof.

Later in the century the precise theorems were formulated. In 1878 F. Georg Frobenius (1849–1917)[24] proved that the only linear associative algebras with real coefficients (of the primary units), with a finite number of primary units, a unit element for multiplication, and obeying the product law are those of the real numbers, the complex numbers and real quaternions. The theorem was proved independently by Charles Sanders Peirce (1839–1914) in an appendix to his father's paper.[25] Another key result, which Weierstrass arrived at in 1861, states that the only linear associative algebras with real or complex coefficients (of the primary units), with a finite number of primary units, obeying the product law and with commutative multiplication are those of the real numbers and the complex numbers. Dedekind obtained the same result about 1870. Weierstrass's result was published in 1884[26] and Dedekind's in the next year.[27]

In 1898 Adolf Hurwitz (1859–1919)[28] showed that the real numbers, complex numbers, real quaternions, and Clifford's biquaternions are the only linear associative algebras satisfying the product law.

These theorems are valuable because they tell us what we can expect in

23. *Amer. Jour. of Math.*, 4, 1881, 97–229.
24. *Jour. für Math.*, 84, 1878, 1–63 = *Ges. Abh.*, 1, 343–405.
25. *Amer. Jour. of Math.*, 4, 1881, 225–29.
26. *Nachrichten König. Ges. der Wiss. zu Gött.*, 1884, 395–410 = *Math. Werke*, 2, 311–32.
27. *Nachrichten König. Ges. der Wiss. zu Gött.*, 1885, 141–59 and 1887, 1–7 = *Werke*, 2, 1–27.
28. *Nachrichten König. Ges. der Wiss. zu Gött.*, 1898, 309–16 = *Math. Werke*, 2, 565–71.

extensions of the complex number system if we wish to preserve at least some of its algebraic properties. Had Hamilton known these theorems he would have saved years of labor in his search for a three-dimensional vector algebra.

The study of linear algebras with a finite and even infinite number of generating (primary) units and with or without division continued to be an active subject well into the twentieth century. Such men as Leonard Eugene Dickson and J. H. M. Wedderburn contributed much to the subject.

Bibliography

Clifford, W. K.: *Collected Mathematical Papers* (1882), Chelsea (reprint), 1968.

Collins, Joseph V.: "An Elementary Exposition of Grassmann's *Ausdehnungslehre*," *Amer. Math. Monthly*, 6, 1899, several parts; and 7, 1900, several parts.

Coolidge, Julian L.: *A History of Geometrical Methods*, Dover (reprint), 1963, pp. 252–64.

Crowe, Michael J.: *A History of Vector Analysis*, University of Notre Dame Press, 1967.

Dickson, Leonard E.: *Linear Algebras*, Cambridge University Press, 1914.

Gibbs, Josiah W., and E. B. Wilson: *Vector Analysis* (1901), Dover (reprint), 1960.

Grassmann, H. G.: *Die lineale Ausdehnungslehre* (1844), Chelsea (reprint), 1969.

——: *Gesammelte mathematische und physikalische Werke*, 3 vols., B. G. Teubner, 1894–1911; Vol. 1, Part I, 1–319 contains *Die lineale Ausdehnungslehre*; Vol. 1, Part II, 1–383 contains *Die Ausdehnungslehre*.

Graves, R. P.: *Life of Sir William Rowan Hamilton*, 3 vols., Longmans Green, 1882–89.

Hamilton, Sir Wm. R.: *Elements of Quaternions*, 2 vols., 1866, 2nd ed., 1899–1901, Chelsea (reprint), 1969.

——: *Mathematical Papers*, Cambridge University Press, 1967, Vol. 3.

——: "Papers in Memory of Sir William R. Hamilton," *Scripta Math.*, 1945; also in *Scripta Math.*, 10, 1944, 9–80.

Heaviside, Oliver: *Electromagnetic Theory*, Dover (reprint), 1950, Vol. 1.

Klein, Felix: *Vorlesungen über die Entwicklung der Mathematik im 19. Jahrhundert*, Chelsea (reprint), 1950, Vol. 1, pp. 167–91; Vol. 2, pp. 2–12.

Maxwell, James Clerk: *The Scientific Papers*, 2 vols., Dover (reprint), 1965.

Peacock, George: "Report on the Recent Progress and Present State of Certain Branches of Analysis," *British Assn. for Advancement of Science Report for 1833*, London, 1834.

——: *A Treatise on Algebra*, 2 vols., 2nd ed., Cambridge University Press, 1845; Scripta Mathematica (reprint), 1940.

Shaw, James B.: *Synopsis of Linear Associative Algebra*, Carnegie Institution of Washington, 1907.

Smith, David Eugene: *A Source Book in Mathematics*, Dover (reprint), 1959, Vol. 2, pp. 677–96.

Study, E.: "Theorie der gemeinen und höheren complexen Grössen," *Encyk. der Math. Wiss.*, B. G. Teubner, 1898, I, 147–83.

33
Determinants and Matrices

> Such is the advantage of a well-constructed language that its
> simplified notation often becomes the source of profound
> theories.
> P. S. LAPLACE

1. *Introduction*

Though determinants and matrices received a great deal of attention in the
nineteenth century and thousands of papers were written on these subjects
they do not constitute great innovations in mathematics. The concept of a
vector, which from the mathematical standpoint is no more than a collection
of ordered triples, nevertheless has a direct physical significance as a force or
velocity, and with it one can write down at once mathematically what the
physics states. The same applies with all the more cogency to gradient,
divergence, and curl. Likewise though mathematically dy/dx is no more than
a symbol for a lengthy expression involving the limit of $\Delta y/\Delta x$, the derivative
is in itself a powerful concept that enables us to think directly and creatively
about physical happenings. Thus though mathematics superficially regarded
is no more than a language or a shorthand, its most vital concepts are those
that supply keys to new realms of thought. By contrast, determinants and
matrices are solely innovations in language. They are shorthand expressions
for ideas which already exist in more expanded form. In themselves they say
nothing directly that is not already said by equations or transformations
albeit in lengthier fashion. Neither determinants nor matrices have influenced
deeply the course of mathematics despite their utility as compact expressions
and despite the suggestiveness of matrices as concrete groups for the discern-
ment of general theorems of group theory. Nevertheless, both concepts have
proved to be highly useful tools and are now part of the apparatus of
mathematics.

2. *Some New Uses of Determinants*

Determinants arose in the solution of systems of linear equations (Chap. 25,
sec. 3). This problem and elimination theory, transformation of coordinates,

795

change of variables in multiple integrals, solution of systems of differential equations arising in planetary motion, reduction of quadratic forms in three or more variables and of pencils of forms (a pencil is the set $A + \lambda B$, where A and B are specific forms and λ is a parameter) to standard forms all gave rise to various new uses of determinants. This nineteenth-century work followed up directly on the work of Cramer, Bezout, Vandermonde, Lagrange, and Laplace.

The word determinant, used by Gauss for the discriminant of the quadratic form $ax^2 + 2bxy + cy^2$, was applied by Cauchy to the determinants that had already appeared in the eighteenth-century work. The arrangement of the elements in a square array and the double subscript notation are also due to him.[1] Thus a third order determinant is written as (the vertical lines were introduced by Cayley in 1841)

$$(1) \qquad \begin{vmatrix} a_{11} & a_{12} & a_{12} \\ a_{21} & a_{22} & a_{23} \\ a_{31} & a_{32} & a_{33} \end{vmatrix}.$$

In this paper Cauchy gave the first systematic and almost modern treatment of determinants. One of the major results is the multiplication theorem for determinants. Lagrange[2] had already given this theorem for third order determinants but because the rows of his determinant were the coordinates of the vertices of a tetrahedron he was not motivated to generalize. With Cauchy the general theorem, expressed in modern notation, states

$$(2) \qquad |a_{ij}| \cdot |b_{ij}| = |c_{ij}|,$$

where $|a_{ij}|$ and $|b_{ij}|$ stand for nth-order determinants and $c_{ij} = \sum_k a_{ik} b_{kj}$. That is, the term in the ith row and jth column of the product is the sum of the products of the corresponding elements in the ith row of $|a_{ij}|$ and the jth column of $|b_{ij}|$. This theorem had been stated but not satisfactorily proved by Jacques P. M. Binet (1786–1856) in 1812.[3] Cauchy also gave an improved statement and a proof of Laplace's expansion theorem for determinants (Chap. 25, sec. 3).

Heinrich F. Scherk (1798–1885) in his *Mathematische Abhandlungen* (1825) gave several new properties of determinants. He formulated the rules for adding two determinants that have a row or column in common and for multiplying a determinant by a constant. He also stated that the determinant of an array which has as a row a linear combination of two or more other rows is zero, and that the value of a triangular determinant (all the elements

1. *Jour. de l'Ecole Poly.*, 10, 1815, 29–112 = *Œuvres*, (2), 1, 91–169.
2. *Nouv. Mém. de l'Acad. de Berlin*, 1773, 85–128 = *Œuvres*, 3, 577–616.
3. *Jour. de l'Ecole Poly.*, 9, 1813, 280–302.

above or below the main diagonal are zero) is the product of the elements on the main diagonal.

One of the consistent workers in determinant theory over a period of more than fifty years was James Joseph Sylvester (1814–97). After winning the second wranglership in the mathematical tripos he was nevertheless barred from teaching at Cambridge University because he was Jewish. From 1841 to 1845 he was a professor at the University of Virginia. He then returned to London and served from 1845 to 1855 as an actuary and a lawyer. The relatively lowly position of professor in a military academy in Woolwich, England, was offered to him and he served there until 1871. A few years of miscellaneous activity were followed by a professorship at Johns Hopkins University where, starting in 1876, he lectured on invariant theory. He initiated research in pure mathematics in the United States and founded the *American Journal of Mathematics*. In 1884 he returned to England, and at the age of seventy became a professor at Oxford University, a post he held until his death.

Sylvester was a lively, sensitive, stimulating, passionate, and even excitable person. His speeches were brilliant and witty and he presented his ideas enthusiastically and with fire. In his papers he used glowing language. He introduced much new terminology and jokingly likened himself to Adam who gave names to the beasts and the flowers. Though he related many diverse fields, such as mechanics and the theory of invariants, he was not given to systematic and thoroughly worked-out theories. In fact he frequently published guesses, and though many of these were brilliant others were incorrect. He ruefully acknowledged that his friends on the Continent "complimented his powers of divination at the expense of his judgment." His chief contributions were to combinatorial ideas and to abstractions from more concrete developments.

One of Sylvester's major accomplishments was an improved method of eliminating x from an nth degree and an mth degree polynomial. He called it the dialytic method.[4] Thus to eliminate x from the equations

$$a_0x^3 + a_1x^2 + a_2x + a_3 = 0$$
$$b_0x^2 + b_1x + b_2 = 0$$

he formed the fifth order determinant

(3)
$$\begin{vmatrix} a_0 & a_1 & a_2 & a_3 & 0 \\ 0 & a_0 & a_1 & a_2 & a_3 \\ b_0 & b_1 & b_2 & 0 & 0 \\ 0 & b_0 & b_1 & b_2 & 0 \\ 0 & 0 & b_0 & b_1 & b_2 \end{vmatrix}.$$

4. *Phil. Mag.*, 16, 1840, 132–35 and 21, 1842, 534–39 = *Coll. Math. Papers*, 1, 54–57 and 86–90.

The vanishing of this determinant is the necessary and sufficient condition that the two equations have a common root. Sylvester gave no proof. The method leads, as Cauchy showed,[5] to the same result as Euler's and Bezout's methods.

The formula for the derivative of a determinant when the elements are functions of t was first given by Jacobi in 1841.[6] If the a_{ij} are functions of t, A_{ij} is the cofactor of a_{ij}, and D is the determinant then

$$\frac{\partial D}{\partial a_{ij}} = A_{ij}$$

and

$$\frac{dD}{dt} = \sum_{i,j} A_{ij} a'_{ij},$$

where the prime denotes differentiation with respect to t.

Determinants were employed in another connection, namely, in the change of variables in a multiple integral. Special results were found first by Jacobi (1832 and 1833). Then Eugène Charles Catalan (1814–94) in 1839[7] gave the result familiar today to students of the calculus. Thus the double integral,

$$(4) \qquad\qquad \iint F(x, y)\, dx\, dy$$

under the change of variables given by

$$(5) \qquad\qquad x = f(u, v), \qquad y = g(u, v)$$

becomes

$$(6) \qquad\qquad \iint G(u, v) \begin{vmatrix} f_u & f_v \\ g_u & g_v \end{vmatrix} du\, dv,$$

where $G(u, v) = F(x(u, v), y(u, v))$. The determinant in (6) is called the Jacobian of x and y with respect to u and v or the functional determinant.

Jacobi devoted a major paper[8] to functional determinants. In this paper Jacobi considers n functions $u_1, \ldots, u_n$ each of which is a function of $x_1, x_2, \ldots, x_n$ and raises the question of when, from such n functions, the x_i can be eliminated so that the u_i's are connected by one equation. If this is not possible the functions u_i are said to be independent. The answer is that

5. *Exercices d'analyse et de physique mathématique*, 1, 1840, 385–422 = *Œuvres*, (2), 11, 466–509.
6. *Jour. für Math.*, 22, 1841, 285–318 = *Werke*, 3, 355–92.
7. *Mémoires couronnés par l'Académie Royale des Sciences et Belles-Lettres de Bruxelles*, 14, 1841.
8. *Jour. für Math.*, 22, 1841, 319–59 = *Werke*, 3, 393–438.

if the Jacobian of the u_i with respect to the x_i vanishes the functions are not independent and conversely. He also gives the product theorem for Jacobians. That is, if the u_i are functions of y_i and the y_i are functions of the x_i, then the Jacobian of the u_i with respect to the x_i is the product of the Jacobian of the u_i with respect to the y_i and the Jacobian of the y_i with respect to the x_i.

3. *Determinants and Quadratic Forms*

The problem of transforming equations of the conic sections and quadric surfaces to simpler forms by choosing coordinate axes which have the directions of the principal axes had been introduced in the eighteenth century. The classification of quadric surfaces in terms of the signs of the second degree terms when the equation is in standard or canonical form, that is, when the principal axes are the coordinate axes, was given by Cauchy in his *Leçons sur les applications du calcul infinitésimal à la géométrie* (1826).[9] However, it was not clear that the same number of positive and negative terms always results from this reduction to standard form. Sylvester answered this question with his law of inertia of quadratic forms in n variables.[10] It was already known that

$$(7) \qquad \sum_{i,j=1}^{n} a_{ij}x_ix_j$$

can always be reduced to a sum of r squares[11]

$$(8) \qquad y_1^2 + \cdots + y_s^2 - y_{s+1}^2 - \cdots - y_{r-s}^2$$

by a real linear transformation

$$x_i = \sum_j b_{ij}y_j, \qquad i = 1, 2, \cdots, n,$$

with non-vanishing determinant. Sylvester's law states that the number s of positive terms and $r - s$ negative ones is always the same no matter what real transformation is used. Regarding the law as self-evident, Sylvester gave no proof.

The law was rediscovered and proved by Jacobi.[12] If a form is positive or zero for all real values of the variables, it is called positive definite. Then all the signs in (8) are positive, and $r = n$. It is called semidefinite if it can take on positive and negative values (in which case $r < n$), and negative

9. *Œuvres*, (2), 5, 244–85.
10. *Phil. Mag.*, (4), 4, 1852, 138–42 = *Coll. Math. Papers*, 1, 378–81.
11. r is the rank of the form, that is, the rank of the matrix of the coefficients. For the notion of rank see section 4.
12. *Jour. für Math.*, 53, 1857, 265–70 = *Werke*, 3, 583–90; see also 593–98.

definite when the form is always negative or zero (and $r = n$). These terms were introduced by Gauss in his *Disquisitiones Arithmeticae* (sec. 271).

The further study of the reduction of quadratic forms involves the notion of the characteristic equation of a quadratic form or of a determinant. A quadratic form in three variables was usually written in the eighteenth century and first half of the nineteenth century as

$$Ax^2 + By^2 + Cz^2 + 2Dxy + 2Exz + 2Fyz$$

and in more recent times as

(9) $a_{11}x_1^2 + a_{22}x_2^2 + a_{33}x_3^2 + 2a_{12}x_1x_2 + 2a_{13}x_1x_3 + 2a_{22}x_2x_3.$

In the latter notation the form has associated with it the determinant

(10) $\begin{vmatrix} a_{11} & a_{12} & a_{13} \\ a_{21} & a_{22} & a_{23} \\ a_{31} & a_{32} & a_{33} \end{vmatrix},$ $a_{ij} = a_{ji}.$

The characteristic equation or latent equation of the form or the determinant is

(11) $\begin{vmatrix} a_{11} - \lambda & a_{12} & a_{13} \\ a_{21} & a_{22} - \lambda & a_{23} \\ a_{31} & a_{32} & a_{33} - \lambda \end{vmatrix} = 0,$

and the values of λ which satisfy this equation are called the characteristic roots or latent roots. From these values of λ the lengths of the principal axes are readily obtained.[13]

The notion of the characteristic equation appears implicitly in the work of Euler[14] on the reduction of quadratic forms in three variables to their principal axes though he failed to prove the reality of the characteristic roots. The notion of characteristic equation first appears explicitly in Lagrange's work on systems of linear differential equations,[15] and in Laplace's work in the same area.[16]

Lagrange, in dealing with the system of differential equations for the motion of the six planets known in his day, was concerned with the secular (long-period) perturbations which these exerted on each other. His characteristic equation (also called the secular equation) was that for a sixth order

13. By a linear transformation $x_i' = \sum_j m_{ij}x_j$, $i, j = 1, 2, 3$, the form (9) can be reduced to $\sum_{j=1}^{3} \lambda_j x_j'^2$ where the λ_j are the characteristic roots of (11). In matrix language the matrix M of the transformation is orthogonal; that is, the transpose of M equals the inverse of M.

14. Chapter 5 of the Appendix to his *Introductio* (1748) = *Opera*, (1), 9, 379–92.

15. *Misc. Taur.*, 3, 1762–65 = *Œuvres*, 1, 520–34, and *Mém. de l'Acad. des Sci., Paris*, 1774 = *Œuvres*, 6, 655–66.

16. *Mém. de l'Acad. des Sci., Paris*, 1772, pub. 1775 = *Œuvres*, 8, 325–66.

determinant and the values of λ determined solutions of the system. He was able to decompose the sixth degree equation and obtain information about the roots. Laplace in his *Mécanique céleste* showed that if the planets all move in the same direction then the six characteristic roots are real and distinct. The reality of the characteristic values for the quadratic problem in three variables was established by Hachette, Monge, and Poisson.[17]

Cauchy recognized the common characteristic value problem in the work of Euler, Lagrange, and Laplace. In his *Leçons* of 1826[18] he took up the problem of the reduction of a quadratic form in three variables and showed that the characteristic equation is invariant for any change of rectangular axes. Three years later in his *Exercices de mathématiques*[19] he took up the problem of the secular inequalities of the planetary paths. In the course of this work he showed that two quadratic forms in n variables

$$A = \sum_{i,j=1}^{n} a_{ij}x_ix_j, \qquad B = \sum_{i,j=1}^{n} b_{ij}x_ix_j$$

(Cauchy's B was a sum of squares) could be reduced simultaneously by a linear transformation

$$x_1 = c_{11}x_1' + \cdots + c_{1n}x_n'$$
$$\cdot \quad \cdot \quad \cdot \quad \cdot \quad \cdot \quad \cdot \quad \cdot$$
$$x_n = c_{n1}x_1' + \cdots + c_{nn}x_n'$$

to a sum of squares. He also solved the problem of finding the principal axes for forms in any number of variables and in this work he again used the notion of characteristic roots.

His work amounts to the following: If A and B are any two given quadratic forms then one can consider the pencil (family, *Schaar*) of forms $uA + vB$, where u and v are arbitrary parameters. The latent roots of the pencil are the values of the ratio $-u/v$ for which the determinant of the pencil $|ua_{ij} + vb_{ij}|$ is zero. Cauchy proved that the latent roots are all real in the special case when one of the forms is positive definite for all real non-zero values of the variables. Since the determinant of $uA + vB$ is symmetric ($d_{ij} = d_{ji}$) and B could be the identity determinant ($b_{ij} = 0$ for $i \neq j$ and $b_{ii} = 1$), Cauchy proved that any real symmetric determinant of any order has real characteristic roots. Cauchy's results, duplicated by Jacobi in 1834,[20] excluded equal latent roots. The term characteristic equation is due to Cauchy.[21]

17. Hachette and Monge: *Jour. de l'Ecole Poly.*, 4, 1801–2, 143–69; Poisson and Hachette, *ibid.*, 170–72.

18. *Œuvres*, (2), 5, 244–85.

19. 4, 1829, 140–60 = *Œuvres*, (2), 9, 174–95.

20. *Jour. für Math.*, 12, 1834, 1–69 = *Werke*, 3, 191–268.

21. *Exercices d'analyse et de physique mathématique*, 1, 1840, 53 = *Œuvres*, (2), 11, 76.

The notion of similar determinants also arose from work on transformations. Two determinants A and B are similar if there exists a non-zero determinant P such that $A = P^{-1}BP$. Similar transformations were considered by Cauchy who showed in his *Leçons* of 1826 [22] that they have the same characteristic values. The importance of similar transformations lies in classifying projective transformations (Chap. 38, sec. 5), a problem that for a long time was treated synthetically. If a figure F is related to a figure G by a linear transformation A and if another such transformation B transforms F into F' and G into G' the transformation C which carries F' into G' will have the same properties as A. The transformation $C = BAB^{-1}$ because B^{-1} carries F' into F, A carries F into G, and B carries G into G'.

In 1858 Weierstrass [23] gave a general method of reducing two quadratic forms simultaneously to sums of squares. He also proved that if one of the forms is positive definite the reduction is possible even when some of the latent roots are equal. Weierstrass's interest in this problem arose from the dynamical problem of small oscillations about a position of equilibrium, and he showed by means of his work on quadratic forms that stability is not destroyed by the presence of equal periods in the system, contrary to the suppositions of Lagrange and Laplace.

Sylvester in 1851,[24] working with the contact and the intersection of curves and surfaces of the second degree was led to consider the classification of pencils of such conics and quadric surfaces. In particular he sought the canonical form of any pencil. Writing a pencil in the form $A + \lambda B$ where

$$A = ax^2 + by^2 + cz^2 + 2dyz + 2ezx + 2fxy$$
$$B = Ax^2 + By^2 + Cz^2 + 2Dyz + 2Ezx + 3Fxy$$

he considered the determinant

(12)
$$\begin{vmatrix} a + \lambda A & f + \lambda F & e + \lambda E \\ f + \lambda F & b + \lambda B & d + \lambda D \\ e + \lambda E & d + \lambda D & c + \lambda C \end{vmatrix}.$$

His method of classification introduced the notion of elementary divisors. The elements of the determinant of $A + \lambda B$ are polynomials in λ. Sylvester proved that if all the minors of any one order of $|A + \lambda B|$ have a factor $\lambda + \varepsilon$ in common, this factor will continue to be common to the same system of minors when A and B are simultaneously transformed by a linear transformation of their variables. He also showed that if all the ith order minors have a factor $(\lambda + \varepsilon)^\alpha$, the $(i + h)$-order minors will contain the factor $(\lambda + \varepsilon)^{(h+1)\alpha}$. The various linear factors to the power to which they

22. *Œuvres*, (2), 5, 244–85.
23. *Monatsber. Berliner Akad.*, 1858, 207–20 = *Werke*, 1, 233–46.
24. *Phil. Mag.*, (4), 1, 1851, 119–40 = *Coll. Math. Papers*, 1, 219–40.

occur in the greatest common divisor $D_i(\lambda)$ of the ith order minors for each i are the *elementary divisors* of $|A + \lambda B|$ or of any general determinant. The quotients of $D_i(\lambda)$ by $D_{i-1}(\lambda)$ for each i are called the invariant factors of $|A + \lambda B|$. Sylvester did not prove that the invariant factors constitute a complete set of invariants for the two quadratic forms.

Weierstrass[25] completed the theory of quadratic forms and extended the theory to bilinear forms, a bilinear form being

$$a_{11}x_1y_1 + a_{12}x_1y_2 + \cdots + a_{nn}x_ny_n.$$

Using Sylvester's notion of elementary divisors Weierstrass obtained the canonical form for a pencil $A + \lambda B$, where A and B are not necessarily symmetric but subject to the condition that $|A + \lambda B|$ is not identically zero. He also proved the converse of a theorem due to Sylvester. The converse states that if the determinant of $A + \lambda B$ agrees in its elementary divisors with the determinant of $A' + \lambda B'$ then a pair of linear transformations can be found which will simultaneously transform A into A' and B into B'.

Among the multitude of theorems on determinants are some concerned with the solution of m linear equations in n unknowns. Henry J. S. Smith (1826–83)[26] introduced the terms augmented array and unaugmented array to discuss the existence and number of solutions of, for example, the system

$$a_1x + a_2y = f$$
$$b_1x + b_2y = g$$
$$c_1x + c_2y = h.$$

The augmented and unaugmented arrays are

$$\left\|\begin{matrix} a_1 & a_2 & f \\ b_1 & b_2 & g \\ c_1 & c_2 & h \end{matrix}\right\| \quad \text{and} \quad \left\|\begin{matrix} a_1 & a_2 \\ b_1 & b_2 \\ c_1 & c_2 \end{matrix}\right\|.$$

A series of results due to many men including Kronecker and Cayley led to the general results now usually stated in terms of matrices but which in the middle of the nineteenth century were stated in terms of augmented and unaugmented determinants. The general results on m equations in n unknowns with m greater than, equal to, or less than n—the equations can be homogeneous (the constant terms are zero) or nonhomogeneous—are stated, for example, in Charles L. Dodgson's (Lewis Carroll, 1832–1898) *An Elementary Theory of Determinants* (1867). In later texts one finds the present condition: In order that a set of m nonhomogeneous linear equations in n unknowns may be consistent it is necessary and sufficient that the highest order non-vanishing minor be of the same order in the unaugmented and augmented arrays, or in terms of matrix language that the ranks of the two matrices be the same.

New results on determinants were obtained throughout the nineteenth century. As an illustration there is the theorem proved by Hadamard in 1893,[27] though known and proved by many others before and after this date. If the elements of the determinant $D = |a_{ij}|$ satisfy the condition $|a_{ij}| \leq A$ then

$$|D| \leq A^n \cdot n^{n/2}.$$

The above theorems on determinants are but a small sample of the multitude that have been established. Beyond a great variety of other theorems on general determinants there are hundreds of others on determinants of special form such as symmetric determinants ($a_{ij} = a_{ji}$), skew-symmetric determinants ($a_{ij} = -a_{ji}$), orthogonants (determinants of orthogonal coordinate transformations), bordered determinants (determinants extended by the addition of rows and columns), compound determinants (the elements are themselves determinants), and many other special types.

4. Matrices

One could say that the subject of matrices was well developed before it was created. Determinants, as we know, were studied from the middle of the eighteenth century onward. A determinant contains an array of numbers and usually one is concerned with the value of that array, given by the definition of the determinant. However, it was apparent from the immense amount of work on determinants that the array itself could be studied and manipulated for many purposes whether or not the value of the determinant came into question. It remained then to recognize that the array as such should be given an identity independent of the determinant. The array itself is called a matrix. The word matrix was first used by Sylvester[28] when in fact he wished to refer to a rectangular array of numbers and could not use the word determinant, though he was at that time concerned only with the determinants that could be formed from the elements of the rectangular array. Later, as we have already noted in the preceding section, augmented arrays were used freely without any mention of matrices. The basic properties of matrices, as we shall see, were also established in the development of determinants.

It is true, as Arthur Cayley insisted (in the 1855 paper referred to below), that logically the idea of a matrix precedes that of a determinant but historically the order was the reverse and this is why the basic properties of

27. *Bull. des Sci. Math.*, (2), 17, 1893, 240–46 = *Œuvres*, 1, 239–45.
28. *Phil. Mag.*, (3), 37, 1850, 363–70 = *Coll. Math. Papers*, 1, 145–51.

matrices were already clear by the time that matrices were introduced. Thus the general impression among mathematicians that matrices were a highly original and independent creation invented by pure mathematicians when they divined the potential usefulness of the idea is erroneous. Because the uses of matrices were well established it occurred to Cayley to introduce them as distinct entities. He says, "I certainly did not get the notion of a matrix in any way through quaternions; it was either directly from that of a determinant or as a convenient way of expression of the equations:

$$x' = ax + by$$
$$y' = cx + dy."$$

And so he introduced the matrix

$$\begin{pmatrix} a & b \\ c & d \end{pmatrix}$$

which represents the essential information about the transformation. Because Cayley was the first to single out the matrix itself and the first to publish a series of articles on the subject he is generally credited with being the creator of the theory of matrices.

Cayley, born in 1821 of an old and talented English family, showed mathematical ability in school. His teachers persuaded his father to send him to Cambridge instead of putting him in the family business. At Cambridge he was senior wrangler in the mathematical tripos and won the Smith prize. He was elected a fellow of Trinity College in Cambridge and assistant tutor, but left after three years because he would have had to take holy orders. He turned to law and spent the next fifteen years in that profession. During this period he managed to devote considerable time to mathematics and published close to 200 papers. It was during this period, too, that he began his long friendship and collaboration with Sylvester.

In 1863, he was appointed to the newly created Sadlerian professorship of mathematics at Cambridge. Except for the year 1882, spent at Johns Hopkins University at the invitation of Sylvester, he remained at Cambridge until his death in 1895. He was a prolific writer and creator in various subjects, notably the analytic geometry of n dimensions, determinant theory, linear transformations, skew surfaces, and matrix theory. Together with Sylvester, he was the founder of the theory of invariants. For these numerous contributions he received many honors.

Unlike Sylvester, Cayley was a man of even temper, sober judgment, and serenity. He was generous in help and encouragement to others. In addition to his fine work in law and prodigious accomplishments in mathematics, he found time to develop interests in literature, travel, painting, and architecture.

It was in connection with the study of invariants under linear transformations (Chap. 39, sec. 2) that Cayley first introduced matrices to simplify
the notation involved.[29] Here he gave some basic notions. This was followed
by his first major paper on the subject, "A Memoir on the Theory of
Matrices."[30]

For brevity we shall state Cayley's definitions for 2 by 2 or 3 by 3
matrices though the definitions apply to n by n matrices and in some cases to
rectangular matrices. Two matrices are equal if their corresponding elements
are equal. Cayley defines the zero matrix and unit matrix as

$$
\begin{pmatrix} 0 & 0 & 0 \\ 0 & 0 & 0 \\ 0 & 0 & 0 \end{pmatrix} \quad \text{and} \quad \begin{pmatrix} 1 & 0 & 0 \\ 0 & 1 & 0 \\ 0 & 0 & 1 \end{pmatrix}.
$$

The sum of two matrices is defined as the matrix whose elements are the
sums of the corresponding elements of the two summands. He notes that this
definition applies to any two m by n matrices and that addition is associative
and convertible (commutative). If m is a scalar and A a matrix then mA is
defined as the matrix whose elements are each m times the corresponding
element of A.

The definition of multiplication of two matrices Cayley took directly
from the representation of the effect of two successive transformations. Thus
if the transformation

$$
\begin{aligned}
x' &= a_{11}x + a_{12}y \\
y' &= a_{21}x + a_{22}y
\end{aligned}
$$

is followed by the transformation

$$
\begin{aligned}
x'' &= b_{11}x' + b_{12}y' \\
y'' &= b_{21}x' + b_{22}y'
\end{aligned}
$$

then the relation between x'', y'' and x, y is given by

$$
\begin{aligned}
x'' &= (b_{11}a_{11} + b_{12}a_{21})x + (b_{11}a_{12} + b_{12}a_{22})y \\
y'' &= (b_{21}a_{11} + b_{22}a_{21})x + (b_{21}a_{12} + b_{22}a_{22})y.
\end{aligned}
$$

Hence Cayley defined the product of the matrices to be

$$
\begin{pmatrix} b_{11}b_{12} \\ b_{21}b_{22} \end{pmatrix} \begin{pmatrix} a_{11}a_{12} \\ a_{21}a_{22} \end{pmatrix} = \begin{pmatrix} b_{11}a_{11} + b_{12}a_{21} & b_{11}a_{12} + b_{12}a_{22} \\ b_{21}a_{11} + b_{22}a_{21} & b_{21}a_{12} + b_{22}a_{22} \end{pmatrix}.
$$

That is, the element c_{ij} in the product is the sum of the products of the
elements in the ith row of the left-hand factor and the corresponding

29. *Jour. für Math.*, 50, 1855, 282–85 = *Coll. Math. Papers*, 2, 185–88.
30. *Phil. Trans.*, 148, 1858, 17–37 = *Coll. Math. Papers*, 2, 475–96.

elements of the *j*th column of the right-hand factor. Multiplication is associative but not generally commutative. Cayley points out that an *m* by *n* matrix can be compounded only with an *n* by *p* matrix.

In this same article, he states that the inverse of

$$
\begin{pmatrix} a, & b, & c \\ a', & b', & c' \\ a'', & b'', & c'' \end{pmatrix} \quad \text{is given by} \quad \frac{1}{\nabla} \begin{pmatrix} \partial_a\nabla, & \partial_{a'}\nabla, & \partial_{a''}\nabla \\ \partial_b\nabla, & \partial_{b'}\nabla, & \partial_{b''}\nabla \\ \partial_c\nabla, & \partial_{c'}\nabla, & \partial_{c''}\nabla \end{pmatrix},
$$

where ∇ is the determinant of the matrix and $\partial_x\nabla$ is the co-factor of x in this determinant, that is, the minor of x with the proper sign. The product of a matrix and its inverse is the unit matrix, denoted by I.

When $\nabla = 0$, the matrix is indeterminate (singular, in modern terminology) and has no inverse. Cayley asserted that the product of two matrices may be zero without either being zero, if either one is indeterminate. Actually, Cayley was wrong; both matrices must be indeterminate. For if $AB = 0$, $A \neq 0$, $B \neq 0$, and only A is indeterminate, then the inverse of B, namely B^{-1}, exists, and $ABB^{-1} = O \cdot B^{-1} = 0$. But $BB^{-1} = I$. Therefore, $AI = 0$ or $A = 0$.

The transverse (transposed or conjugate) matrix is defined as the one in which rows and columns are interchanged. The statement is made (without proof) that $(LMN)' = N'M'L'$, where prime denotes transpose. If $M' = M$ then M is called symmetric and if $M' = -M$ then M is skew-symmetric (or alternating). Any matrix can be expressed as the sum of a symmetrical matrix and a skew-symmetrical one.

Another concept carried over from determinant theory is the characteristic equation of a square matrix. For the matrix M it is defined to be

$$
|M - xI| = 0,
$$

where $|M - xI|$ is the determinant of the matrix $M - xI$ and I is the unit matrix. Thus if

$$
M = \begin{pmatrix} a & b \\ c & d \end{pmatrix},
$$

the characteristic equation (Cayley does not use the term though it was introduced for determinants by Cauchy [sec. 3]) is

(13) $$x^2 - (a + d)x + (ad - bc) = 0.$$

The roots of this equation are the characteristic roots (eigenvalues) of the matrix.

In the 1858 paper Cayley announced what is now called the Cayley-Hamilton theorem for square matrices of any order. The theorem states that if M is substituted for x in (13) the resulting matrix is the zero matrix.

Cayley states that he has verified the theorem for the 3 by 3 case and that further proof is not necessary. Hamilton's association with this theorem rests on the fact that in introducing in his *Lectures on Quaternions*[31] the notion of a linear vector function $\mathbf{r}'$ of another vector $\mathbf{r}$, a linear transformation from x, y, and z to x', y', and z' is involved. He proved that the matrix of this transformation satisfied the characteristic equation of that matrix, though he did not think formally in terms of matrices.

Other mathematicians found special properties of the characteristic roots of classes of matrices. Hermite[32] showed that if the matrix $M = M*$, where $M*$ is the transpose of the matrix formed by replacing each element of M by its complex conjugate (such M's are now called Hermitian), then the characteristic roots are real. In 1861 Clebsch[33] deduced from Hermite's theorem that the non-zero characteristic roots of a real skew-symmetric matrix are pure imaginaries. Then Arthur Buchheim[34] (1859–88) demonstrated that if M is symmetric and the elements are real, the characteristic roots are real, though this result was already established for determinants by Cauchy.[35] Henry Taber (1860–?) in another paper[36] asserted as evident that if

$$x^n - m_1 x^{n-1} + m_2 x^{n-2} - \cdots \pm m_n = 0$$

is the characteristic equation of any square matrix M, then the determinant of M is m_n and if by a principal minor of a matrix we understand the *determinant* of a minor whose diagonal is part of the main diagonal of M, then m_i is the sum of the principal i-rowed minors. In particular, then, m_1, which is also the sum of the characteristic roots, is the sum of the elements along the main diagonal. This sum is called the trace of the matrix. The proofs of Taber's assertions were given by William Henry Metzler (1863–?).[37]

Frobenius[38] raised a question related to the characteristic equation. He sought the minimal polynomial—the polynomial of lowest degree—which the matrix satisfies. He stated that it is formed from the factors of the characteristic polynomial and is unique. It was not until 1904[39] that Kurt Hensel (1861–1941) proved Frobenius's statement of uniqueness. In the same article, Hensel also proved that if $f(x)$ is the minimal polynomial of a matrix M and $g(x)$ is any other polynomial satisfied by M, then $f(x)$ divides $g(x)$.

31. 1853, p. 566.
32. *Comp. Rend.*, 41, 1855, 181–83 = *Œuvres*, 1, 479–81.
33. *Jour. für Math.*, 62, 1863, 232–45.
34. *Messenger of Math.*, (2), 14, 1885, 143–44.
35. *Œuvres*, (2), 9, 174–91.
36. *Amer. Jour. of Math.*, 12, 1890, 337–96.
37. *Amer. Jour. of Math.*, 14, 1891/92, 326–77.
38. *Jour. für Math.*, 84, 1878, 1–63 = *Ges. Ahb.*, 1, 343–405.
39. *Jour. für Math.*, 127, 1904, 116–66.

The notion of the rank of a matrix was introduced by Frobenius[40] in 1879 though in connection with determinants. A matrix A with m rows and n columns (of order m by n) has k-rowed minors of all orders from 1 (the elements of A themselves) to the smaller of the two integers m and n inclusive. A matrix has rank r if and only if it has at least one r-rowed minor whose determinant is not zero while the determinants of all minors of order greater than r are zero.

Two matrices A and B can be related in various ways. They are equivalent if there exist two non-singular matrices U and V such that $A = UBV$. Sylvester had shown in his work on determinants[41] that the greatest common divisor d'_i of the i-rowed minor determinants of B equals the greatest common divisor d_i of the i-rowed minor determinants of A. Then H. J. S. Smith, working with matrices with integral elements, showed[42] that every matrix A of rank ρ is equivalent to a diagonal matrix with elements $h_1, h_2, \ldots, h_\rho$ down the main diagonal and such that h_i divides $h_i + 1$. The quotients $h_1 = d_1$, $h_2 = d_2/d_1$, ... are called the invariant factors of A. Further if

$$h_i = p_1^{l_{i1}} p_2^{l_{i2}} \cdots p_k^{l_{ik}}$$

(where the p_i are primes) these various powers $p_i^{l_{ij}}$ are the elementary divisors of A. The invariant factors determine the elementary divisors and conversely.

These ideas on invariant factors and elementary divisors, which stem from Sylvester's and Weierstrass's work on determinants (as noted earlier), were carried over to matrices by Frobenius in his 1878 paper. The significance of the invariant factors and elementary divisors for matrices is that the matrix A is equivalent to the matrix B if and only if A and B have the same elementary divisors or invariant factors.

Frobenius did further work with invariant factors in his paper of 1878 and then organized the theory of invariant factors and elementary divisors in logical form.[43] The work in the 1878 paper enabled Frobenius to give the first general proof of the Cayley-Hamilton theorem and to modify the theorem when some of the latent roots (characteristic roots) of the matrix are equal. In this paper he also showed that when $AB^{-1} = B^{-1}A$, in which case there is an unambiguous quotient A/B, then $(A/B)^{-1} = B/A$ and that $(A^{-1})^T = (A^T)^{-1}$, where A^T is the transpose of A.

The subject of orthogonal matrices has received considerable attention. Although the term was used by Hermite in 1854,[44] it was not until 1878 that the formal definition was published by Frobenius (see ref. above). A matrix M is orthogonal if it is equal to the inverse of its transpose, that is, if

40. *Jour. für Math.*, 86, 1879, 146–208 = *Ges. Abh.*, 1, 482–544.
41. *Phil. Mag.*, (4), 1, 1851, 119–40 = *Coll. Math. Papers*, 1, 219–40.
42. *Phil. Trans.*, 151, 1861–62, 293–326 = *Coll. Math. Papers*, 1, 367–409.
43. *Sitzungsber. Akad. Wiss. zu Berlin*, 1894, 31–44 = *Ges. Abh.*, 1, 577–90.
44. *Cambridge and Dublin Math. Jour.*, 9, 1854, 63–67 = *Œuvres*, 1, 290–95.

$M = (M^T)^{-1}$. In addition to the definition, Frobenius proved that if S is a symmetric matrix and T, a skew-symmetric one, an orthogonal matrix can always be written in the form $(S - T)/(S + T)$ or more simply $(I - T)/(I + T)$.

The notion of similar matrices, like many other notions of matrix theory, came from earlier work on determinants as far back as Cauchy's. Two square matrices A and B are similar if there exists a non-singular matrix P such that $B = P^{-1}AP$. The characteristic equations of two similar matrices A and B are the same and the matrices have the same invariant factors and the same elementary divisors. For matrices with complex elements Weierstrass proved this result in his 1868 paper (though he worked with determinants). Since a matrix represents a linear homogeneous transformation similar matrices can be thought of as representing the same transformation but referred to two different coordinate systems.

Using the notion of similar matrices and the characteristic equation, Jordan [45] showed that a matrix can be transformed to a canonical form. If the characteristic equation of a matrix J is

$$f(\lambda) = \lambda^n + b_1\lambda^{n-1} + \cdots + b_n = 0$$

and if

$$f(\lambda) = (\lambda - \lambda_1)^{l_1} \cdots (\lambda - \lambda_k)^{l_k},$$

where the λ_i are distinct, then let

$$J_i = \begin{pmatrix} \lambda_i & 1 & 0 & 0 & \cdots & 0 \\ 0 & \lambda_i & 1 & 0 & \cdots & 0 \\ \cdot & \cdot & \cdot & \cdot & \cdot & \cdot \\ 0 & 0 & 0 & 0 & \cdots & \lambda_i \end{pmatrix}$$

denote an l_ith order matrix. Jordan showed that J can be transformed to a similar matrix having the form

$$\begin{pmatrix} J_1 & 0 & 0 & \cdots & 0 \\ 0 & J_2 & 0 & \cdots & 0 \\ \cdot & \cdot & \cdot & \cdot & \cdot \\ 0 & 0 & 0 & \cdots & J_k \end{pmatrix}.$$

This is the Jordan canonical, or normal, form of a matrix.

The similarity transformation from A to B was also treated by Frobenius under the name of contragredient transformation in his 1878 paper. In the

45. *Traité des substitutions*, 1870, Book II, 88–249.

same discussion he treated the notion of congruent matrices or cogredient transformations. This tells us that if $A = P^T BP$ then A is congruent with B, written $A \overset{c}{=} B$. For example, the transformation of matrix A that results in the simultaneous interchange of the same rows and columns of A is a congruence transformation. Also, a symmetric matrix A of rank r can be reduced by a congruent transformation to a diagonal matrix of the same rank; that is,

$$
P^T AP = \begin{bmatrix}
d_{11} & 0 & \cdots & 0 & \cdots & 0 \\
0 & d_{22} & \cdots & 0 & \cdots & 0 \\
\cdot & \cdot & \cdot & \cdot & \cdot & \cdot \\
0 & 0 & \cdots & d_{rr} & \cdots & 0 \\
\cdot & \cdot & \cdot & \cdot & \cdot & \cdot \\
0 & 0 & \cdots & 0 & \cdots & 0
\end{bmatrix}.
$$

There are many basic theorems on congruent matrices. For example, if S is symmetric and S_1 is congruent to S, then S_1 is symmetric and if S is skew then S_1 is skew.

In his 1892 paper in the *American Journal of Mathematics* Metzler introduced transcendental functions of a matrix, writing each as a power series in a matrix. He established series for e^M, e^{-M}, $\log M$, $\sin M$, and $\sin^{-1} M$. Thus

$$
e^M = \sum_{n=0}^{\infty} M^n / n!.
$$

The ramifications of the theory of matrices are numerous. Matrices have been used to represent quadratic and bilinear forms. The reduction of such forms to simple canonical forms is the core of the work on the invariants of matrices. They are intimately connected with hypernumbers and Cayley in his 1858 paper developed the idea of treating hypernumbers as matrices.

Both determinants and matrices have been extended to infinite order. Infinite determinants were involved in Fourier's work of determining the coefficients of a Fourier series expansion of a function (Chap. 28, sec. 2) and in Hill's work on the solution of ordinary differential equations (Chap. 29, sec. 7). Isolated papers on infinite determinants were written between these two outstanding nineteenth-century researches but the major activity postdates Hill's.

Infinite matrices were implicitly and explicitly involved in the work of Fourier, Hill, and Poincaré, who completed Hill's work. However, the great impetus to the study of infinite matrices came from the theory of integral equations (Chap. 45). We cannot devote space to the theory of determinants and matrices of infinite order.[46]

46. See the reference to Bernkopf in the bibliography at end of chapter.

In the elementary work on matrices the elements are ordinary real numbers though a great deal of what was done on behalf of the theory of numbers was limited to integral elements. However, they can be complex numbers and indeed most any other quantities. Naturally, the properties the matrices themselves possess depend upon the properties of the elements. Much late nineteenth- and early twentieth-century research has been devoted to the properties of matrices whose elements are members of an abstract field. The importance of matrix theory in the mathematical machinery of modern physics cannot be treated here, but in this connection a prophetic statement made by Tait is of interest. "Cayley is forging the weapons for future generations of physicists."

Bibliography

Bernkopf, Michael: "A History of Infinite Matrices," *Archive for History of Exact Sciences*, 4, 1968, 308–58.

Cayley, Arthur: *The Collected Mathematical Papers*, 13 vols., Cambridge University Press (1889–97), Johnson Reprint Corp., 1963.

Feldman, Richard W., Jr.: (Six articles on matrices with various titles), *The Mathematics Teacher*, 55, 1962, 482–84, 589–90, 657–59; 56, 1963, 37–38, 101–2, 163–64.

Frobenius, F. G.: *Gesammelte Abhandlungen*, 3 vols., Springer-Verlag, 1968.

Jacobi, C. G. J.: *Gesammelte Werke*, Georg Reimer, 1884, Vol. 3.

MacDuffee, C. C.: *The Theory of Matrices*, Chelsea, 1946.

Muir, Thomas: *The Theory of Determinants in the Historical Order of Development* (1906–23), 4 vols., Dover (reprint), 1960.

————: List of writings on the theory of matrices, *Amer. Jour. of Math.*, 20, 1898, 225–28.

Sylvester, James Joseph: *The Collected Mathematical Papers*, 4 vols., Cambridge University Press, 1904–12.

Weierstrass, Karl: *Mathematische Werke*, Mayer und Müller, 1895, Vol. 2.

Abbreviations

Journals whose titles have been written out in full in the text are not listed here.

Abh. der Bayer. Akad. der Wiss. Abhandlungen der Königlich Bayerischen Akademie der Wissenschaften (München)

Abh. der Ges. der Wiss. zu Gött. Abhandlungen der Königlichen Gesellschaft der Wissenschaften zu Göttingen

Abh. König. Akad. der Wiss., Berlin Abhandlungen der Königlich Preussischen Akademie der Wissenschaften zu Berlin

Abh. Königlich Böhm. Ges. der Wiss. Abhandlungen der Königlichen Böhmischen Gesellschaft der Wissenschaften

Abh. Math. Seminar der Hamburger Univ. Abhandlungen aus dem Mathematischen Seminar Hamburgischen Universität

Acta Acad. Sci. Petrop. Acta Academiae Scientiarum Petropolitanae

Acta Erud. Acta Eruditorum

Acta Math. Acta Mathematica

Acta Soc. Fennicae Acta Societatis Scientiarum Fennicae

Amer. Jour. of Math. American Journal of Mathematics

Amer. Math. Monthly American Mathematical Monthly

Amer. Math. Soc. Bull. American Mathematical Society, Bulletin

Amer. Math. Soc. Trans. American Mathematical Society, Transactions

Ann. de l'Ecole Norm. Sup. Annales Scientifiques de l'Ecole Normale Supérieure

Ann. de Math. Annales de Mathématiques Pures et Appliquées

Ann. Fac. Sci. de Toulouse Annales de la Faculté des Sciences de Toulouse

Ann. Soc. Sci. Bruxelles Annales de la Société Scientifique de Bruxelles

Annali di Mat. Annali di Matematica Pura ed Applicata

Annuals of Math. Annals of Mathematics

Astronom. Nach. Astronomische Nachrichten

Atti Accad. Torino Atti della Reale Accademia delle Scienze di Torino

Atti della Accad. dei Lincei, Rendiconti Atti della Reale Accademia dei Lincei, Rendiconti

Brit. Assn. for Adv. of Sci. British Association for the Advancement of Science

Bull. des Sci. Math. Bulletin des Sciences Mathématiques

Bull. Soc. Math. de France Bulletin de la Société Mathématique de France

Cambridge and Dublin Math. Jour. Cambridge and Dublin Mathematical Journal

Comm. Acad. Sci. Petrop. Commentarii Academiae Scientiarum Petropolitanae

Comm. Soc. Gott. Commentationes Societatis Regiae Scientiarum Gottingensis Recentiores

Comp. Rend. Comptes Rendus

Corresp. sur l'Ecole Poly. Correspondance sur l'Ecole Polytechnique

Encyk. der Math. Wiss. Encyklopädie der Mathematischen Wissenschaften

Gior. di Mat. Giornale di Matematiche

Hist. de l'Acad. de Berlin Histoire de l'Académie Royale des Sciences et des Belles-Lettres de Berlin

Hist. de l'Acad. des Sci., Paris Histoire de l'Académie Royale des Sciences avec les Mémoires de Mathématique et de Physique

Jahres, der Deut. Math.-Verein. Jahresbericht der Deutschen Mathematiker-Vereinigung

Jour. de l'Ecole Poly. Journal de l'Ecole Polytechnique

Jour. de Math. Journal de Mathématiques Pures et Appliquées

Jour. des Sçavans Journal des Sçavans

Jour. für Math. Journal für die Reine und Angewandte Mathematik

Jour. Lon. Math. Soc. Journal of the London Mathematical Society

Königlich Sächsischen Ges. der Wiss. zu Leipzig Berichte über die Verhandlungen der Königlich Sächsischen Gesellschaft der Wissenschaften zu Leipzig

Math. Ann. Mathematische Annalen

Mém. de l'Acad. de Berlin See *Hist. de l'Acad. de Berlin*

Mém. de l'Acad. des Sci., Paris See *Hist. de l'Acad. des Sci., Paris;* after 1795, Mémoires de l'Academie des Sciences de l'Institut de France

Mém. de l'Acad. Sci. de St. Peters. Mémoires de l'Académie Impériale des Sciences de Saint-Petersbourg

Mém. des sav. étrangers Mémoires de Mathématique et de Physique Présentés à l'Académie Royal des Sciences, par Divers Sçavans, et Lus dans ses Assemblées

Mém. divers Savans See *Mém. des sav. étrangers*

Misc. Berolin. Miscellanea Berolinensia; also as *Hist. de l'Acad. de Berlin (q.v.)*

Misc. Taur. Miscellanea Philosophica-Mathematica Societatis Privatae Taurinensis (published by Accademia della Scienze di Torino)

Monatsber. Berliner Akad. Monatsberichte der Königlich Preussischen Akademie der Wissenschaften zu Berlin

N.Y. Math. Soc. Bull. New York Mathematical Society, Bulletin

Nachrichten König. Ges. der Wiss. zu Gött. Nachrichten von der Königlichen Gesellschaft der Wissenschaften zu Göttingen

Nou. Mém. de l'Acad. Roy. des Sci., Bruxelles Nouveaux Mémoires de l'Académie Royale des Sciences, des Lettres, et des Beaux-Arts de Belgique

Nouv. Bull. de la Soc. Philo. Nouveau Bulletin de la Société Philomatique de Paris

Nouv. Mém. de l'Acad. de Berlin Nouveaux Mémoires de l'Académie Royale des Sciences et des Belles-Lettres de Berlin

Nova Acta Acad. Sci. Petrop. Nova Acta Academiae Scientiarum Petropolitanae

Nova Acta Erud. Nova Acta Eruditorum

Novi Comm. Acad. Sci. Petrop. Novi Commentarii Academiae Scientiarum Petropolitanae

Phil. Mag. The Philosophical Magazine

Philo. Trans. Philosophical Transactions of the Royal Society of London

Proc. Camb. Phil. Soc. Cambridge Philosophical Society, Proceedings

Proc. Edinburgh Math. Soc. Edinburgh Mathematical Society, Proceedings

Proc. London Math. Soc. Proceedings of the London Mathematical Society

Proc. Roy. Soc. Proceedings of the Royal Society of London

Proc. Royal Irish Academy Proceedings of the Royal Irish Academy

Quart. Jour. of Math. Quarterly Journal of Mathematics

Scripta Math. Scripta Mathematica

Sitzungsber. Akad. Wiss zu Berlin Sitzungsberichte der Königlich Preussischen Akademie der Wissenschaften zu Berlin

Sitzungsber. der Akad. der Wiss., Wien Sitzungsberichte der Kaiserlichen Akademie der Wissenschaften zu Wien. Mathematisch-Naturwissenschaftlichen Klasse

Trans. Camb. Phil. Soc. Cambridge Philosophical Society, Transactions

Trans. Royal Irish Academy Transactions of the Royal Irish Academy

Zeit. für Math. und Phys. Zeitschrift für Mathematik und Physik

Zeit. für Physik Zeitschrift für Physik

Name Index

iii

Subject Index

xiii